T0235054

Theory of Superconductivity

Revised Printing

Theory of Superconductivity

Revised Printing

J. R. Schrieffer
Florida State University
Tallahassee, Florida

CRC Press
Taylor & Francis Group
Boca Raton London New York

CRC Press is an imprint of the
Taylor & Francis Group, an informa business

ABP

Advanced Book Program

The publisher is pleased to acknowledge the assistance of Paul Orban, who produced the illustrations for the original edition. The revised edition (1983) includes an Appendix containing Nobel Lectures, December 11, 1972, by J. R. Schrieffer, Leon N. Cooper, and John Bardeen, reprinted with permission of the Nobel Foundation.
Copyright © 1973 by the Nobel Foundation.

First published 1964 by Westview Press

Published 2018 by CRC Press
Taylor & Francis Group
6000 Broken Sound Parkway NW, Suite 300
Boca Raton, FL 33487-2742

Copyright © 1999, 1964 by J. R. Schrieffer
CRC Press is an imprint of Taylor & Francis Group, an Informa business

No claim to original U.S. Government works

This book contains information obtained from authentic and highly regarded sources. Reasonable efforts have been made to publish reliable data and information, but the author and publisher cannot assume responsibility for the validity of all materials or the consequences of their use. The authors and publishers have attempted to trace the copyright holders of all material reproduced in this publication and apologize to copyright holders if permission to publish in this form has not been obtained. If any copyright material has not been acknowledged please write and let us know so we may rectify in any future reprint.

Except as permitted under U.S. Copyright Law, no part of this book may be reprinted, reproduced, transmitted, or utilized in any form by any electronic, mechanical, or other means, now known or hereafter invented, including photocopying, microfilming, and recording, or in any information storage or retrieval system, without written permission from the publishers.

Trademark Notice: Product or corporate names may be trademarks or registered trademarks, and are used only for identification and explanation without intent to infringe.

Visit the Taylor & Francis Web site at
http://www.taylorandfrancis.com

and the CRC Press Web site at
http://www.crcpress.com

Library of Congress Catalog Card Number: 99-60035

ISBN 13: 978-0-7382-0120-7 (pbk)

Cover design by Suzanne Heiser

Advanced Book Classics

Anderson: Basic Notions of Condensed Matter Physics, ABC ppbk,
ISBN 0-201-32830-5

Atiyah: K-Theory, ABC ppbk, ISBN 0-201-40792-2

Bethe: Intermediate Quantum Mechanics, ABC ppbk, ISBN 0-201-32831-3

Clemmow: Electrodynamics of Particles and Plasmas, ABC ppbk,
ISBN 0-20147986-9

Davidson: Physics of Nonneutral Plasmas, ABC ppbk
ISBN 0-201-57830-1

DeGennes: Superconductivity of Metals and Alloys, ABC ppbk,
ISBN 0-7382-0101-4

d'Espagnat: Conceptual Foundations Quantum Mechanics, ABC ppbk,
ISBN 0-7382-0104-9

Feynman: Photon-Hadron Interactions, ABC ppbk, ISBN 0-201-36074-8

Feynman: Quantum Electrodynamics, ABC ppbk, ISBN 0-201-36075-4

Feynman: Statistical Mechanics, ABC ppbk, ISBN 0-201-36076-4

Feynman: Theory of Fundamental Processes, ABC ppbk, ISBN 0-201-36077-2

Forster: Hydrodynamic Fluctuations, Broken Symmetry, and Correlation Functions,
ABC ppbk, ISBN 0-201-41049-4

Gell-Mann/Ne'eman: The Eightfold Way, ABC ppbk, ISBN 0-7382-0299-1

Gottfried: Quantum Mechanics, ABC ppbk, ISBN 0-201-40633-0

Kadanoff/Baym: Quantum Statistical Mechanics, ABC ppbk, ISBN 0-201-41046-X

Khalatnikov: An Intro to the Theory of Superfluidity, ABC ppbk,
ISBN 0-7382-0300-9

Ma: Modern Theory of Critical Phenomena, ABC ppbk, ISBN 0-7382-0301-7

Migdal: Qualitative Methods in Quantum Theory, ABC ppbk, ISBN 0-7382-0302-5

Negele/Orland: Quantum Many-Particle Systems, ABC ppbk, ISBN 0-7382-0052-2

Nozieres/Pines: Theory of Quantum Liquids, ABC ppbk, ISBN 0-7382-0229-0

Nozieres: Theory of Interacting Fermi Systems, ABC ppbk, ISBN 0-201-32824-0

Parisi: Statistical Field Theory, ABC ppbk, ISBN 0-7382-0051-4

Pines: Elementary Excitations in Solids, ABC ppbk, ISBN 0-7382-0115-4

Pines: The Many-Body Problem, ABC ppbk, ISBN 0-201-32834-8

Quigg: Gauge Theories of the Strong, Weak, and Electromagnetic Interactions, ABC ppbk, ISBN 0-201-32832-1

Richardson: Experimental Techniques in Condensed Matter Physics at Low Temperatures, ABC ppbk ISBN 0-201-36078-0

Rohrlich: Classical Charges Particles, ABC ppbk ISBN 0-201-48300-9

Schrieffer: Theory of Superconductivity, ABC ppbk ISBN 0-7382-0120-0

Schwinger: Particles, Sources, and Fields Vol. 1, ABC ppbk ISBN 0-7382-0053-0

Schwinger: Particles, Sources, and Fields Vol. 2, ABC ppbk ISBN 0-7382-0054-9

Schwinger: Particles, Sources, and Fields Vol. 3, ABC ppbk ISBN 0-7382-0055-7

Schwinger: Quantum Kinematics and Dynamics, ABC ppbk, ISBN 0-7382-0303-3

Thom: Structural Stability and Morphogenesis, ABC ppbk, ISBN 0-201-40685-3

Wyld: Mathematical Methods for Physics, ABC ppbk, ISBN 0-7382-0125-1

CONTENTS

PREFACE

The material presented here is an outgrowth of a series of lectures I gave at the University of Pennsylvania during the fall of 1962. I have stressed the fundamentals of the microscopic theory of superconductivity rather than attempting to give a broad survey of the field as a whole. As a result, a number of highly interesting and important areas are not discussed here; an example is the application of the microscopic theory to type II (or "hard") superconductors. The material presented here is primarily intended to serve as a background for reading the literature in which detailed applications of the microscopic theory are made to specific problems.

A variety of formal techniques have been used in the literature to describe the pairing correlations basic to superconductivity. For this reason I have developed a number of these techniques in the text and it is hoped that the inelegance of this approach will be justified by the usefulness of the material.

A brief review of the simple experimental facts and several phenomenological theories of superconductivity is given in the first chapter. This is followed in Chapter 2 by an account of the original pairing theory advanced by Bardeen, Cooper, and the author. A number of applications of this theory are worked out in Chapter 3. This first portion of the book uses only the techniques of quantum mechanics which are covered in a standard graduate course on quantum theory. While the notation of second quantization is used as a convenient shorthand, this formalism is reviewed in the appendix.

In Chapters 4 and 5 the many-body aspects of the coupled electron–ion system are developed with a view to treating in a more realistic manner the effective interaction between electrons which brings about superconductivity. In addition, the basis for treating strong quasi-particle damping effects important in strong coupling superconductors is developed. In Chapter 6 a discussion of elementary excitations in normal metals is given, which lays the ground work for the field-theoretic treatment of the superconducting state given in Chapter 7. There, the noninstantaneous nature of the interaction bringing about superconductivity is treated as well as the breakdown of the quasi-particle approximation and the resolution of this difficulty. In the final chapter the electromagnetic properties of superconductors are treated, as well as the collective excitations of the system.

I should like to thank Drs. P. W. Anderson, J. Bardeen, L. P. Kadanoff, D. J. Scalapino, Y. Wada, and J. W. Wilkins for many helpful discussions during the preparation of this manuscript. I am also indebted to Drs. F. Bassani and J. E. Robinson, who prepared a set of notes covering a lecture series I gave at Argonne National Laboratory during the spring of 1961. Much of the material in Chapters 4 and 5 and in the appendix is related to their notes. In addition, I would like to express my sincere appreciation to Mrs. Dorothea Hofford for the speed and accuracy with which she typed the manuscript. Finally, I should like to thank my wife for her considerable help in preparing this book.

J. R. Schrieffer

Philadelphia, Pennsylvania
July 1964

PREFACE TO THE REVISED PRINTING

Since the first appearance of this book in 1964, the field of superconductivity has undergone dramatic growth in scope and level of activity. The two-volume treatise, Superconductivity, edited by R. D. Parks, gave an account of the field at the close of the 60s. Collections of papers describing current research on superconductivity can be found in proceedings of conferences, such as the International Conference on Low Temperature Physics and the Applied Superconductivity Conference.

Since 1964 many significant areas of research have developed including type II superconductivity and the Abrikosov vortex lattice. This fundamental understanding led to high field magnets having important technological applications. Josephson tunnel junctions proved to exhibit a wide variety of interesting phenomena which have led to a spectrum of devices of use in high precision measurements and in computers. The long sought after organic superconductor has been discovered, and major advances in fabricating new superconducting materials have been made. ^3He was discovered to be a spin triplet superconductor, with many remarkable properties.

On the theoretical side, there was early recognition that the pairing correlations of metallic superconductors also played a fundamental role in the structure of atomic nuclei. The theory of superconductivity also forms the basis for understanding the structure of neutron stars, despite their enormously high temperature. Finally, the broken symmetry concept inherent in the pairing theory has been helpful in setting up gauge theories of elementary particles. Possibly other developments stemming from the pairing theory lie ahead.

This revised printing does not attempt to give an account of the above topics since this would be impossible in a single volume. Rather, we note that the fundamentals of the theory of superconductivity as discussed have remained unchanged over the past two decades, and the recent developments have been built on that theoretical foundation. Hopefully, the reader will find the text of continuing value as an introduction to this fascinating and active field.

J. ROBERT SCHRIEFFER
Santa Barbara, 1983

THEORY OF SUPERCONDUCTIVITY

Revised Printing

CHAPTER 1

INTRODUCTION

The phenomenon of superconductivity is a remarkable example of quantum effects operating on a truly macroscopic scale.[1] In a superconducting material, a finite fraction of the electrons are in a real sense condensed into a "macromolecule" (or "superfluid") which extends over the entire volume of the system and is capable of motion as a whole. At zero temperature the condensation is complete and all the electrons participate in forming this superfluid, although only those electrons near the Fermi surface have their motion appreciably affected by the condensation. As the temperature is increased, a fraction of the electrons evaporate from the condensate and form a weakly interacting gas of excitations (or "normal fluid"), which also extends throughout the entire volume of the system, interpenetrating the superfluid.[2] As the temperature approaches a critical value T_c, the fraction of electrons remaining in the superfluid tends to zero and the system undergoes a second-order phase transition from the superconducting to the normal state. This two-fluid picture of a superconductor is formally analogous to that which characterizes superfluid He^4, although there are important differences between these systems.[1,3]

The amazing properties of superconductors (e.g., perfect diamagnetism, zero d-c electrical resistance, etc.[4]) are related to the peculiar excitation spectrum of the superfluid. As we shall see, the superfluid can carry out potential (or irrotational) flow with little change of its "internal energy" (i.e., energy associated with forces binding the superfluid together). On the other hand, the superfluid *cannot* support rotational flow. In analogy with superfluid He^4, if one tries to force the superfluid into motion having vorticity (i.e., a nonvanishing curl of its linear momentum), a fraction of the superfluid is necessarily converted into normal fluid. Since the normal fluid does not take advantage of the forces binding the superfluid together, there is in general a large increase in energy associated with creating this vorticity. It is reasonable, therefore, that the superfluid possesses a rigidity or stiffness with respect to perturbations which, like the magnetic field, tends to impart vorticity (i.e., angular momentum) to the system. On the basis of this assumed rigidity, London[1,5] was able to account theoretically for the perfect diamagnetism of bulk superconductors in weak magnetic fields (the Meissner effect[6]) and for the apparent lack of d-c electrical resistance, as first observed by Kamerlingh Onnes in 1911.[7]

As we shall see, the microscopic theory of superconductivity proposed by Bardeen, Cooper, and the author[8] can be thought of in terms of this sort of two-fluid picture.[9] In the lowest approximation the superfluid is formed from pairs of electrons which are bound together by lattice polarization forces. The pairs *greatly overlap* with each other in space, and it is the strong *pair–pair correlations* in addition to correlations between mates of a pair which are ultimately responsible for the rigidity of the superfluid wave function discussed above. More generally, these correlations are responsible for an energy gap in the elementary excitation spectrum of a superconductor from which many properties of the superconductor (in addition to its electromagnetic behavior) follow as a consequence. In the theory, the normal fluid is composed of the gas of elementary excitations of the system.

It is perhaps not surprising that the microscopic theory of superconductivity followed Onnes' remarkable discovery of this phenomenon by almost fifty years, considering the physical and mathematical complications of the problem. It was not until 1950 that the basic forces responsible for the condensation were recognized, through the insight of Fröhlich.[10] He suggested that an effective interaction between electrons arising from their interaction with crystal lattice vibrations (phonons) was of primary importance in bringing about the condensation. At this time, independent experiments on the isotope effect in superconductors were being carried out by Reynolds *et al.*[11] and by Maxwell[12] which gave experimental support to Fröhlich's point of view. Early theories of Fröhlich[10] and Bardeen[13] based on a perturbation treatment of the electron–phonon interaction ran into mathematical difficulties. The significance of these difficulties was emphasized by Schafroth's[14] proof that the Meissner effect *cannot* be obtained in any *finite* order of perturbation theory, beginning with the uncoupled system. Later, Migdal[15] showed that there is no energy gap in the electronic excitation spectrum within the perturbation theory. In the BCS theory, the electron–phonon coupling constant g enters in the nonanalytic fashion e^{-1/g^2}, in agreement with Schafroth's and Migdal's results.

The microscopic theory explains essentially all of the general features of superconductivity. In addition to this qualitative explanation, it is in remarkably good quantitative agreement with experiment considering the crudeness of the approximations necessitated by our uncertainties regarding electronic and phononic band structure, electron–phonon matrix elements, etc., in real metals.

In this book we shall attempt to give an account of the underlying physical ideas of the theory. While some of the discussion is couched in the language of the many-body problem, much of this formalism is developed in the text. In general, we shall not give a detailed discussion of the relation between theory and experiment and the reader is referred to a number of books and review articles[9, 16] covering this area. We list below a few of the most important simple experimental facts about superconductors. One conventionally distinguishes between the behavior of type I (or soft) superconductors and type II (or hard) superconductors.

1-1 SIMPLE EXPERIMENTAL FACTS

Electromagnetic Properties

The d-c electrical resistivity of materials in the soft superconducting state is zero. This fact is established to better than one part in 10^{15} of the resistance of the normal state at the corresponding temperature.[16f] At $T = 0$ the resistivity of a superconductor ideally remains zero up to a critical frequency $\hbar\omega_g \sim 3.5k_BT_c$ (presumably the threshold for creating excitations out of the condensate). In practice, the edge of the gap is smeared and a precursor electromagnetic absorption is observed below the edge of the gap in certain cases. At finite temperatures, there is a finite a-c resistivity for all $\omega > 0$ (presumably because of absorption by the thermally excited normal fluid if $\omega < \omega_g$). For $\omega \gg \omega_g$, the resistivities of the normal and superconducting states are essentially equal, independent of temperature.

In 1933, Meissner and Oschenfeld[6] discovered that a bulk superconductor is a perfect diamagnet. Thus the magnetic field **B** penetrates only to a depth $\lambda \simeq 500$ Å and is excluded from the main body of the material. If one (incorrectly) argues that the vanishing *zero-frequency* electrical resistance implies that there can be no electric field (of any frequency) in a superconductor, Maxwell's equation

$$\nabla \times \mathbf{E} = -\frac{1}{c}\frac{\partial \mathbf{B}}{\partial t} \tag{1-1}$$

shows that the magnetic field present in the normal metal will be "frozen in" when the metal becomes superconducting. This is contrary to the Meissner effect, which states that the field is expelled in the superconducting phase. The point is that the superfluid gives rise to a purely inductive impedance which vanishes only at zero frequency.[9] It is this nonzero impedance which permits the expulsion of **B**. This point is discussed further below.

The perfect exclusion of the magnetic flux in bulk soft superconductors increases the free energy per unit volume of the superconductor by $H^2/8\pi$,[5] if H is the externally applied field. Since there is only a finite amount of energy reduction due to the condensation into the superconducting phase, there must be a critical field $H_c(T)$ at which the total free energy of the superconducting and normal states are equal. The critical field is a maximum H_0 at $T = 0$ and falls to zero at $T = T_c$ as shown in Figure 1-1. For typical "soft" superconductors like Al, Sn, In, Pb, etc., H_0 is of order a few hundred gauss. In "hard" superconductors like Nb_3Sn, superconductivity can persist to an "upper" critical field H_{c2} of order 10^5 gauss presumably due to the magnetic flux penetrating into the bulk of the material for H larger than a "lower" critical field H_{c1}.[17, 18] Thus, as opposed to a soft superconductor, a perfect Meissner effect does not exist above H_{c1} in a hard superconductor.

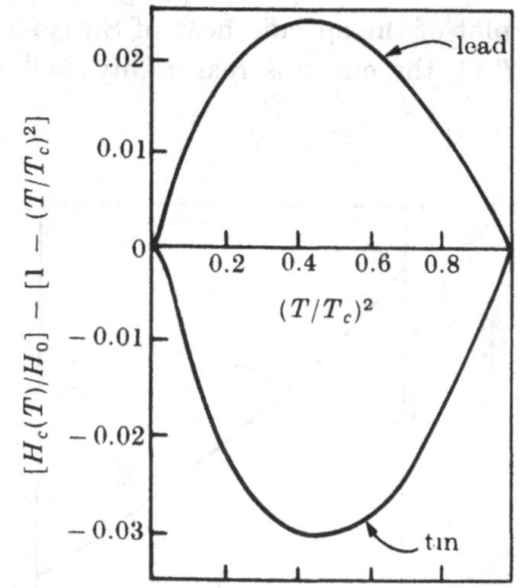

FIGURE 1-1 The deviations of the critical field from Tuyn's law $H_c(T) = H_0[1 - (T/T_c)^2]$, i.e., the prediction of the Gorter–Casimir model.

If one has a multiply connected superconductor, e.g., a hollow cylinder, the flux passing through the hole cannot have an arbitrary value, but is quantized to multiples of $hc/2e \simeq 4 \times 10^{-7}$ gauss cm^2. Quantization of flux in units twice this size was predicted by London[1] while the experimental observation of the effect and the establishment of the correct unit of flux was carried out independently by Deaver and Fairbank[20a] and by Doll and Näbauer.[20b]

Thermodynamic Properties

In zero magnetic field, there is a second-order phase transition at T_c.[21] The jump in specific heat is generally about three times the electronic specific heat γT_c in the normal state just above the transition. In well-annealed pure specimens the width of the transition can be as small as 10^{-4} °K although this is not believed to be the intrinsic width of the transition.[22] As $T/T_c \to 0$, the electronic specific heat generally falls as $ae^{-b/T}$, presumably due to the energy gap for creating elementary excitations. The ratio of the energy gap $2\Delta(0)$ at $T = 0$ to $k_B T_c$ is usually of order 3.5, the ratio being larger for stronger coupling superconductors like Pb and Hg. A plot of the specific heat of Sn is shown in Figure 1-2. For $T \geqslant T_c/2$, the curve is reasonably well fitted by αT^3.

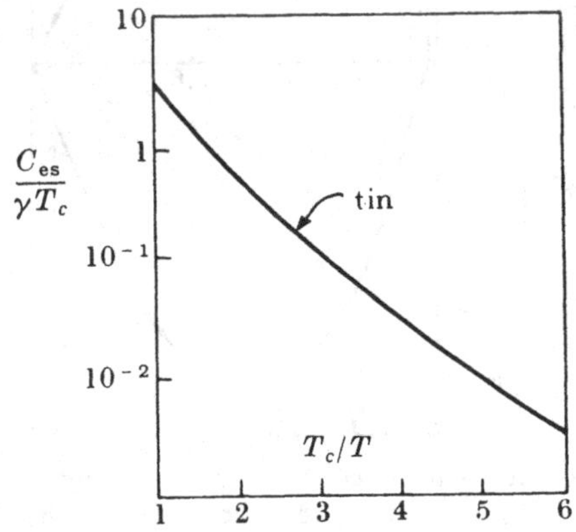

FIGURE 1-2 The electronic specific heat of Sn.

In the presence of a magnetic field the N–S-phase transition for a bulk specimen is first order, i.e., a latent heat is involved.[4]

Isotope Effect

As we discussed above, the isotope effect shows that lattice vibrations play an essential role in bringing about superconductivity. In particular, one finds that the critical field at zero temperature H_0 and the transition temperature T_c vary as

$$T_c \sim \frac{1}{M^\alpha} \sim H_0 \qquad (\alpha \simeq \tfrac{1}{2}) \tag{1-2}$$

when the isotopic mass M of the material is varied. Thus, T_c and H_0 are larger for lighter isotopes. If lattice vibrations were not important in the phenomenon there is no reason why T_c should change as neutrons are added to the nuclei since their main effect is to change the mass of the ions. While the value $\alpha = 0.45$ to 0.50 is approximately correct for many superconductors there are a number of notable exceptions, for example Ru, Mo, Nb_3Sn, and Os[23] which have small or vanishing isotope effects. As Garland[24] has shown, this does not preclude the phonons from causing the transition. Although the actual mechanism in these materials is not firmly established at present, it is not unlikely that the electron–phonon interaction is the appropriate mechanism even in these exceptional cases.

Energy Gap

There are several direct ways of observing the energy gap in the elementary excitation spectrum of superconductors.[16d, g] As we mentioned above, the threshold for absorbing electromagnetic radiation gives a value for the energy gap.[25] An even simpler method[26] (due to Giaever) is to measure the electron tunneling current between two films of a superconducting material separated by a thin (~20 Å) oxide layer. As $T \to 0$, no current flows until the applied voltage (times the electronic charge) V exceeds the energy gap 2Δ. As the temperature is increased, a finite current flows for $V < 2\Delta(T)$; however, a break in the curve persists for

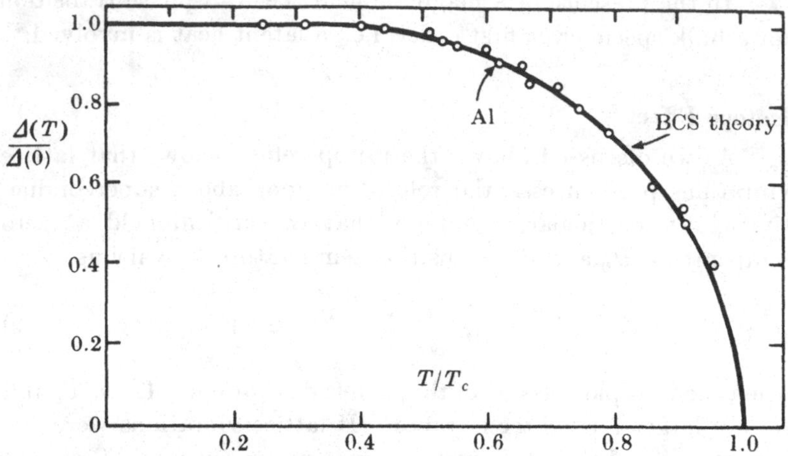

FIGURE 1-3 The temperature-dependent energy gap of Al as determined by electron tunneling.

$V = 2\Delta(T)$. The temperature-dependent energy gap observed in this way is plotted in Figure 1-3. The temperature dependence of the energy gap can also be simply determined from the rate of attenuation of sound waves,[27] the rate of decay of nuclear magnetization,[28, 29] and the impurity-limited electronic thermal conductivity.[30] All of these methods give essentially the same results.

Coherence Effects

If one attempts to account for the rate of electromagnetic and acoustic absorption as well as the rate of nuclear spin relaxation in superconductors on the basis of a simple two-fluid energy-gap model, one quickly discovers inconsistencies. Experimentally the rate of acoustic absorption decreases monotonically as T decreases below T_c,[27] while the nuclear spin relaxation rate initially rises, passing through a peak before dropping to zero at low temperature.[28] If one takes the same matrix elements as in the normal state for the coupling of the excitations with phonons and with the nuclear spins, the two processes should have identical temperature dependences. Therefore, at least some of these matrix elements differ from those in the normal metal. As we shall see, the matrix elements appropriate to the superconducting state are linear combinations of those in the normal state.[8] Since the coefficients in the linear combination depend upon the nature of the coupling (scalar, vector, spin), the square of the matrix elements in the superconducting state differ for coupling the excitations to acoustic (scalar) and electromagnetic (vector) or nuclear magnetization (spin) variables.

1-2 PHENOMENOLOGICAL THEORIES

Gorter–Casimir Model

In 1934, Gorter and Casimir[2] advanced a two-fluid model along the lines which we discussed above. If x represents the fraction of electrons which are in the "normal" fluid and $(1 - x)$ the fraction condensed into the superfluid, they assumed the free energy for the electrons is of the form

$$F(x, T) = x^{1/2} f_n(T) + (1 - x) f_s(T) \tag{1-3}$$

where f_n and f_s were chosen to be

$$f_n(T) = -\tfrac{1}{2}\gamma T^2 \tag{1-4}$$

and

$$f_s(T) = -\beta = \text{const.} \tag{1-5}$$

In the normal metal the electronic free energy is just (1-4) so that the free energy of the S- and N-phases agree when $(1 - x) \rightarrow 0$, i.e., at T_c. The energy $-\beta$ represents the condensation energy associated with the superfluid. By minimizing $F(x, T)$ with respect to x for fixed T, one finds that the fraction x of "normal" electrons at a temperature T is given by

$$x = \left(\frac{T}{T_c}\right)^4 \tag{1-6}$$

From the thermodynamic relation

$$\frac{H_c^2(T)}{8\pi} = F_n(T) - F_s(T) \tag{1-7}$$

one finds from (1-3)–(1-6) the expression

$$H_c(T) = H_0\left[1 - \left(\frac{T}{T_c}\right)^2\right] \tag{1-8}$$

for the temperature-dependent critical field. Thus, H_c is predicted to be a parabolic function of (T/T_c), in rough agreement with experiment. In addition, the free energy gives the electronic specific heat in the S-phase as

$$C_{es}(T) = 3\gamma T_c\left(\frac{T}{T_c}\right)^3 \tag{1-9}$$

so that the relative jump in the electronic specific heat at T_c is 3, again in general agreement with experiment. This agreement is not completely surprising since the theory was constructed in what appears to be a rather artificial manner in order to obtain agreement with experiment. In particular, one would expect the exponent r to be unity rather than one-half if x represents the fraction of electrons which are normal. Furthermore, the condensation energy β would be expected to increase as more particles condense into the ordered phase. Nonetheless, the Gorter–Casimir theory leads to nontrivial predictions which, when combined with the London theory, are in reasonably good agreement with experiment. Unfortunately, there is little detailed connection between the expression (1-3) and that given by the microscopic theory.

The London Theory

In the year following Gorter and Casimir's work, F. and H. London advanced a phenomenological theory of the electromagnetic behavior of superconductors.[1, 31] Their scheme is based on a two-fluid type concept with superfluid and normal fluid densities n_s and n_n plus the associated velocities $\mathbf{v}_s$ and $\mathbf{v}_n$. Owing to local charge neutrality, the densities are restricted by

$n_s + n_n = n$, where n is the average number of electrons per unit volume. The super and normal fluid current densities are postulated to satisfy

$$\frac{d\mathbf{J}_s}{dt} = \frac{n_s e^2}{m}\mathbf{E} \qquad (\mathbf{J}_s = -en_s\mathbf{v}_s) \tag{1-10a}$$

$$\mathbf{J}_n = \sigma_n \mathbf{E} \qquad (\mathbf{J}_n = -en_n\mathbf{v}_n) \tag{1-10b}$$

The first of these equations is nothing more than $\mathbf{F} = m\mathbf{a}$ applied to a set of free particles of charge $-e$ and density n_s. Apparently the superfluid is unaffected by the usual scattering mechanisms which produce the finite conductivity σ_n associated with the normal fluid.

The second (and most famous) equation of the London theory is

$$\nabla \times \mathbf{J}_s = -\frac{n_s e^2}{mc}\mathbf{B} \tag{1-11}$$

This latter equation leads to the Meissner effect. One can see this by considering the curl of one of Maxwell's equations:

$$\nabla \times \nabla \times \mathbf{B} = \frac{4\pi}{c}\nabla \times J_s \tag{1-12}$$

where we have neglected the displacement current and the normal fluid current J_n since we are interested in the static Meissner effect. On combining (1-11) and (1-12) one has

$$\nabla^2\mathbf{B} = \frac{4\pi n_s e^2}{mc^2}\mathbf{B} = \frac{1}{\lambda_L{}^2}\mathbf{B} \tag{1-13}$$

where London's penetration depth λ_L is defined by

$$\lambda_L = \left(\frac{mc^2}{4\pi n_s e^2}\right)^{1/2} \tag{1-14}$$

If (1-13) is applied to a plane boundary located at $x = 0$, the magnetic field (parallel to the surface) decreases into the superconductor according to

$$B(x) = B(0)e^{-x/\lambda_L} \tag{1-15}$$

Therefore the magnetic field vanishes in the bulk of the material and one obtains perfect diamagnetism, as required. For solutions of London's equations for other geometries see London's book.[1]

To understand the relation between London's two equations, we notice that the curl of (1-10a) is the time derivative of (1-11). Therefore, outside of a constant of integration, the Meissner effect follows from the "perfect" conductivity of the superfluid, i.e., (1-10a). By postulating (1-11), the Londons added the all-important restriction that $\mathbf{B} = 0$ inside the superconductor *regardless* of its history, which is the essence of the Meissner effect.

If one combines the result (1-6) of the Gorter–Casimir model

$$(1 - x) = 1 - \left(\frac{T}{T_c}\right)^4 = \frac{n_s(T)}{n} \tag{1-16}$$

for the temperature dependence of the superfluid density, with London's expression (1-14) for the penetration depth, one finds

$$\lambda(T) = \frac{\lambda(0)}{[1 - (T/T_c)^4]^{1/2}} \tag{1-17}$$

Thus, for $T = T_c$, $\lambda = \infty$ so that no flux is excluded at T_c. As T drops infinitesimally below T_c, λ decreases rapidly, thereby establishing the Meissner effect in bulk specimens for all $T < T_c$. This temperature dependence is surprisingly close to that observed experimentally although the results of the microscopic theory are in somewhat better agreement with experiment than is (1-17).

The fact that the supercurrents are uniquely determined by the magnetic-field configuration (according to the Meissner effect) guarantees that one can apply reversible thermodynamics to quasi-static processes in superconductors, an important fact.[4]

If we introduce the vector potential $\mathbf{A}$, the second London equation (1-11) can be written as

$$\nabla \times \mathbf{J}_s = -\frac{n_s e^2}{mc} \nabla \times \mathbf{A} \tag{1-18}$$

As London pointed out, this equation can be satisfied by taking

$$\mathbf{J}_s = -\frac{n_s e^2}{mc} \mathbf{A} \tag{1-19}$$

if **A** is properly gauged. For the supercurrent to be conserved we must require

$$\nabla \cdot \mathbf{A} = 0 \tag{1-20}$$

We are, however, still free to add to **A** the gradient of any function χ which satisfies Laplace's equation $\nabla^2\chi = 0$. For an isolated simply connected body, the normal component $J_{s\perp}$ must vanish on the surface. Therefore, $A_\perp$ must also vanish on the surface. This condition determines $(\nabla\chi)_\perp$ over the entire surface and therefore determines χ to within an additive constant (which of course cannot contribute to **A** or $\mathbf{J}_s$). For a massive body these conditions ensure that $\mathbf{A} = 0$ in the bulk of the material. If current is flowing through the boundaries (e.g., when a superconductor is an element in a circuit), the current on the boundary uniquely specifies **A**. Therefore, while (1-19) does not appear to be gauge-invariant, the theory is in fact gauge-invariant since one is told to throw away any part of **A** which does not satisfy the London gauge conditions. In this way, physical predictions are independent of the choice of gauge.

In a multiply connected body, the restrictions

$$\nabla^2\chi' = 0 \qquad \left.\frac{\partial\chi'}{\partial n}\right|_{\text{surface}} = 0 \tag{1-21}$$

no longer require that an added gauge potential $\nabla\chi'$ be zero. Therefore **A** is *not* uniquely determined by the boundary condition $A_\perp = 0$, in this case. If we form the line integral of **A** around a loop surrounding a hole in the multiply connected body, Stokes's theorem gives

$$\oint \mathbf{A} \cdot d\mathbf{l} = \int \mathbf{B} \cdot d\mathbf{S} = \Phi \tag{1-22}$$

for the flux through the loop. If the path is taken within the interior of the superconductor where $\mathbf{B} = 0$, then $\nabla \times \mathbf{A} = 0$ and we may write **A** as the gradient of a scalar

$$\mathbf{A} = \nabla\chi \tag{1-23}$$

While $\nabla \chi$ must be single valued, χ will not be single valued in general, since

$$\oint \mathbf{A} \cdot d\mathbf{l} = \oint \nabla \chi \cdot d\mathbf{l} = \Delta \chi = \Phi \tag{1-24}$$

where $\Delta \chi$ is the change of χ as one travels once around the hole. By specifying the flux Φ through each hole, one can uniquely determine $\mathbf{A}$, as required.

F. London's Justification of the London Theory

F. London[5] pointed out that the equation (1-19)

$$\mathbf{J}_s = -\frac{n_s e^2}{mc} \mathbf{A} \tag{1-25}$$

could be deduced from first principles if one *assumed* that the many-body wave function Ψ_s describing the superfluid is rigid with respect to perturbations due to a transverse vector potential ($\nabla \cdot \mathbf{A} = 0$). One can see this as follows. The current density $\mathbf{J}_{s0}$ in the absence of $\mathbf{A}$

$$\mathbf{J}_{s0}(\mathbf{r}) = -\frac{e\hbar}{2mi} \sum_{j=1}^{n_s} \int (\Psi_s^* \nabla_j \Psi_s - \Psi_s \nabla_j \Psi_s^*)\, \delta(\mathbf{r}_j - \mathbf{r})\, d^3 r_1 \cdots d^3 r_{n_s} \tag{1-26}$$

clearly vanishes. If a weak magnetic field is applied to the system and Ψ_s is unaffected to first order by this perturbation, the paramagnetic current (1-26) continues to vanish, while the diamagnetic current is given by

$$\begin{aligned} \mathbf{J}_s(r) &= -\sum_{j=1}^{n_s} \frac{e^2}{mc} \mathbf{A}(\mathbf{r}) \int \Psi_s^* \Psi_s\, \delta(\mathbf{r}_j - \mathbf{r})\, d^3 r_1 \cdots d^3 r_{n_s} \\ &= -\frac{n_s e^2}{mc} \mathbf{A}(\mathbf{r}) \end{aligned} \tag{1-27}$$

in agreement with (1-25). More accurately, it is assumed that in the long wavelength limit, the paramagnetic and diamagnetic currents of the normal fluid exactly cancel each other (as they do in the Landau diamagnetism of the normal state) while the paramagnetic current of the superfluid vanishes, leaving the diamagnetic supercurrent. We have suggested above that the origin of the London "rigidity" is the energy gap in the excitation spectrum of the system. This somewhat imprecise statement is not in conflict with the fact that insulators *also* possess an energy gap in their excitation spectra. This follows since interband matrix elements of the magnetic perturbation are large in this case so that the paramagnetic current is nonzero and just cancels the diamagnetic current (see Chapter 8).

As we shall see below, the microscopic theory reduces exactly to the form (1-27) in the limit of fields which vary slowly in space.

On the basis of London's quantum interpretation of the London equations,[1] he concluded that the flux Φ trapped through a hole of a multiply connected superconducting body must be an integral multiple of $hc/e \simeq 4 \times 10^{-7}$ gauss cm^2. To understand this result, consider two concentric superconducting cylinders, as shown in Figure 1-4. Suppose that the thickness of the cylinders is large compared to the penetration depth λ and that a flux Φ is trapped within the hole of the inner cylinder. Furthermore, assume that there is no magnetic field in the region between the

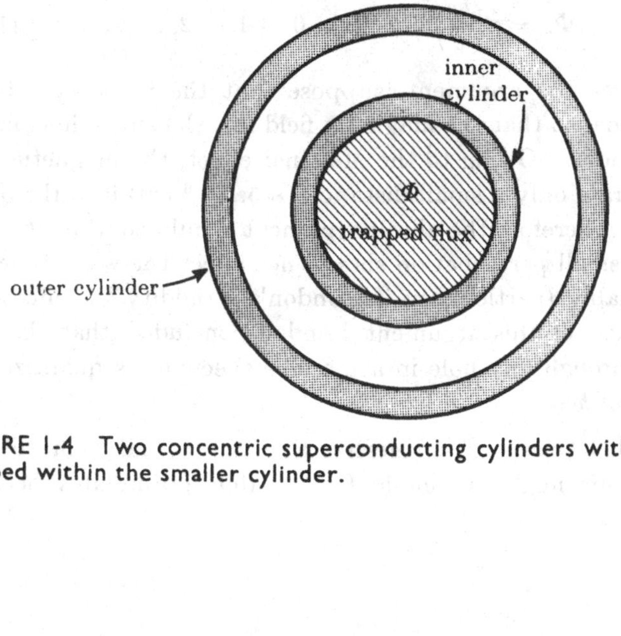

FIGURE 1-4 Two concentric superconducting cylinders with a flux Φ trapped within the smaller cylinder.

inner and outer cylinders, so that the flux through the hole of the outer cylinder is also equal to Φ. The inner cylinder acts only as a shield to ensure that no *magnetic field* touches the physically interesting outer cylinder. Let Ψ_0 be the wave function for the outer cylinder when there is no flux trapped, $\Phi = 0$. To determine the wave function Ψ_Φ in the case $\Phi \neq 0$, we note that the vector potential in the outer ring is in the θ direction and has the value

$$A_\theta(r) = \frac{\Phi}{2\pi r} = \frac{1}{r}\frac{\partial}{\partial\theta}\left(\frac{\Phi\theta}{2\pi}\right) = \nabla_\theta\left(\frac{\Phi\theta}{2\pi}\right) \tag{1-28}$$

Since **A** in the outer cylinder is the gradient of the scalar $(\Phi\theta/2\pi)$, it follows that Ψ_0 and Ψ_Φ are related by the gauge transformation

$$\Psi_\Phi = e^{-ie\Phi\sum_j \theta_j/hc}\Psi_0 \tag{1-29}$$

where θ_j is the azimuthal coordinate of the jth electron. If Ψ_Φ and Ψ_0 are to be single-valued functions of the coordinates θ_j, one must have

$$\frac{e\Phi}{hc} = \text{integer} \tag{1-30}$$

or Φ is quantized to the London values

$$\Phi_n = n\left(\frac{hc}{e}\right) \qquad (n = 0, \pm 1, \pm 2, \ldots) \tag{1-31}$$

To complete the argument, suppose that the inner cylinder is made normal so that the magnetic field fills the entire hole in the outer cylinder. Owing to the Meissner effect, the magnetic field will penetrate only a small distance ($\sim 5.10^{-6}$ cm) into the outer cylinder. Therefore the above argument should continue to hold since this small perturbation should not affect the wave function Ψ appreciably (particularly if London's "rigidity" is effective). On the basis of this argument London concluded that the flux trapped through any hole in a massive specimen is quantized to multiples of hc/e.

In 1953 Onsager[32] suggested that the actual value of the flux quantum might be one-half this value, presumably because of the effective charge of the entities making up the superfluid being $2e$. In a beautiful set of experiments carried out by Deaver and Fairbank[20a] and independently by Doll and Näbauer,[20b] Onsager's suggestion was verified. In essence the difficulty in London's argument is that there is another series of low-lying states which are distinct from London's state Ψ_n and *cannot* be generated from the ground state Ψ_0 by a gauge transformation. This second series of states, first discussed by Byers and Yang,[19] leads to the quantized flux values

$$\Phi_n = \left(n + \frac{1}{2}\right) \frac{hc}{e} \qquad (n = 0, \pm 1, +2, \ldots) \qquad (1\text{-}32)$$

By taking the London series (1-31) and the Byers–Yang series (1-32) together, one obtains the result suggested by Onsager

$$\Phi_n = n\left(\frac{hc}{2e}\right) \qquad (n = 0, \pm 1, \pm 2, \ldots) \qquad (1\text{-}33)$$

in agreement with experiment. The fact that these are the *only* allowed values for Φ follows from the BCS pairing theory since other values of Φ lead to an extraordinarily high energy of the electron system and are therefore unstable. The problem of flux quantization is discussed further in Chapter 8.

Pippard's Nonlocal Generalization of the London Theory

The basic equations (1-10) and (1-11) of the London theory are "local" in the sense that they relate the current densities and the electromagnetic potentials at the *same* point in space. On the basis of numerous experimental results, Pippard[33] concluded that these local relations must be replaced by nonlocal relations giving the currents at a given point in space as a space average of the field strengths taken over a region of extent $\xi_0 \sim 10^{-4}$ cm about the point in question. One of the most compelling arguments for this generalization is that the penetration depth λ increases appreciably if a sufficient amount of impurity is introduced into the material. This effect sets in when the mean free path l of electrons in the normal state falls below a distance ξ_0, known as Pippard's "coherence" length. As we shall see, ξ_0 is a measure of the size of the pair bound state from which the superfluid wave function is constructed. In the microscopic theory it is related to the energy gap 2Δ by $\xi_0 = \hbar v_F/\pi\Delta$, where v_F is the Fermi velocity. On the other hand, in the London theory λ is not expected to be appreciably affected by impurities, particularly near $T = 0$, where all of the electrons are condensed. In choosing a form for the nonlocal relations, Pippard was guided by Chamber's nonlocal expression,[34] relating the current density and electric field strength in the normal metal

$$\mathbf{J}(\mathbf{r}) = \frac{3\sigma}{4\pi l}\int \frac{\mathbf{R}[\mathbf{R}\cdot\mathbf{E}(\mathbf{r}')]}{R^4}\, e^{-R/l}\, d^3r' \qquad \mathbf{R} \equiv \mathbf{r} - \mathbf{r}' \tag{1-34}$$

where σ is the long wavelength electrical conductivity. Chamber's expression is a solution of Boltzmann's transport equation if the scattering mechanism is characterized by a mean free path l. For fields varying slowly over a mean free path l, (1-34) reduces to Ohm's law $\mathbf{J} = \sigma\mathbf{E}$. With Chamber's expression in mind, Pippard assumed that London's equation

$$\mathbf{J}_s(\mathbf{r}) = -\frac{1}{c\Lambda(T)}\mathbf{A}(\mathbf{r}) \qquad \frac{1}{\Lambda(T)} \equiv \frac{n_s(T)e^2}{m} \tag{1-35}$$

should be replaced by

$$\mathbf{J}_s(\mathbf{r}) = -\frac{3}{4\pi\xi_0 c\Lambda}\int \frac{\mathbf{R}[\mathbf{R}\cdot\mathbf{A}(\mathbf{r}')]}{R^4}\, e^{-R/\xi}\, d^3r' \tag{1-36}$$

The effective coherence length ξ is given by

$$\frac{1}{\xi} = \frac{1}{\xi_0} + \frac{1}{\alpha l} \tag{1-37}$$

where α is an empirical constant of order unity and ξ_0 is a length characteristic of the material. For a pure material, Pippard's equation reduces to London's equation if $\mathbf{A}(\mathbf{r})$ varies slowly over a coherence length. For an impure material, Pippard's equation leads to an extra factor $\xi/\xi_0 < 1$ multiplying $(1/c\Lambda)$ in London's equation in this long wavelength limit, thereby increasing the effective penetration depth. In most cases distances of order $\lambda \ll \xi$ are of importance in penetration phenomena so that the full reduction ξ/ξ_0 is not effective. In highly impure specimens λ is of order or greater than ξ and one has $\lambda \sim (\xi_0/l)^{1/2}$.

That the effective coherence length ξ should be bounded by the mean free path l is certainly reasonable from a physical point of view. It is a tribute to Pippard's insight into the physics of the problem that his equation is almost identical to that given by the microscopic theory.[8]

A good deal of the qualitative aspects concerning the electromagnetic properties of superconductors can be understood on the basis of a simple energy-gap model. Prior to the BCS theory, Bardeen[16c] gave a theoretical derivation of the nonlocal electrodynamics. He assumed that the single-particle matrix elements of the magnetic perturbation were unaltered by the condensation and that the single-particle excitation spectrum was altered only by adding a constant to the excitation energy, thereby creating an energy gap. Subsequent to the work of BCS, Ferrell, Glover, and Tinkham[35] employed the Kramers–Kronig relation to give quite a general discussion of how the electrodynamic behavior of a superconductor comes about, because of its energy gap. For a review of their arguments, the reader is referred to Tinkham's review article.[16f]

Ginsburg–Landau Theory

In 1950 Ginsburg and Landau[36] proposed an extension of the London theory which takes into account the possibility of the superfluid density n_s varying in space. They phrased the theory in terms of an effective wave function $\Psi(\mathbf{r})$ which we normalize such that the local density of condensed electrons is given by

$$|\Psi(\mathbf{r})|^2 = \frac{n_s(\mathbf{r})}{n} \tag{1-38}$$

where n is the total number of electrons per unit volume. Roughly speaking, $\Psi(\mathbf{r})$ corresponds to the center-of-mass wave function of the BCS pairs. Ginsburg and Landau treated $\Psi(\mathbf{r})$ as an order parameter which is to be determined at each point in space by minimizing the free-energy functional $F(\Psi, T)$ of the system. The problem is then one of guessing an appropriate form for F.

Suppose that $f(\Psi, T)$ is the difference of free energy per unit volume between the S- and N-phases when Ψ is uniform. Then it is natural to include in F the term

$$\int f[\Psi(\mathbf{r}), T]\, d^3r \tag{1-39}$$

While $f(\Psi, T)$ is not known *a priori*, Ginsburg and Landau determined this function for small Ψ (which is all that is needed when T is near T_c) by expanding f as a power series in $|\Psi|^2$ and retaining the first two nonvanishing terms; thus

$$f(\Psi, T) \cong a(T)|\Psi|^2 + \tfrac{1}{2}b(T)|\Psi|^4 \tag{1-40}$$

for $|\Psi|^2 \ll 1$. The equilibrium value $|\Psi_e|^2$ is determined by minimizing f:

$$\frac{\partial f}{\partial|\Psi|^2} = 0 = a(T) + b(T)|\Psi_e|^2 \tag{1-41}$$

and therefore

$$|\Psi_e|^2 = -\frac{a(T)}{b(T)} \tag{1-42}$$

From (1-40) and (1-42) one finds the (zero-field) free-energy difference per unit volume between the S- and N-phases is given by

$$f_S(T) - f_N(T) \equiv f(T) = -\frac{1}{2}\frac{a^2(T)}{b(T)} = -\frac{H_c{}^2(T)}{8\pi} \tag{1-43}$$

where we have used the thermodynamic relation between the critical field and the N–S free-energy difference. If we use the fact that in the London theory $\lambda^2(T) \sim 1/n_s(T)$, we obtain a second relation between $a(T)$ and $b(T)$:

$$\frac{\lambda^2(0)}{\lambda^2(T)} = \frac{|\Psi_e(T)|^2}{|\Psi_e(0)|^2} = |\Psi_e(T)|^2 = -\frac{a(T)}{b(T)} \tag{1-44}$$

From (1-43) and (1-44) we find

$$a(T) = -\frac{H_c^2(T)}{4\pi}\frac{\lambda^2(T)}{\lambda^2(0)}$$
$$b(T) = \frac{H_c^2(T)}{4\pi}\frac{\lambda^4(T)}{\lambda^4(0)} \tag{1-45}$$

and therefore $f(\Psi, T)$ given by (1-40) can be expressed in terms of experimentally measurable quantities.

If $\Psi(r)$ is not uniform in space, Ginsburg and Landau argue that extra terms should be included in F which involve the rate of change of Ψ in space. Presumably these terms would come from (a) the kinetic energy associated with extra wiggles in the many-body wave function describing n_s and/or v_s changing in space and (b) the interaction energy density being influenced by the variations of the superfluid density in a region surrounding the point in question. If $|\Psi|^2$ varies slowly in space, it should be sufficient to keep the leading term in $|\text{grad } \Psi|^2$. On the basis of gauge invariance, one would expect that this term, when combined with the effect of a vector potential $\mathbf{A}(\mathbf{r})$, would lead to a free-energy contribution of the form

$$\int \frac{n^*}{2m^*}\left|\frac{\hbar}{i}\nabla\Psi(\mathbf{r}) + \frac{e^*}{c}\mathbf{A}(\mathbf{r})\Psi(\mathbf{r})\right|^2 d^3r \tag{1-46}$$

where e^* is the effective charge of the "entities" forming the superfluid. (As we shall see, $2n^* = n$, $e^* = 2e$, and $m^* = 2m$, consistent with the pairing theory.)

By minimizing the total free-energy difference

$$F(\Psi, T) = \int \frac{n^*}{2m^*}\left|\frac{\hbar}{i}\nabla\Psi(\mathbf{r}) + \frac{e^*}{c}A(\mathbf{r})\Psi(\mathbf{r})\right|^2 d^3r + \int\left[a(T)|\Psi(\mathbf{r})|^2 + \frac{1}{2}b(T)|\Psi(\mathbf{r})|^4\right] d^3r + \int \frac{H(\mathbf{r})^2}{8\pi} d^3r \tag{1-47}$$

with respect to $\Psi(r)$, one finds the constitutive equation of the Ginsburg–Landau theory

$$\frac{\hbar^2}{2m^*}\left[\nabla + \frac{ie^*}{\hbar c}A(\mathbf{r})\right]^2 \Psi(\mathbf{r}) + \frac{H_c^2(T)}{4\pi n^*}\frac{\lambda^2(T)}{\lambda^2(0)}\left[1 - \frac{\lambda^2(T)}{\lambda^2(0)}|\Psi(\mathbf{r})|^2\right]\Psi(\mathbf{r}) = 0 \tag{1-48}$$

The current density is given by

$$\mathbf{J}_s(\mathbf{r}) = -\frac{n^*|\Psi(\mathbf{r})|^2}{m^*c} e^{*2}\mathbf{A}(\mathbf{r}) - \frac{n^*e^*\hbar}{2m^*i}\{\Psi^*(\mathbf{r})\nabla\Psi(\mathbf{r}) - \Psi(\mathbf{r})\nabla\Psi^*(\mathbf{r})\} \tag{1-49}$$

with our normalization of Ψ. As in the London theory one is to use the gauge $\nabla \cdot \mathbf{A} = 0$. Therefore, (1-48) and (1-49) together with Maxwell's equation $\nabla \times \nabla \times \mathbf{A} = 4\pi\mathbf{J}/c$ lead to two nonlinear differential equations which determine the functions $\Psi(\mathbf{r})$ and $\mathbf{A}(\mathbf{r})$.

We note that if $\mathbf{A} = 0$ and Ψ is uniform in space, (1-48) reduces to the condition

$$1 - \frac{\lambda^2(T)|\Psi|^2}{\lambda^2(0)} = 0 \tag{1-50}$$

which states that Ψ is equal to its equilibrium value (1-44), as required. If Ψ is perturbed slightly from its equilibrium value at some point, say $\mathbf{r} = 0$, then the linearized Ginsburg–Landau equation

$$\frac{\hbar^2\nabla^2}{2m^*}\tilde{\Psi}(\mathbf{r}) - \frac{H_c^2(T)}{2\pi n^*}\frac{\lambda^2(T)}{\lambda^2(0)}\tilde{\Psi} = 0 \tag{1-51}$$

for the deviation $\tilde{\Psi}(r)$ leads to

$$\tilde{\Psi} \sim \frac{e^{-r/d}}{r} \tag{1-52}$$

Thus the perturbation dies away exponentially, with the characteristic length

$$d = \left[\frac{\pi n^*\hbar^2}{m^*H_c^2(T)}\right]^{1/2}\frac{\lambda(0)}{\lambda(T)} \sim \frac{\xi_0}{[1 - T/T_c]^{1/2}} \tag{1-53}$$

where the last estimate uses the microscopic theory to relate H_0 and ξ_0. We see that even though the relation between $\mathbf{J}_s$ and $\mathbf{A}$ is approximated by a local expression, the Ginsburg–Landau theory definitely includes nonlocal effects and the coherence length appears in a natural way.

Gor'kov[37] has given a derivation of the Ginsburg–Landau theory starting from the microscopic theory. He finds that the GL wave function Ψ is proportional to the local value of the energy-gap parameter Δ. His derivation is outlined in Chapter 8.

The GL theory is particularly useful in calculations where one cannot treat the magnetic field by perturbation theory. Typical examples of such situations include thin films in strong magnetic fields, N–S phase boundaries, the intermediate state, etc. One can give a simple derivation of flux quantization on the basis of the current equation (1-49), and one finds the flux quantum to be hc/e^*. The experimentally observed value $hc/2e$ leads to the value $e^* = 2e$, as mentioned above. The GL theory has recently played an important role in explaining the magnetic behavior of so-called "hard" superconductors, which are particularly interesting materials, due to their high critical fields ($\sim 10^5$ gauss). The fundamental theoretical work in this area is due to Abrikosov,[17] who established the vortex picture to account for this new magnetic behavior. Each vortex carries one quantum of flux.

Unfortunately, the original Ginsburg–Landau theory is restricted to the temperature range $(T_c - T)/T_c \ll 1$, although it has recently been extended to all temperatures under suitable conditions by Werthamer and by Tewordt.[38]

CHAPTER 2

THE PAIRING THEORY OF SUPERCONDUCTIVITY

In analogy with a free electron gas, normal (N) metals exhibit a single-particle excitation spectrum which, in the limit of a large system, is a continuum starting at zero energy. The degeneracy associated with this spectrum leads to the linear temperature dependence of the electronic specific heat near 0°K, and to the large electrical and thermal conductivities of these materials. In the superconducting (S) phase, the single-particle excitation is radically different from that of normal metals. In superconductors a minimum energy 2Δ, called the energy gap, is required to make a single-particle excitation from the ground state.

2-1 PHYSICAL NATURE OF THE SUPERCONDUCTING STATE

This qualitative difference in the excitation spectra is paralleled by a qualitative difference in the wave functions of the N- and S-phases of metals. In the N-phase, the probability that two single-particle states i and j are simultaneously occupied is a smoothly varying function of the quantum numbers i and j. For example, in a pure single crystal, the expectation value

$$P_{kk'}{}^N = \langle N|n_{k\uparrow}n_{k'\downarrow}|N\rangle \tag{2-1}$$

is a smoothly varying function of **k** and **k**′ (so long as one does not cross the Fermi surface in varying **k** or **k**′). Here $|N\rangle$ represents a typical state in the normal phase and $n_{k\uparrow}$ is the operator which measures the number of electrons in **k** ↑, etc. (see the Appendix). In the superconducting phase,[8] the corresponding probability

$$P_{kk'}{}^S = \langle S|n_{k\uparrow}n_{k'\downarrow}|S\rangle \tag{2-2}$$

is also a smoothly varying function of **k** and **k**′ except when **k** and **k**′ are related by the "pairing" condition. This condition states that for a given state **k**, there exists a single mate $\bar{\mathbf{k}}$ such that the probability $P_{k\bar{k}}{}^S$ is larger than $P_{kk'}{}^S$ by a finite amount, for all states **k**′ in the vicinity of $\bar{\mathbf{k}}$. This singular behavior of the two-particle correlation function, which has been stressed by Yang,[39] is no doubt the sort of picture F. London had in mind when he suggested that superconductivity is due to a condensation of the electrons in momentum space.[1] When proper account is taken of residual interactions conventionally neglected in the description of the normal state, these "pairing correlations" leading to superconductivity emerge in a natural manner. Above the superconducting transition temperature, the pairing correlations are broken up by thermal fluctuations and play no important role in the normal phase.

It is essential to realize at the outset that the lowering in energy of the S-phase due to interactions between mates (say **k** ↑ and $\bar{\mathbf{k}}$ ↓) of a given pair depends critically on the choice of mates (**k**′ ↑ and $\bar{\mathbf{k}}'$↓) for other pairs. In fact, the energy gap and most of the observed properties of the superconducting phase would be *absent* were it not for strong correlations *between the pairs*. The reason for the simple BCS model working so well is that in real metals these pair–pair correlations are almost entirely due to Pauli principle restrictions rather than correlations due to true dynamical interactions between the pairs. This fact allows one to treat the system in lowest order as if dynamical interactions existed *only between mates* of a pair. The pair–pair correlations would then be accounted for by working out this reduced problem consistently with Fermi–Dirac statistics so as to include the crucial Pauli principle correlations between the pairs. We shall call this scheme the pairing (or BCS) approximation.

For a translationally invariant system, we shall see that the pairing ($\mathbf{k}\uparrow, -\mathbf{k}\downarrow$) of Bloch states leads to the lowest energy of the system. Supercurrent-carrying states are generated by translating this state of the system by an amount $\mathbf{q}/2$ in $\mathbf{k}$-space. The pairing would then be ($\mathbf{k} + \mathbf{q}/2\uparrow, -\mathbf{k} + \mathbf{q}/2\downarrow$) and the electrons would have a net drift velocity $\mathbf{v}_d = \hbar\mathbf{q}/2m$. More generally, corresponding to each physical system and each state of that system there is a choice of pairing of single-particle states which minimizes the energy (or free energy, at finite temperature). For example, in a superconductor with nonmagnetic impurities present one should pair one-electron states φ_n which include the impurity scattering potential, as Anderson[40] first pointed out. He showed that one should pair a state φ_n and its time-reversed mate $\varphi_n{}^*$ to form the ground state of the system in this case. For a uniform hollow cylinder in the absence of a magnetic field, one would pair the state (n, m, k) with its time-reversed partner $(n, -m, -k)$, where n and m are the radial and azimuthal quantum numbers, respectively, and k is the wave number for motion along the axis of the cylinder. In the presence of a magnetic field, the best pairing depends on the thickness of the cylinder and the strength of the field. For a thickness $d \gg \lambda$ (the penetration depth) one would pair $(n, m + \nu, k)$ with $(n, -m + \nu, k)$ or $(n, m + \nu, k)$ with $(n, -m + \nu + 1, k)$ depending on whether the flux trapped in the hole is an even or odd multiple of the flux quantum $hc/2e$, that is, $\nu hc/e$ or $(\nu + \frac{1}{2})hc/e$. As we shall see, these are the only allowed values of the flux trapped within a thick-walled superconducting cylinder. We shall study in detail these various possible pairings in later chapters when we apply the basic theory to physical problems.

While the above "pairing" approximation gives a good account of the single-particle excitation spectrum, there exist collective modes, such as the plasmons, arising from residual interactions neglected within this approximation. In addition there may be small momentum exciton-like collective modes which lie within the gap. For larger momentum, the exciton states rise above the gap edge and pass into the continuum, thereby becoming heavily damped. The nature of the collective states and their effect on system properties is discussed in Chapter 8.

Also neglected within the simplest pairing approximation are damping effects. In the strong-coupling superconductors, such as lead and mercury, it is essential to include these effects on the same footing as the pairing correlations to obtain a reasonable description of these "bad actors."

The above discussion suggests that the excited states of a superconductor can be represented by a two-fluid model, one for the condensed electrons and one for the excitations. As we mentioned in the introduction, phenomenological two-fluid models (most notably the Gorter–Casimir model[2] and the Ginsburg–Landau model[36]) have played an extremely important role in laying the ground work for our present understanding of superconductivity. While there are important differences between the predictions of the pairing theory and the earlier two-fluid models, the theories share the basic idea that the superfluid electrons (i.e., the strongly correlated pairs in our case) can be described by a local density $\rho_s(\mathbf{r})$ and a local flow velocity $\mathbf{v}_s(\mathbf{r})$. The excited electrons then form an interpenetrating normal fluid which in local thermal equilibrium can be described by the local quantities $\rho_n(\mathbf{r})$ and $\mathbf{v}_n(\mathbf{r})$. As we shall see, the superfluid can only carry out potential flow, that is curl $\mathbf{v}_s(\mathbf{r}) = 0$, a condition emphasized by F. London.[1] (No such restriction holds for the normal fluid.) Many of the observed properties of superconductors can be understood in terms of a two-fluid model having a temperature-dependent energy gap for creating normal fluid (excitations) from the superfluid component.[16c] As we shall see, one can often interpret the results of the microscopic theory in terms of such a model.[9]

2-2 THE ONE-PAIR PROBLEM

To understand the origin and consequence of pairing correlations, it is helpful to consider the problem, first studied by Cooper,[41] of a pair of electrons interacting above a noninteracting Fermi sea of electrons via a velocity-dependent nonretarded two-body potential V. Thus, all but two of the electrons are assumed to be noninteracting. The background electrons enter the total problem only through the Pauli principle by blocking states below the Fermi surface from participating in the remaining two-particle problem. If we measure the kinetic energy ϵ_k relative to its value at the Fermi surface, only states with $\epsilon_k > 0$ are available to the interacting pair of electrons. Since the system is assumed to be translationally invariant and one neglects spin-dependent forces, the center-of-mass momentum $\hbar\mathbf{q}$ of the pair and the total spin S are constants of motion. The orbital wave function of the pair can then be written as

$$\psi(\mathbf{r}_1, \mathbf{r}_2) = \varphi_q(\boldsymbol{\rho})e^{i\mathbf{q}\cdot\mathbf{R}} \tag{2-3}$$

where the relative and center-of-mass coordinates are defined by $\boldsymbol{\rho} = \mathbf{r}_1 - \mathbf{r}_2$ and $\mathbf{R} = (\mathbf{r}_1 + \mathbf{r}_2)/2$, respectively. The relative coordinate wave function is symmetric for the singlet spin state ($S = 0$) and antisymmetric for the triplet states ($S = 1$). In the limit $\mathbf{q} \to 0$ the relative coordinate problem is spherically symmetric so that $\varphi(\boldsymbol{\rho})$ is an eigenfunction of angular momentum and can be labeled by the angular momentum quantum numbers l and m. For $\mathbf{q} \neq 0$, the component of angular momentum along $\mathbf{q}$ and parity remain good quantum numbers but l is no longer sharp.

For simplicity, we first consider the zero momentum states $\mathbf{q} = 0$, so that ψ can be expanded as

$$\psi(\mathbf{r}_1, \mathbf{r}_2) = \varphi(\boldsymbol{\rho}) = \sum_k a_k e^{i\mathbf{k}\cdot\boldsymbol{\rho}} = \sum_k a_k e^{i\mathbf{k}\cdot\mathbf{r}_1} e^{-i\mathbf{k}\cdot\mathbf{r}_2} \tag{2-4}$$

In (2-4), the sum is restricted to the available states ($\epsilon_k > 0$). Since the factors $e^{i\mathbf{k}\cdot\mathbf{r}_1}$ and $e^{-i\mathbf{k}\cdot\mathbf{r}_2}$ can be thought of as single-particle states of momentum $\mathbf{k}$ and $-\mathbf{k}$, we see that the pair wave function is a superposition of configurations in each of which a definite pair state $(\mathbf{k}, -\mathbf{k})$ is occupied.

To find the zero-spin eigenstates of the pair, we write Schrödinger's equation as

$$(W - H_0)\psi = V\psi \tag{2-5a}$$

and from (2-4) one has

$$(W - 2\epsilon_k)a_{\mathbf{k}} = \sum_{k'} V_{\mathbf{kk'}} a_{\mathbf{k'}} \tag{2-5b}$$

where the matrix element $V_{\mathbf{kk'}}$ is defined by

$$V_{\mathbf{kk'}} = \langle \mathbf{k}, -\mathbf{k}|V|\mathbf{k'}, -\mathbf{k'}\rangle \tag{2-6}$$

In Figure 2-1 a typical scattering process caused by V is illustrated.

While the Schrödinger equation (2-5b) cannot be analytically solved in general, the solution is immediate if $V_{\mathbf{kk'}}$ is taken to be a factorizable potential $V_{\mathbf{kk'}} = \lambda w_{\mathbf{k'}}^* w_{\mathbf{k}}$. More generally, if the system is isotropic $V_{\mathbf{kk'}}$ can be expanded into its partial wave components

$$V_{\mathbf{kk'}} = \sum_{l=0}^{\infty} \sum_{m=-l}^{l} V_l(|\mathbf{k}|, |\mathbf{k'}|) Y_l^m(\Omega_{\mathbf{k}}) Y_l^{-m}(\Omega_{\mathbf{k'}}) \tag{2-7}$$

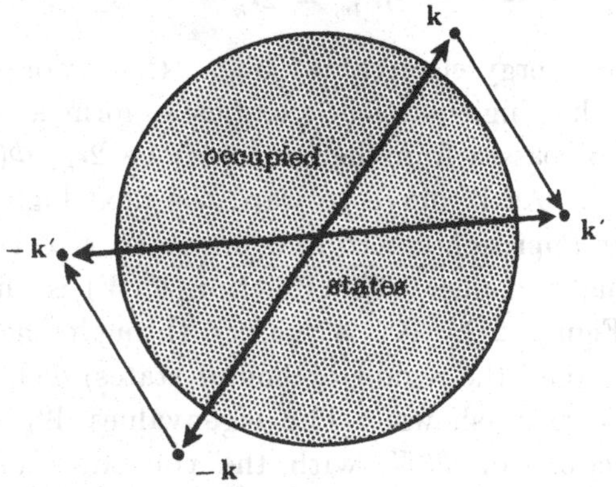

FIGURE 2-1 A typical transition occurring in Cooper's problem in which one pair of electrons interacts above a quiescent Fermi sea. The center-of-mass momentum of the pair is chosen to be zero in this drawing.

and the (l, m) eigenstates of the pair can be determined if V_l is taken to be factorizable,

$$V_l(|\mathbf{k}|, |\mathbf{k}'|) = \lambda_l w_k{}^l w_{k'}{}^{l*} \tag{2-8}$$

In this case we have from (2-5b)

$$(W_{lm} - 2\epsilon_k)a_k = \lambda_l w_k{}^l \sum_{k'} w_{k'}{}^{l*} a_{k'} \tag{2-9a}$$

where

$$a_{\mathbf{k}} = a_k Y_l{}^m(\Omega_{\mathbf{k}}) \tag{2-9b}$$

Equation (2-9a) can be written as

$$a_k = \frac{\lambda_l w_k{}^l C}{W_{lm} - 2\epsilon_k} \tag{2-10a}$$

where the constant C is defined as

$$C = \sum_{\mathbf{k}'} w_{k'}{}^{l*} a_{k'} \tag{2-10b}$$

By substituting (2-10a) into the definition (2-10b) one obtains the equation

$$1 = \lambda_l \sum_k |w_k{}^l|^2 \frac{1}{W_{lm} - 2\epsilon_k} \equiv \lambda_l \Phi(W_{lm}) \tag{2-11}$$

determining the energy eigenvalues W_{lm}. If we work in a large but finite box the single-particle energies ϵ_k form a discrete set so that when W passes from below to above $2\epsilon_k$, $\Phi(W)$ jumps from $-\infty$ to ∞. As W moves toward the next higher value of $2\epsilon_k$, $\Phi(W)$ again approaches $-\infty$ and jumps to $+\infty$ as W passes through this higher value. The function $\Phi(W)$ is shown schematically in Figure 2-2. As W passes through the origin to negative values (i.e., the region of bound states) $\Phi(W)$ increases from $-\infty$ to zero as shown. The eigenvalues W_{lm} are given by the intersections of $\Phi(W)$ with the constant function $1/\lambda_l$, as shown for both positive (repulsive) and negative (attractive) λ_l. While the eigenvalues in the continuum are trapped between the unperturbed energies $2\epsilon_k$ and approach the unperturbed energies as the size of the box goes to infinity, a state is bound split off from the continuum for an attractive l-wave potential, as shown in the figure. For the simple case

$$w_k{}^l = \begin{cases} 1 & 0 < \epsilon_k < \omega_c \\ 0 & \text{otherwise} \end{cases} \tag{2-12a}$$

and $\lambda_l < 0$, the binding energy $|W_{lm}|$ of the pair in the split-off state is given by

$$\frac{1}{|\lambda_l|} = \frac{N(0)}{2} \log \left[\frac{|W_{lm}| + 2\omega_c}{|W_{lm}|} \right] \tag{2-12b}$$

or

$$|W_{lm}| = \frac{2\omega_c}{\exp\left[\dfrac{2}{N(0)|\lambda_l|}\right] - 1} \tag{2-12c}$$

We have assumed the density of states $N(\epsilon_k)$ is slowly varying in the interval $0 < \epsilon_k < \omega_c$ and have approximated it by $N(0)$, the density of single-electron states of one-spin orientation, evaluated at the Fermi surface. From (2-12c) one has for weak coupling $[N(0)|\lambda_l| \ll 1]$

$$|W_{lm}| \cong 2\omega_c \exp\left[-\frac{2}{N(0)|\lambda_l|}\right] \tag{2-13a}$$

while for strong coupling $[N(0)|\lambda_l| \gg 1]$ one obtains

$$|W_{lm}| \simeq N(0)|\lambda_l|\omega_c \tag{2-13b}$$

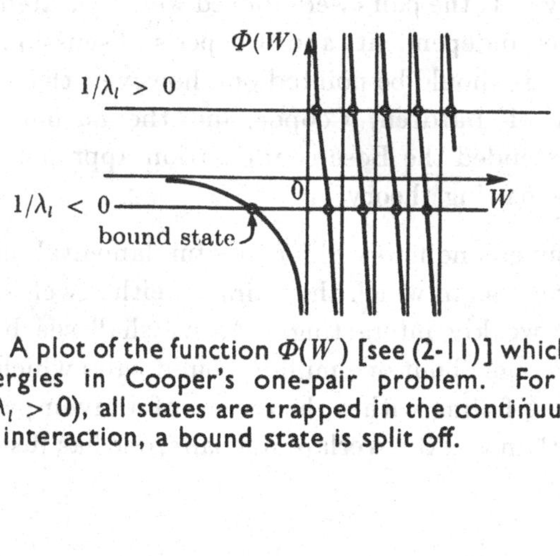

FIGURE 2-2 A plot of the function $\Phi(W)$ [see (2-11)] which determines the eigenenergies in Cooper's one-pair problem. For a repulsive interaction ($\lambda_l > 0$), all states are trapped in the continuum, while for an attractive interaction, a bound state is split off.

From (5-13a) we see that the binding energy is an extremely sensitive function of the coupling strength for weak coupling; however, a bound state exists for arbitrarily weak coupling so long as the potential is attractive near the Fermi surface. This important result was discovered by Cooper,[41] who suggested that the instability of the normal phase, because of pairs of electrons entering this type of bound state, was associated with the occurrence of the superconducting phase.

Earlier work of Schafroth, Blatt, and Butler (SBB)[42] is closely related to Cooper's discussion. Schafroth[43] had suggested that the superconducting state might correspond to a Bose–Einstein condensation of pairs of electrons into localized bound states. An attempt to develop a theory along these lines was made by SBB using what they call the quasi-chemical equilibrium approach for evaluating the partition function of the system. Owing to mathematical difficulties, they were not able to carry out calculations based on their general formulation for any model which exhibited superconducting properties. For a qualitative picture they suggested a model with localized pairs such that the size of the pair bound state is small compared to the average spacing between pairs. The bound pairs would presumably be capable of translational motion relative to the other pairs and one would obtain a continuum of Bose–Einstein excitations above the ground state without an energy gap, in contrast with the pairing theory. If the pairs were indeed well separated they could be treated as independent and Cooper's discussion would be appropriate. It should be pointed out, however, that, subsequent to the work of Bardeen, Cooper, and the author, Blatt and Matsubara extended the Bose condensation approach to give the results of the pairing theory.

Actual superconductors differ in a fundamental manner from a bound pair model in which the pairs are either well separated in space and/or weakly interacting. As we shall see below, there are on the average about one million bound pairs which have their centers of mass falling within the extent of a given pair function. Thus, rather than weakly overlapping pairs, one has just the reverse limit, very *strongly* overlapping pairs. As we mentioned above, it is surprising that one can meaningfully treat such a system in zero order by including only dynamical interactions between mates of a pair and neglect all but the Pauli principle restrictions when treating interactions between the pairs. It is intended that the discussion below and in the following chapters will help clarify this point.

Returning to the one-pair problem of Cooper, it is interesting to see how the energy of the bound state varies with the center-of-mass momentum $\hbar\mathbf{q}$. If we assume only the s-wave ($l = 0$) part of V is important (as appears to be the case except for the crystalline anisotropy effects), one finds that the binding energy $W(\mathbf{q})$ satisfies

$$1 = |\lambda_0| \sum_{\mathbf{k}} \frac{1}{|W_q| - \epsilon_{\mathbf{k}+\mathbf{q}/2} - \epsilon_{\mathbf{k}-\mathbf{q}/2}} \tag{2-14}$$

where $|\mathbf{k} + \mathbf{q}/2|$ and $|\mathbf{k} - \mathbf{q}/2|$ are required to be greater than k_F, and the sum is to be cut off at $\epsilon_k = \omega_c$. For small q one finds

$$|W_q| = |W_0| - \frac{v_F \hbar q}{2} \tag{2-15}$$

where

$$|W_0| = \frac{2\omega_c}{\left\{\exp\left[\dfrac{2}{N(0)|\lambda_l|}\right] - 1\right\}}$$

as above. Thus, the pair energy increases linearly with the center-of-mass momentum in the limit $\mathbf{q} \to 0$, rather than as q^2, as one might expect. As Cooper pointed out, the drift of the pair with respect to the noninteracting Fermi sea strongly reduces the binding energy of the pair due to the reduced density of low-energy states available to the pair. This effect dominates the q^2 increase of kinetic energy for small q.

If $|W_0|$ is imagined to be of order kT_c, (2-15) shows that the pair would have lost most of its binding energy when

$$q \sim \frac{k_B T_c}{\hbar v_F} \sim \frac{k_B T_c}{E_F} k_F \sim 10^{-4} k_F \sim 10^4 \text{ cm}^{-1} \tag{2-16}$$

This number is roughly numerically equal to the reciprocal of

Pippard's coherence length $\xi_0 \sim 10^{-4}$ cm,[33] about which we shall have more to say. If one calculates the pair function $\varphi(\boldsymbol{\rho})$ from (2-4) and (2-10a) one finds the size of the bound state is of order ξ_0. Thus, one would be required to have an extremely small density of bound pairs if an isolated pair model were to be appropriate. In fact, the density would be so small that the predicted N–S energy difference at zero temperature would be many orders of magnitude too small to agree with experiment.

Thus far we have considered only the singlet spin state of the bound pair. If there is a strong attractive odd l potential, the triplet state will have the largest binding energy and one might expect the pairing in the superconducting state to be in a triplet state. There is no experimental evidence to support other than singlet pairing at present.

In closing this section we emphasize that the single-pair model exhibits a continuous spectrum above the ground state, without an energy gap.

2-3 LANDAU'S THEORY OF A FERMI LIQUID

Looking back at Cooper's argument, one might raise several objections to the conclusion that the bound state in the two-particle problem has anything to do with the occurrence of superconductivity. For example, one knows that Coulomb and phonon interactions between electrons in the normal state lead to a correlation energy of order one electron-volt per electron,[44] compared with the negligibly small binding energy $W \simeq 10^{-4}$ ev of a bound pair. Is it not possible that the strong correlations between all the electrons in the normal state will necessarily lead to fluctuations which break up the weakly bound state of a given pair of electrons? In addition, even if such a bound state could exist in a metal, would not the strong overlap of the pairs required to fit the observed condensation energy lead to interactions which would destroy the concept of bound pairs?

In answering the first objection, it is important to recognize that Landau's theory of a Fermi liquid[45] gives a good account of the low-lying single-particle excitations of the normal state. In this theory, the excited states of a normal metal are placed in one-to-one correspondence with those of a free-electron gas. In Landau's theory, one asserts that the essential effect of the interactions between the electrons in the normal state of a metal is to shift the effective mass of an electron (now called a "quasi-particle") by an amount which is observed to be of order 10 to 50 per cent. An important feature of the theory is that a quasi-particle, as opposed to a "bare" electron, is a stable excitation in the immediate vicinity of the Fermi surface (at sufficiently low temperature). There is, however, a coupling between quasi-particles which arises from interactions neglected within this Fermi liquid approximation. This residual coupling leads to superconductivity.

The basis for Landau's theory has been extensively investigated and one knows that the theory is correct in all orders of perturbation theory[46] starting from the noninteracting system. The theory no doubt has a wider range of validity than that of the perturbation series itself, although its exact limitations are not known at present. Empirically, Landau's theory works very well in the normal state.

The remarkably small energy difference between the normal and superconducting states of a metal (10^{-8} ev per electron) strongly suggests that there is only a subtle shift of the electron–electron correlations between the two states. Since Landau's theory gives a good account of the normal state, it is reasonable to use the complete set of wave functions given by this theory as a basis for constructing the wave functions of the superconducting state. This procedure is particularly appealing because the superconducting wave functions primarily involve normal state configurations in which quasi-particle excitations are present only near the Fermi surface. However, these are just the configurations which are best described by the Landau theory. Therefore, Cooper's result is to be understood in the sense that his two-particle problem is actually a two-quasi-particle problem.

A difficulty with the above approach is that one knows little about the interaction between quasi-particles in the normal phase from an experimental point of view. While the quasi-particle's effective mass involves the forward scattering amplitude of two quasi-particles, the effective mass also involves band-structure effects which are difficult to estimate accurately. More important is the fact that one needs the interaction for finite momentum and energy transfer so that one is forced to estimate the interaction theoretically. While this problem is not completely settled at present, it appears that one understands the essential features of the interaction, the remaining complications arising primarily from detailed crystalline effects (see Chapter 7).

In regard to the second objection mentioned above, it is true that the simple picture of bound pairs of electrons forming the ground state of the superconductor is impaired by their great overlap. Nevertheless there remains the strongly correlated occupancy of a given quasi-particle state (say $\mathbf{k}\uparrow$) with its mate (say $-\mathbf{k}\downarrow$) as in Cooper's problem. Interactions between quasi-particles which tend to break up this correlated occupancy are presumably already included in Landau's description of the normal state. Thus a simplified model in which one includes pairing correlations between otherwise noninteracting quasi-particles is not at the outset an unreasonable starting point. It is this point of view Bardeen, Cooper, and the author took in constructing the microscopic theory of superconductivity.

2-4 THE PAIRING APPROXIMATION

We saw above that for a translationally invariant normal system carrying no current, the $\mathbf{q} = 0$ pair state is the most unstable, in the sense that it is the pair with the largest binding energy W. In Chapter 7 a time-dependent treatment of the normal state instability is given and one finds the greatest growth rate is for $\mathbf{q} = 0$ pairs, in agreement with Cooper's result. Thus it is natural to solve the reduced problem in which interactions are considered only between electrons of opposite momentum. One hopes that the resolution of the strongest instability will modify the system so as to remove the $\mathbf{q} \neq 0$ pair instabilities as well. This is exactly what happens. We restrict our attention to pairing electrons of opposite spin orientation.

We shall use the formalism of second quantization to describe the interacting electron system; this scheme is reviewed in the Appendix. The creation and destruction operators for electrons of wavevector **k** and z-component of spin s ($\uparrow$ or $\downarrow$) are denoted by $c_{ks}{}^{+}$ and c_{ks}, respectively. They satisfy the usual Fermi anticommutation rules. The Hamiltonian for the reduced problem of the $\mathbf{q} = 0$ pairs is

$$H_{\text{red}} = \sum_{ks} \epsilon_k n_{ks} + \sum_{kk'} V_{k'k} b_{k'}{}^{+} b_k \tag{2-17}$$

where the pairing matrix element $V_{k'k}$ is given by

$$V_{k'k} = \langle k', -k' | V | k, -k \rangle \tag{2-18}$$

and the operator $b_k{}^+$ creates a pair of electrons in the single-particle states $k\uparrow$ and $-k\downarrow$, that is,

$$\begin{aligned} b_k{}^{+} &= c_{k\uparrow}{}^{+} c_{-k\downarrow}{}^{+} \\ b_k &= c_{-k\downarrow} c_{k\uparrow} \end{aligned} \tag{2-19}$$

This type of Hamiltonian forms the basis of the theory of superconductivity proposed by Bardeen, Cooper, and the author.[8] Further argument for concentrating on these particular interactions in describing the ground state and the low-lying excited states are given in the original paper by Bardeen, Cooper, and the author and in a review article by Bardeen and the author.[9] There it is argued, on the basis of phase-space considerations as well as effects due to the antisymmetry of the wave functions, that the $\mathbf{q} = 0$ pair state should be macroscopically occupied in cases where the superfluid momentum density is zero (although $\mathbf{v}_s$ need not be zero in the presence of magnetic fields). We expect the matrix element $V_{k'k}$ to be predominantly negative near the Fermi surface for superconductivity to occur. As we shall see in Chapter 7, this attraction is due to the ions overscreening the Coulomb repulsion, thereby reversing the sign of the effective interaction. While (2-17) is written in terms of bare single-particle operators c_{ks}, the reduced Hamiltonian can be formally viewed as a model Hamiltonian describing residual interactions ($V_{k'k}$) between the quasi-particles in the normal phase as discussed above. Since the pairing correlations constitute a fractional change of only $\sim 10^{-8}$ in the total correlation energy of a metal, it is clear that this more liberal interpretation of H_{red} must be adopted. When the quasi-particle picture of the normal phase excitations is insufficient, as in lead and mercury, other techniques must be used in treating the superconducting phase (see Chapters 6 and 7). Note that the form of H_{red} depends upon the choice of pairing, as discussed in Section 2-1.

Since the pairing interaction maintains the pairing condition, it follows that the eigenstates of H_{red} can be labeled by those states **k**, s which are occupied, their mates $-\mathbf{k}$, $-\mathbf{s}$ being unoccupied. This labeling of states leads to a one-to-one correspondence of the eigenstates of H_{red} with the eigenstates of a noninteracting Fermi gas (or the normal state). As we shall see, if excitations happen to be in states $\mathbf{k}\uparrow$ *and* $-\mathbf{k}\downarrow$, special care must be taken so that these configurations are properly orthogonal to the ground-state wave function.

It is clear that if V is attractive, the ground state of H_{red} has no pair state ($\mathbf{k}\uparrow$, $-\mathbf{k}\downarrow$) occupied by a single electron. In this case the operator $n_{k\uparrow} + n_{-k\downarrow}$ can be replaced by $2b_k^+ b_k$ that is, twice the pair occupation number. The reduced Hamiltonian is then

$$H_{\text{red}}{}^0 = \sum_k 2\epsilon_k b_k{}^+ b_k + \sum_{k,k'} V_{k'k} b_{k'}{}^+ b_k \tag{2-20}$$

It might be argued that eigenstates of $H_{\text{red}}{}^0$ follow immediately by forming new operators B_n which are linear combinations of the b_k's such that H_{red} is of the form $\sum_n \tilde{\epsilon}_n B_n{}^+ B_n$. This argument is incorrect. If the operators b_k and $b_k{}^+$ described true Bose particles (rather than pairs of fermions) the B_n's and $B_n{}^+$'s would also describe Bose particles and the ground state would be formed by placing all the bosons in the lowest state. Rather, one finds by direct calculation the commutation relations

$$[b_k, b_{k'}{}^+] = 0 \qquad \text{for } \mathbf{k} \neq \mathbf{k}' \tag{2-21a}$$

$$[b_k, b_{k'}{}^+] = 1 - (n_{k\uparrow} + n_{-k\downarrow}) \qquad \text{for } \mathbf{k} = \mathbf{k}' \tag{2-21b}$$

and

$$[b_k, b_{k'}] = 0 = [b_k{}^+, b_{k'}{}^+] \tag{2-21c}$$

These are not of the form required by Bose–Einstein statistics. The factor $(n_{k\uparrow} + n_{-k\downarrow})$ in (2-21b) represents the effect of the

Pauli principle acting on the individual electrons forming the pair. Perhaps it is simplest to view the "pairon" operators b_k and b_k^+ as satisfying Bose–Einstein statistics for $\mathbf{k} \neq \mathbf{k}'$ and satisfying the Pauli principle $b_k^{+2} = 0 = b_k^2$ for $\mathbf{k} = \mathbf{k}'$. It is the latter relation which ruins a simple Bose gas picture.[47]

In attempting to find a variational estimate of the ground-state energy and wave function of H_{red}, the author tried to adapt the intermediate coupling approximation of Tomonaga, familiar in the coupled meson–nucleon problem and the polaron problem.[48, 49] In these problems one assumes that successive bosons (mesons or phonons) are emitted into the *same* orbital state (about the proton or the electron, respectively). The form of the orbital state φ and the weight A_N of the $N/2$ boson state are then determined by minimizing the system energy. Lee, Low, and Pines[49a] simplified the procedure for the polaron by assuming what is equivalent to a parameterized form for the weights A_N. Their wave function, after a canonical transformation has been made to eliminate the electron's coordinate from the problem, is

$$|\psi_0\rangle \alpha \prod_k e^{g_k(a_k^+ + a_{-k})}|0\rangle \tag{2-22}$$

where the a^+'s are phonon creation operators. The function g_k is essentially the Fourier transform of the orbital wave function φ of the phonons surrounding the electron.

The application of this physical idea to superconductivity is complicated by several features. First, the "pairon" operators do not truly satisfy Bose statistics and, second, the number of electrons is a definite number N_0 in our system, rather than being a probability distribution $|A_N|^2$ of finite width about the value N_0. The author tried to describe the ground state of H_{red} by

$$|\psi_0\rangle \alpha \prod_k e^{g_k b_k^+} |0\rangle = \prod_k (1 + g_k b_k^+)|0\rangle \tag{2-23}$$

where we have used the fact that $b_k^{+2} = 0$ in expanding out the exponential. The normalization integral is easily seen to be

$$\langle\psi_0|\psi_0\rangle = \prod_k (1 + |g_k|^2) \tag{2-24}$$

so that

$$|\psi_0\rangle = \prod_k \frac{1 + g_k b_k^+}{(1 + |g_k|^2)^{1/2}} |0\rangle \tag{2-25}$$

is a properly normalized state. By expanding out the infinite product one sees that $|\psi_0\rangle$ has a nonvanishing amplitude for all even numbers of electrons, 0, 2, 4, ... However, by choosing g_k appropriately the mean number of particles described by $|\psi_0\rangle$ can be adjusted to be the required number N_0. As in the grand canonical ensemble, one can show that the width of the distribution is of order $N_0^{1/2}$ so that particle number fluctuations cause no difficulty in a large system.

Since we want to minimize the ground-state energy subject to the constraint

$$\langle\psi_0|N_{op}|\psi_0\rangle \equiv \langle\psi_0| \sum_{ks} n_{ks}|\psi_0\rangle = N_0 \tag{2-26}$$

we use the Lagrange multiplier scheme and minimize

$$\delta W = \delta\langle\psi_0|H_{\text{red}} - \mu N_{op}|\psi_0\rangle = 0 \tag{2-27}$$

On combining (2-20) and (2-27) one finds the quantity to be minimized is[8]

$$W = \sum_k 2(\epsilon_k - \mu)v_k^2 + \sum_{k,\,k'} V_{k'k} u_k v_k u_{k'} v_{k'} \tag{2-28}$$

where u_k and v_k are defined by

$$u_k = \frac{1}{(1 + g_k^2)^{1/2}} \tag{2-29a}$$

and

$$v_k = \frac{g_k}{(1 + g_k^2)^{1/2}} \tag{2-29b}$$

thus

$$u_k^2 + v_k^2 = 1 \tag{2-29c}$$

We have assumed phases are chosen so that $V_{k'k}$ and g_k are real quantities. On minimizing W one finds

$$u_k^2 = \frac{1}{2}\left(1 + \frac{\epsilon_k - \mu}{E_k}\right) \tag{2-30a}$$

$$v_k^2 = \frac{1}{2}\left(1 - \frac{\epsilon_k - \mu}{E_k}\right) \tag{2-30b}$$

and

$$u_k v_k = \frac{\Delta_k}{2E_k} \tag{2-30c}$$

where E_k is defined by

$$E_k = +[(\epsilon_k - \mu)^2 + \Delta_k{}^2]^{1/2} \tag{2-30d}$$

As we shall see, E_k turns out to be the energy required to create a quasi-particle of momentum **k** in the superconducting state. The "energy-gap" parameter Δ_k satisfies the integral equation

$$\Delta_k = -\sum_{k'} V_{kk'} \frac{\Delta_{k'}}{2E_{k'}} \tag{2-30e}$$

One must simultaneously solve (2-30e) and the constraint condition

$$\langle\psi_0|N_{op}|\psi_0\rangle = 2\sum_k v_k{}^2 = N_0 \tag{2-31}$$

to determine Δ_k and μ. If the single-particle energy ϵ_k is measured relative to the Fermi energy in the normal state, μ is just the shift of the chemical potential between the normal and superconducting states. For a system possessing particle-hole symmetry in the vicinity of the Fermi surface, one finds $\mu = 0$. In general, this is an excellent approximation and we shall assume $\mu = 0$ in solving for Δ_k. Once the energy-gap equation (2-30e) is solved, one can obtain the energy difference $W_N - W_S$ between the N- and S-states by inserting the expressions for u_k and v_k back into (2-28). An explicit solution of (2-30e) is easily obtained if $V_{kk'}$ is approximated by the s-wave potential

$$V_{kk'} \equiv \begin{cases} -V < 0 & \text{for } |\epsilon_k| \text{ and } |\epsilon_{k'}| < \omega_c \\ 0 & \text{otherwise} \end{cases} \tag{2-32}$$

so that $V_{kk'}$ is attractive in a shell of width $2\omega_c$ centered at the Fermi surface. In this case one finds

$$\Delta_k = \begin{cases} \Delta_0 & \text{for } |\epsilon_k| < \omega_c \\ 0 & \text{otherwise} \end{cases} \tag{2-33}$$

where

$$\Delta_0 = \frac{\omega_c}{\sinh\left[\dfrac{1}{N(0)V}\right]} \simeq 2\omega_c \exp\left[-\frac{1}{N(0)V}\right] \tag{2-34}$$

The right-hand equality holds in the weak-coupling limit $N(0)V \gtrsim \frac{1}{4}$. By substituting this result into the expression for the ground-state energy (2-28) and subtracting the ground-state energy of the normal phase (i.e., the unperturbed Fermi sea in this model) one finds the condensation energy

$$W_N - W_S = \frac{1}{2} N(0)\,\Delta_0{}^2 \cong 2N(0)\omega_c{}^2 \exp\left[-\frac{2}{N(0)V}\right] \tag{2-35}$$

Since thermodynamics[4] gives the relation

$$W_N - W_S = \frac{H_0{}^2}{8\pi} \tag{2-36}$$

where H_0 is the critical magnetic field for destroying superconductivity at zero temperature, we find

$$H_0 = 2[\pi N(0)]^{1/2}\,\Delta_0 \tag{2-37}$$

By using experimental values of $N(0)$ and Δ_0 one finds values of H_0 which are in reasonably good agreement with experiment.[9, 16]

We note that the condensation energy (2-35) is not an analytic function of the coupling constant $N(0)V$ so that a perturbation treatment starting from the normal phase could not give this result unless one sums an infinite number of graphs of a selected class.[50]

Returning to the ground-state wave function (2-23), we would expect on the basis of the Tomonaga scheme that the projection of $|\psi_0\rangle$ onto the N-particle space would lead to a function (in the coordinate representation) of the form

$$\begin{aligned} \langle r_1, s_1; r_2, s_2; \cdots r_N, s_N|\psi_0\rangle &\equiv \psi_{0N} \\ &= \mathscr{A}\varphi(\mathbf{r}_1 - \mathbf{r}_2)\chi_{12}\varphi(\mathbf{r}_3 - \mathbf{r}_4)\chi_{34}\cdots\varphi(\mathbf{r}_{N-1} - \mathbf{r}_N)\chi_{N-1,N} \end{aligned} \tag{2-38}$$

The function φ is the relative coordinate wave function of a pair (*the same function for all pairs*) and χ_{ij} is corresponding spin function $\uparrow(i)\downarrow(j)$. Thus within the pairing approximation all pairs are in the *same* state in the ground-state wave function. The operator $\mathscr{A}$ in (2-38) antisymmetrizes the entire function.

This result, first noted by Dyson, can be obtained by taking the inner product of $|\psi_0\rangle$ with the basis vector

$$|\mathbf{r}_1, s_1; \mathbf{r}_2, s_2; \cdots \mathbf{r}_N, s_N\rangle = \psi_{s_1}{}^+(\mathbf{r}_1)\psi_{s_2}{}^+(\mathbf{r}_2)\cdots\psi_{s_N}{}^+(\mathbf{r}_N)|0\rangle$$

and expanding the ψ^+'s in terms of the creation operators $c_{ks_i}{}^+$ by

$$\psi_{s_i}{}^+(\mathbf{r}_i) = \sum_k e^{-i\mathbf{k}\cdot\mathbf{r}_i} c_{ks_i}{}^+ \tag{2-39}$$

The N-particle state (2-38) has been discussed by Blatt[51a] in the case where φ is a general function of $\mathbf{r}_1$ and $\mathbf{r}_2$. This generalization corresponds to considering pairings between states other than $\mathbf{k}\uparrow$ and $-\mathbf{k}\downarrow$.

The orbital function φ in (2-38) is given by

$$\varphi(\boldsymbol{\rho}) = \sum_k g_k e^{i\mathbf{k}\cdot\boldsymbol{\rho}} \tag{2-40}$$

thus g_k is the Fourier transform of $\varphi(\boldsymbol{\rho})$, as stated above. While there is a formal similarity between (2-38) and the wave function for a condensed Bose–Einstein gas of pairs of electrons with opposite spin, the antisymmetrization operator $\mathscr{A}$ is all important in real superconductors. In fact, the (unnormalized) ground state of the noninteracting Fermi gas can be written in this form with

$$g_k = \begin{cases} 1 & |k| < k_f \\ 0 & |k| < k_f \end{cases} \tag{2-41}$$

so that antisymmetrization removes the correlations between opposite spin electrons implied by φ in this case. In the superconducting state g_k differs from (2-41) only for values of $\mathbf{k}$ in the immediate vicinity of the Fermi surface. This difference is reflected in a long-range tail of $\varphi(\boldsymbol{\rho})$ which increases the probability of two antiparallel spin electrons being near each other out to a range $\xi_0 \equiv \hbar v_F/\pi\Delta_0 \sim 10^{-4}$ cm, that is, Pippard's coherence length. As we mentioned above, this quasi-bound state has such a long range in space that on the average about 10^6 other pairs have their centers of mass in this region. [In this estimate electrons deep within the Fermi sea have not been counted since they behave essentially as if the material were in the normal phase.] Thus an isolated pair picture has little meaning here.

2-5 QUASI-PARTICLE EXCITATIONS

To find the excited states of the BCS reduced Hamiltonian (2-17) we consider adding an electron to the system in the state $\mathbf{p}\uparrow$ (its mate $-\mathbf{p}\downarrow$ being empty). The only effect of this process is to block the pair state $(\mathbf{p}\uparrow, -\mathbf{p}\downarrow)$ from participating in the pairing interaction (due to the Pauli principle). Since $-\mathbf{p}\downarrow$ is assumed to be empty, the electron in $\mathbf{p}\uparrow$ cannot be scattered out of this state, due to the form of the pairing interaction (2-20). Of course, residual interactions not explicitly included in H_{red} will allow this process to take place; however, these interactions appear to have a small effect on the excitation spectrum (since they are implicitly included in the quasi-particles of the normal state).

The quasi-particle energy is defined to be the total excitation energy of the system when the extra electron is added to the system. From (2-28) we see that by deleting the pair state $(\mathbf{p}\uparrow, -\mathbf{p}\downarrow)$, the energy of the interacting pairs is increased by

$$-2\epsilon_p v_p^2 - 2\left[\sum_{k'} V_{pk'} u_{k'} v_{k'}\right] u_p v_p \tag{2-42}$$

To this we must add the single-particle energy ϵ_p of the added electron. The total excitation energy is given by

$$\epsilon_p[1 - 2v_p^2] + 2\Delta_p u_p v_p \tag{2-43}$$

where we have used the gap equation (2-30e) to simplify the interaction energy term. If we use the results (2-30) for u_k and v_k (with $\mu = 0$), we find the excitation energy

$$W_{p\uparrow} - W_0 = \frac{\epsilon_p^2}{E_p} + \frac{\Delta_p^2}{E_p} = E_p \tag{2-44}$$

Thus, the parameter E_p defined by (2-30c) is just the energy required to create a quasi-particle in state $\mathbf{p}\uparrow$. A plot of E_p vs. $\mathbf{p}$ is given in Figure 2-3. The minimum energy required to add an electron to the system is $\Delta_{k_f} \equiv \Delta_0 \sim 10^{-3} - 10^{-4}$ ev. In principle the chemical potential μ should be shifted a small amount to ensure $\langle N\rangle = N_0 + 1$ in the excited state; however, this correction has negligible effect in a large system.

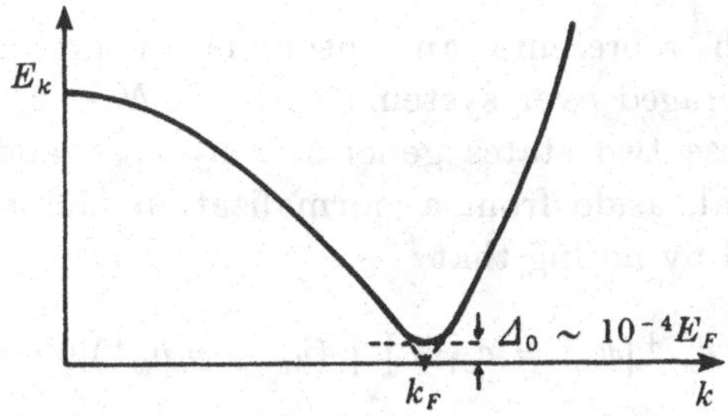

FIGURE 2-3 The quasi-particle energy E_k in the superconducting state plotted as a function of the wavevector **k**. The energy $E_k = (\epsilon_k{}^2 + \Delta_k{}^2)^{1/2}$ differs from the corresponding energy $|\epsilon_k|$ of the normal state only in the vicinity of the Fermi surface. The energy gap observed in experiments which do not inject or withdraw electrons from the system is $2\Delta_0$, a minimum energy Δ_0 being required to create each quasi-particle produced in a one-electron transition. Note: All energies are measured with respect to the Fermi energy.

In the above calculation, nothing has been said about **p** being above (or below) the Fermi surface. Since the pairing interactions smooth out the jump in the single-particle occupation numbers $\langle n_k \rangle$ in the normal phase at the Fermi surface, as shown in Figure 2-4, there is a finite probability of being able to add an electron to the system in a state p below the Fermi surface. The excitation energy is positive in this case as it is for $|\mathbf{p}| > p_f$. In an analogous way one can calculate the energy required to remove an electron in the state $\mathbf{p}\uparrow$ from the ground state. One again finds the energy $E_p(>0)$ regardless of whether $|\mathbf{p}|$ is greater or less than the Fermi momentum.

Therefore, the minimum energy required to create a single-particle-like excitation from the superconducting ground state is $2\Delta_0$, Δ_0 for removing an electron from one state and Δ_0 for placing it in another state.

It is important to realize that the states created by *adding* an electron to $|\psi_{0N}\rangle$ in state $\mathbf{p}\uparrow$ or *removing* an electron from state $-\mathbf{p}\downarrow$ in $|\psi_{0N}\rangle$ are *identical* within the pairing approximation except that the number of superfluid pairs in the two states differs by unity. If instead of $|\psi_{0N}\rangle$ we work with the state $|\psi_0\rangle$

(2-23), which represents an ensemble of ground-state wave functions averaged over systems with $\cdots N - 2, N, N + 2 \cdots$ particles, these two states generated by $c_{p\uparrow}{}^+$ and by $c_{-p\downarrow}$ are truly identical, aside from a normalization factor. This result is established by noting that

$$\begin{aligned} c_{p\uparrow}{}^+|\psi_0\rangle &= c_{p\uparrow}{}^+ \prod_k (u_k + v_k b_k{}^+)|0\rangle \\ &= u_p c_{p\uparrow}{}^+ \prod_{k \neq p} (u_k + v_k b_n{}^+)|0\rangle \\ &\equiv u_p|\psi_{p\uparrow}\rangle \end{aligned} \tag{2-45a}$$

and

$$\begin{aligned} c_{-p\downarrow}|\psi_0\rangle &= c_{-p\downarrow} \prod_k (u_k + v_k b_k{}^+)|0\rangle \\ &= -v_p c_{p\uparrow}{}^+ \prod_{k \neq p} (u_k + v_k b_k{}^+)|0\rangle \\ &= -v_p|\psi_{p\uparrow}\rangle \end{aligned} \tag{2-45b}$$

where $|\psi_{p\uparrow}\rangle$ is the normalized one-quasi-particle state

$$|\psi_{p\uparrow}\rangle = c_{p\uparrow}{}^+ \prod_{k \neq p} (u_k + v_k b_k{}^+)|0\rangle \tag{2-46}$$

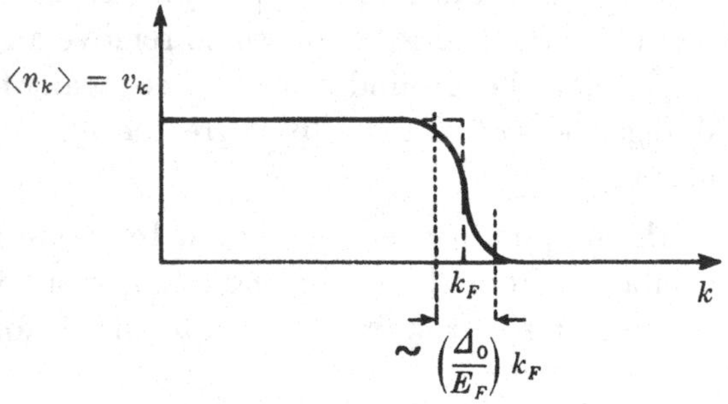

FIGURE 2-4 A plot of the average occupation number $\langle n_k \rangle$ of the Bloch states in the superconducting state if one makes a single-particle model for the normal state. The occupation number in the normal state, shown here as 1 for $k < k_F$ and 0 for $k > k_F$, is also rounded due to normal state interactions, although a discontinuity of $\langle n_k \rangle$ presumably remains. The "smearing" of the Fermi surface by the pairing correlations occurs only over a range $\sim 10^{-4} k_F$ about the Fermi surface.

An important mathematical simplification occurs if one considers the linear combination

$$\gamma_{p\uparrow}^{+} = u_p c_{p\uparrow}^{+} - v_p c_{-p\downarrow} \tag{2-47a}$$

of the two equivalent operators. From (2-45) we see that $\gamma_{p\uparrow}^{+}$ applied to $|\psi_0\rangle$ creates the normalized state $|\psi_{p\uparrow}\rangle$,

$$\gamma_{p\uparrow}^{+}|\psi_0\rangle = |\psi_{p\uparrow}\rangle \tag{2-47b}$$

The orthogonal combination

$$\gamma_{-p\downarrow} = u_p c_{-p\downarrow} + v_p c_{p\uparrow}^{+} \tag{2-48a}$$

when applied to $|\psi_0\rangle$ leads to the null-state vector (not to be confused with the vacuum $|0\rangle$)

$$\gamma_{-p\downarrow}|\psi_0\rangle = 0 \tag{2-48b}$$

The relations (2-47a) and (2-48a), and their Hermitian conjugates

$$\gamma_{p\uparrow} = u_p c_{p\uparrow} - v_p c_{-p\downarrow}^{+} \tag{2-49a}$$

$$\gamma_{-p\downarrow}^{+} = u_p c_{-p\downarrow}^{+} + v_p c_{p\uparrow} \tag{2-49b}$$

were introduced independently by Bogoliubov[52] and by Valatin.[53] These relations are known as the B–V transformation. As the notation suggests $\gamma_{p\uparrow}^{+}$ and $\gamma_{-p\downarrow}^{+}$ create quasi-particles in states $p\uparrow$ and $-p\downarrow$, respectively, while $\gamma_{p\uparrow}$ and $\gamma_{-p\downarrow}$ destroy quasi-particles in these states. Thus one has the relations

$$\gamma_{p\uparrow}^{+}|\psi_0\rangle = |\psi_{p\uparrow}\rangle \tag{2-50a}$$

$$\gamma_{-p\downarrow}^{+}|\psi_0\rangle = |\psi_{-p\downarrow}\rangle \tag{2-50b}$$

$$\gamma_{p\uparrow}|\psi_0\rangle = 0 \tag{2-50c}$$

$$\gamma_{-p\downarrow}|\psi_0\rangle = 0 \tag{2-50d}$$

The last two relations are equivalent to the statement that $|\psi_0\rangle$ is the vacuum state for quasi-particles. It follows by direct computation that the quasi-particle operators satisfy Fermi–Dirac statistics:

$$\{\gamma_{ps}, \gamma_{p's'}^{+}\} = \delta_{pp'}\,\delta_{ss'} \tag{2-51a}$$

$$\{\gamma_{ps}, \gamma_{p's'}\} = \{\gamma_{ps}^{+}, \gamma_{p's'}^{+}\} = 0 \tag{2-51b}$$

and can be thought of as leading to excitations that form a weakly interacting Fermi gas.

It is important to remember that the γ and γ^+'s must operate on the ensemble-averaged states, e.g., $|\psi_0\rangle$, not on the N-particle projection of these states $|\psi_{0N}\rangle$. If one thinks of $\gamma_{p\uparrow}{}^+$ as acting on an N-particle state, it would appear that a quasi-particle is a linear combination of a particle and a hole. This is *not* correct. In the N-particle system a quasi-particle of momentum $\mathbf{p}$ and spin s is nothing more than an electron definitely occupying the state $\mathbf{p}$, s with its mate $-\mathbf{p}$, $-s$ being definitely empty. In configuration space, the $N + 1$ particle wave function corresponding to $|\psi_{p\uparrow}\rangle$ is

$$\psi_{p\uparrow}(\mathbf{r}_1, s_1; \cdots \mathbf{r}_{N+1}, s_{N+1}) = \mathscr{A}\varphi'(\mathbf{r}_1 - \mathbf{r}_2)\chi_{12}\varphi'(\mathbf{r}_3 - \mathbf{r}_4)\chi_{34}\cdots \varphi'(\mathbf{r}_{N-1} - \mathbf{r}_N)\chi_{N-1,N}\exp(i\mathbf{p}\cdot\mathbf{r}_{N+1})\uparrow_{N+1} \tag{2-52}$$

where $\varphi'(\boldsymbol{\rho})$ is given by (2-40) with the term $\mathbf{k} = \mathbf{p}$ deleted. For some purposes it is convenient to discuss the excitation in terms of the empty state $-\mathbf{p}$, $-s$ and call the excitation a "hole." In other cases one prefers to concentrate on the occupied state $\mathbf{p}$, s. The wave function for the pair state ($\mathbf{p}\uparrow$, $-\mathbf{p}\downarrow$) is the same in either case regardless of the words used to describe it. However, one must keep in mind that the number of superfluid pairs differs by unity in the two descriptions of the same state.

An excited state having quasi-particles in $\mathbf{k}_1 s_1$, $\mathbf{k}_2 s_2 \cdots \mathbf{k}_n s_n$ is given by

$$|\psi_{k_1 s_1, k_2 s_2, \cdots k_n s_n}\rangle = \gamma_{k_1 s_1}{}^+\gamma_{k_2 s_2}{}^+\cdots\gamma_{k_n s_n}{}^+|0\rangle \tag{2-53}$$

The excitation energy is $E_{k_1 s_1} + E_{k_2 s_2} + \cdots E_{k_n s_n}$. The Bogoliubov–Valatin operators have the important property that the excited state

$$|\psi_{p\uparrow, -p\downarrow}\rangle = \gamma_{p\uparrow}{}^+\gamma_{-p\downarrow}{}^+|\psi_0\rangle \tag{2-54}$$

is orthogonal to the ground state. This is not true if one generates the excitations by applying $c_{p\uparrow}{}^+c_{p\downarrow}{}^+$ to $|\psi_0\rangle$. In the original BCS treatment these doubly excited pair states (called "real" pairs as opposed to the "virtual" pairs occurring in the ground state) were treated separately and were represented by the factor

$$(v_p - u_p b_p{}^+) \tag{2-55}$$

in the wave function. However, this is just the factor one obtains by expressing the γ^+'s in (2-54) in terms of the c and c^+'s and simplifying the factors involving $b_p{}^+$ in the wave function.

It has been shown by several groups[54] that for H_{red} the variational solutions for $|\psi_0\rangle$ and the quasi-particle spectrum are exact in the limit of a large system so long as the number of excitations present is small compared to the number of electrons participating in the pairing interactions. In the next section we shall see how these results are generalized to finite temperature where the latter condition is not satisfied.

2-6 LINEARIZED EQUATIONS OF MOTION

In the original work of Bardeen, Cooper, and the author a complete discussion of the thermodynamic properties of the superconducting state was carried out within the pairing approximation. As for $T = 0$, their treatment of the system described by H_{red} is exact in the limit of large volume. In agreement with experiment, they obtained a second-order phase transition at T_c and an exponentially vanishing electronic specific heat for $T \lesssim \frac{1}{2}T_c$. For a discussion of this work and its comparison with experiment, the reader is referred to the original BCS paper[8] and to a review article by Bardeen and the author.[9]

Rather than repeating the BCS finite-temperature treatment, we would like to illustrate an alternative procedure based on a linearization of the equations of motion for the single-particle operators $c_{p\uparrow}{}^+$ and $c_{-p\downarrow}$. The discussion follows closely a treatment given by Valatin,[55] and leads to results identical to those of BCS.

To fix ideas, we begin with the reduced Hamiltonian (2-24)

$$H_{\text{red}} = \sum_{k,s'} \epsilon_k n_{ks} + \sum_{k,k'} V_{k'k} b_{k'}{}^+ b_k \tag{2-17}$$

(although, owing to our approximations, the scheme gives essentially the same results if the full two-body interaction is considered). The basic idea is to find eigenoperators, say $\mu_\alpha{}^+$ and μ_β, which satisfy

$$[H_{\text{red}}, \mu_\alpha{}^+] = \Omega_\alpha \mu_\alpha{}^+ \tag{2-56a}$$

and the Hermitian conjugate relation

$$[H_{\text{red}}, \mu_\beta] = -\Omega_\beta \mu_\beta \tag{2-56b}$$

where the Ω's are positive quantities. It follows that the eigen-operators μ_α^+ and μ_β create and destroy excitations of the system since by applying the operator equation (2-56b) to the ground state $|\psi_0\rangle$ of H_{red} we find

$$[H_{\text{red}}, \mu_\alpha^+]|\psi_0\rangle = (H_{\text{red}} - W_0)\mu_\alpha^+|\psi_0\rangle = \Omega_\alpha \mu_\alpha^+|\psi_0\rangle \tag{2-57}$$

Thus $|\psi_\alpha\rangle \equiv \mu_\alpha^+|\psi_0\rangle$ is an eigenstate of H_{red} with excitation energy Ω_α. In an analogous manner one finds that μ_β lowers the energy of the system by Ω_β, from which it follows that the ground state (or the excitation vacuum) satisfies

$$\mu_\beta|\psi_0\rangle = 0 \tag{2-58}$$

for all β. Operators which approximately satisfy (2-56) presumably give an approximate description of the excitations. Except in extremely simple systems, the exact operators μ_α^+ can neither be found nor are of great interest since physically interesting probes (i.e., external fields, injected particles, etc.) create complicated superpositions of such excitations (see Chapter 5).

Suppose we try to find an operator which adds a quasi-particle of momentum $\mathbf{p}$ and spin $\uparrow$ to the ground state of H_{red}. The simplest fermion operator which will add this momentum and spin to the system is $c_{p\uparrow}^+$, so we try

$$[H_{\text{red}}, c_{p\uparrow}^+] = \epsilon_p c_{p\uparrow}^+ + \sum_{k'} V_{k'p} b_{k'}^+ c_{-p\downarrow} \tag{2-59}$$

In the absence of the interaction $c_{p\uparrow}^+$ satisfies (2-56a) with the excitation energy $\Omega_\alpha = \epsilon_p$. The "excitation" energy is just the energy required to add an electron to the system in state $\mathbf{p}$. In the presence of V, $c_{p\uparrow}^+$ is no longer an eigenoperator. In fact we must go out of the operator subspace of $c_{k\uparrow}^+$ and include products of the forms c^+cc in constructing μ_α^+. When this more complicated guess for μ_α^+ is commuted with H_{red}, still higher order polynomials in c^+ and c appear. In most cases the series continues on to infinite order, just as the series of equations determining the Green's functions, which are discussed in Chapter 5. To obtain a tractable problem we must cut off the chain at a certain order by approximating the commutators. Whether the termination is meaningful clearly depends on the physics of the problem. Fortunately, the interactions in H_{red} are sufficiently simple that one can cut off the series by including in a sense only a linear combination of c and c^+.

By taking matrix elements of (2-59) between the N-particle ground state $|0, N\rangle$ and the $N + 1$ particle state $|p\uparrow, N + 1\rangle$ with one quasi-particle present in state $p\uparrow$, we have

$$\begin{aligned}(\Omega_{p\uparrow} - \epsilon_p)\langle p\uparrow, N + 1|c_{p\uparrow}^+|0, N\rangle \\ = \sum_{\alpha, k'} V_{k'p}\langle p\uparrow, N + 1|c_{-p\downarrow}|\alpha, N + 2\rangle\,\langle\alpha, N + 2|b_{k'}^+|0, N\rangle\end{aligned} \tag{2-60}$$

where the sum is over the eigenstates of the $N + 2$ particle system. If we measure all energies relative to the chemical potential, $\mu = \lim_{n/N\to 0} (W_{0,N+n} - W_{0,N})/n$, where $n \gg 1$, $\Omega_{p\uparrow}$ is the energy required to add a quasi-particle in $\mathbf{p}\uparrow$ to $|0, N\rangle$. We argue that for a large system the intermediate state sum is given by retaining only the $N + 2$ particle ground state. It is not that the matrix elements of $b_{k'}^+$ for all other α are small compared to the one for $\alpha = 0$, but as we shall see the matrix element of $c_{-p\downarrow}$ entering the equation is small when the $\alpha \neq 0$ matrix element of $b_{k'}^+$ is large, and therefore the product is negligible. Thus (2-60) becomes

$$(\Omega_p - \epsilon_p)F_p = \sum_k V_{kp}B_kG_p \tag{2-61}$$

where

$$F_p = \langle p\uparrow, N + 1|c_{p\uparrow}^+|0, N\rangle \tag{2-62a}$$

$$G_p = \langle p\uparrow, N + 1|c_{-p\downarrow}|0, N + 2\rangle \tag{2-62b}$$

$$B_k = \langle 0, N + 2|b_k^+|0, N\rangle \tag{2-62c}$$

Another relation between F and G can be obtained by taking the matrix element of $(H_{\text{red}}, c_{-p\downarrow})$ between the states $|0, N + 2\rangle$ and $|p\uparrow, N + 1\rangle$. If the intermediate state sum is again replaced by the single term $\alpha = 0$, one finds

$$(\Omega_p + \epsilon_p)G_p = \sum_k V_{kp}B_kF_p \tag{2-63}$$

where we have chosen phases so that all quantities are real. The secular equation for (2-61) and (2-63) is

$$\begin{vmatrix} \Omega_p - \epsilon_p & \Delta_p \\ \Delta_p & \Omega_p + \epsilon_p \end{vmatrix} = \Omega_p^2 - \epsilon_p^2 - \Delta_p^2 = 0 \tag{2-64}$$

where the parameter Δ_p is defined by

$$\Delta_p = -\sum_k V_{kp} B_k \tag{2-65}$$

(The matrix element B_k is still to be determined.) From equation (2-64) we find the (positive) excitation energy Ω_p of the state $|p\uparrow, N+1\rangle$ is given by

$$\Omega_p = +(\epsilon_p^2 + \Delta_p^2)^{1/2} \equiv E_p \tag{2-66}$$

The negative energy root $-E_p$ corresponds to the process in which a quasi-particle in the time-reversed state $-\mathbf{p}\downarrow$ is destroyed. The eigenoperators $\mu_\alpha^+ \equiv \gamma_{p\uparrow}^+$ and $\mu_\beta \equiv \gamma_{-p\downarrow}$ corresponding to the positive and negative energy solutions are of the form

$$\gamma_{p\uparrow}^+ = u_p c_{p\uparrow}^+ - v_p c_{-p\downarrow} R^+ \tag{2-67a}$$

and

$$\gamma_{-p\downarrow} = u_p c_{-p\downarrow} + v_p R c_{p\uparrow}^+ \tag{2-67b}$$

The operator R^+ transforms a given state in an N-particle system into the corresponding state in the $N+2$ particle system; thus

$$R^+|0, N\rangle = |0, N+2\rangle \tag{2-68}$$

and

$$R^+|k, s; N\rangle = |k, s; N+2\rangle$$

while

$$R|0, N+2\rangle = |0, N\rangle, \text{ etc.}$$

By inserting the eigenvalues back into (2-61) or (2-63) and requiring that the γ^+ and γ's satisfy Fermi anticommutation relations it follows that u_p and v_p are given by

$$u_p^2 = \frac{1}{2}\left(1 + \frac{\epsilon_p}{E_p}\right) \tag{2-69a}$$

$$v_p^2 = \frac{1}{2}\left(1 - \frac{\epsilon_p}{E_p}\right) \tag{2-69b}$$

$$u_p v_p = \frac{\Delta_p}{2E_p} \tag{2-69c}$$

The formal similarity between these results and those of the last section is complete if we require that $|0, N\rangle$ be the ground state of the system, that is,

$$\gamma_{p\uparrow}|0, N\rangle = 0 \tag{2-70a}$$

and

$$\gamma_{-p\downarrow}|0, N\rangle = 0 \tag{2-70b}$$

Thus, the γ^+'s create noninteracting fermion excitations from the "vacuum state" $|0, N\rangle$.

By inverting (2-67a), (2-67b), and their Hermitian conjugates to solve for the c-operators in terms of the γ's, one finds from the definition of B_k (2-62c) the relation

$$B_k = u_k v_k = \frac{\Delta_k}{2E_k} \tag{2-71}$$

On combining this result with the definition of Δ_p (2-65) we find an equation determining the parameter Δ_p:

$$\Delta_p = -\sum_k V_{kp} \frac{\Delta_k}{2E_k} \tag{2-72}$$

which is just the energy-gap equation (2-30e). Thus the excitation energies are identical in the two approaches and the quasi-particle operators differ only by the presence of R.[56]

As in the BCS treatment, it is straightforward to generalize these results to finite temperature. The only change is that instead of the ground state $|0, N\rangle$ appearing one has a typical state $|T, N\rangle$ excited at the temperature T. All goes through as above except for the relation (2-71). For $T \neq 0$ one has

$$B_k = \langle T, N + 2|b_k^+|T, N\rangle = \frac{\Delta_k}{2E_k}(1 - f_{k\uparrow} - f_{-k\downarrow}) \tag{2-73}$$

where f_{ks} is the expectation value of the quasi-particle occupation number $\gamma_{ks}^+\gamma_{ks}$ in the state $|T, N\rangle$. Since the quasi-particles are essentially independent fermions (whose properties change slowly with temperature) f_{ks} is given by the Fermi distribution function

$$f_{ks} = \frac{1}{e^{\beta E_k} + 1} \qquad (E_k > 0) \tag{2-74}$$

and

$$B_k = \frac{\Delta_k}{2E_k} \tanh \frac{\beta E_k}{2} \tag{2-75}$$

By inserting this result into (2-65) we obtain the finite-temperature BCS gap equation

$$\Delta_p = -\sum_k V_{kp} \frac{\Delta_k}{2E_k} \tanh \frac{\beta E_k}{2} \tag{2-76}$$

This finite-temperature treatment of the pairing theory is entirely equivalent to the BCS treatment, which, as we mentioned above, gives an exact account of the system described by the reduced Hamiltonian (in the limit of large volume). If V_{kp} is approximated by (2-32), Δ_k is again of the form (2-33) where $\Delta_0(\beta)$ satisfies

$$\frac{1}{N(0)V} = \int_0^{\omega_c} \frac{d\epsilon}{(\epsilon^2 + \Delta_0^2)^{1/2}} \tanh \left[\frac{\beta}{2} (\epsilon^2 + \Delta_0^2)^{1/2}\right] \tag{2-77}$$

As T increases from zero, Δ_0 decreases as shown in Figure 2-5, vanishing at the transition temperature T_c. Thus, T_c is given by

$$\frac{1}{N(0)V} = \int_0^{\omega_c} \frac{d\epsilon}{\epsilon} \tanh \left[\frac{\epsilon}{2k_B T_c}\right] \tag{2-78}$$

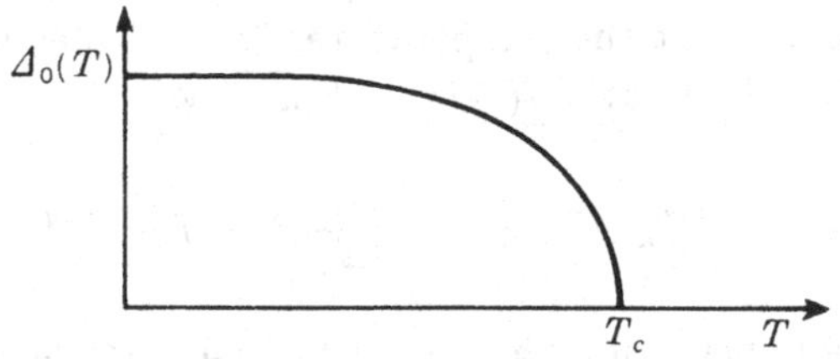

FIGURE 2-5 A plot of the temperature dependence of the energy-gap parameter $\Delta_0(T)$. Note that Δ_0 vanishes with infinite slope as $T \to T_c$, leading to the second-order phase transition.

In the weak-coupling limit this gives

$$k_B T_c = 1.14\omega_c \exp\left[-\frac{1}{N(0)V}\right] \tag{2-79}$$

so that the ratio $2\Delta_0(T = 0)/k_B T_c$ is 3.52 in this limit. While this ratio is in reasonably good agreement with experiment for weak-coupling superconductors,[9, 16] the ratio is too small to account for the observed ratio for lead and mercury. It now appears that temperature-dependent damping effects account for the discrepancy.[57]

The free energy of the superconducting state

$$F_s = W_s - TS \tag{2-80}$$

can be obtained by calculating the expectation value of H_{red} with respect to the typical state $|T, N\rangle$ and using the standard expression

$$S = -2k_B \sum_k \{f_k \log f_k + (1 - f_k)\log(1 - f_k)\} \tag{2-81}$$

for the entropy of the quasi-particle (normal) fluid, where f_k is given by (2-74). The energy W_s is easily seen to be[8]

$$W_s = 2\sum_k |\epsilon_k|\left[f_k + \frac{1}{2}\left(1 - \frac{|\epsilon_k|}{E_k}\right)\tanh\frac{\beta E_k}{2}\right] + \sum_k \frac{\Delta_k^2}{2E_k}\tanh\frac{\beta E_k}{2} + 2\sum_{|k|<k_F}\epsilon_k \tag{2-82}$$

The bulk critical magnetic field $H_c(T)$ is given by

$$\frac{H_c^2}{8\pi} = F_N(T) - F_S(T) \tag{2-83}$$

where the free energy of the normal state $F_N(T)$ is given by (2-80), (2-81), and (2-82) with $\Delta_k = 0$. H_0 is plotted in Figure 2-6 for the potential (2-32). The electronic specific heat can be calculated from

$$c_{es} = 2k_B\beta^2 \sum_k f_k(1 - f_k)\left[E_k^2 + \frac{\beta}{2}\frac{d\Delta_k^2}{d\beta}\right] \tag{2-84}$$

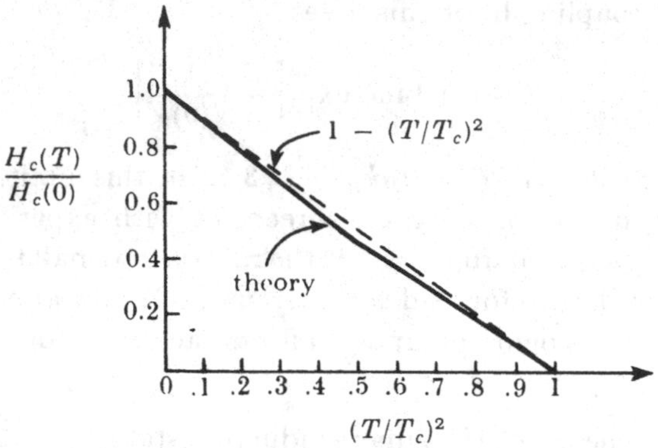

FIGURE 2-6 A plot of the critical magnetic field versus temperature.

and is plotted in Figure 2-7. The jump of the electronic-specific heat at T_c is due to Δ^2 being proportional to $(T_c - T)$ near T_c so that the derivative in (2-84) is discontinuous at T_c.

The reader is referred to the literature[9, 16] for a detailed discussion of the thermodynamic properties of the system.

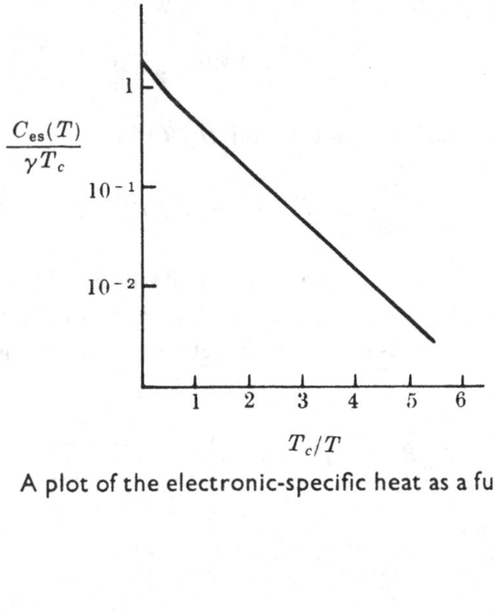

FIGURE 2-7 A plot of the electronic-specific heat as a function of T_c/T.

2-7 CONCLUDING REMARKS

Before closing this chapter we would like to make a few remarks.

1. While there is a formal similarity between the ground state (2-38) of H_{red} and a condensed Bose–Einstein gas, the analogy must be used with care due to the strong overlap of the pair functions. As a result of this overlap the excitation spectrum in real metals exhibits an energy gap rather than a continuous spectrum characteristic of a Bose gas. If the treatment is extended to include the interactions neglected in H_{red} and one assumes all interactions to be of short range, a continuous boson spectrum, starting at zero energy appears in the energy gap corresponding to density fluctuations in the electron system. In real metals these low-lying boson modes are pushed up to the plasmon energy ($\sim 10^4 \times 2\Delta_0$) due to the long-range Coulomb interaction between electrons so that there are no low-lying boson modes except for dressed lattice vibrations (phonons) in cases of physical interest.[58]

2. The discussion in this chapter has emphasized states for which the momentum density of the superfluid is zero, i.e., the pairing $(k\uparrow, -k\downarrow)$ was treated. If the Hamiltonian of the electron system were Galilean-invariant, states with finite superfluid momentum could be formed by translating the zero-momentum eigenfunction by an amount $\mathbf{q}/2$ in momentum space. The transformed wave function would be

$$\exp\left(\frac{i}{2}\sum_{j=1}^{N}\mathbf{q}\cdot\mathbf{r}_j\right)\psi_0(\mathbf{r}_1s_1, \mathbf{r}_2s_2, \ldots \mathbf{r}_Ns_N)$$
$$= \mathscr{A}\varphi(\mathbf{r}_1 - \mathbf{r}_2)\exp\left[\frac{i\mathbf{q}\cdot(\mathbf{r}_1+\mathbf{r}_2)}{2}\right]\uparrow_1\downarrow_2$$
$$\times\cdots\varphi(\mathbf{r}_{N-1} - \mathbf{r}_N)\exp\left[\frac{i\mathbf{q}\cdot(\mathbf{r}_{N-1}+\mathbf{r}_N)}{2}\right]\uparrow_{N-1}\downarrow_N \qquad (2\text{-}85)$$

so that the "center-of-mass" wave function of a pair would go from the $\mathbf{q} = 0$ state to the plane wave state of momentum $\mathbf{q}$. For states involving mass flow of this sort, a condensed Bose gas picture may be helpful; however, one must be cautious in using this picture in detailed calculations. In particular since the fixed ions will create a magnetic field in the moving frame, the function φ will change its form. Also, if the wavevector $\mathbf{q}$ varies appreciably over a coherence length ξ_0 the separation into center-of-mass and relative coordinates is of questionable value.

3. We have concentrated on singlet spin pairing with φ being an s-state in the absence of crystal anisotropy. The problems of triplet spin pairing and $l \neq 0$ orbital states have been treated by a number of authors[59] and we refer the reader to the literature for a discussion of these questions. In addition, pairing of states other than Bloch functions can be easily handled, since the basic scheme does not rely on the form of the states being paired.

4. In Section 2-2 we saw that not only the $\mathbf{q} = 0$ pairs are unstable if we consider fluctuations about the normal state, but the $q \neq 0$ pairs are also unstable. In Chapter 8 we investigate the stability of the ground state $|\psi_0\rangle$ given by the pairing approximation. As mentioned above, there are no unstable pair fluctuations about this state. This result is due to the finite energy required to create from the superfluid the quasi-particles which one tries to bind together by the residual interactions.

5. In Section 2-6 we stated that the matrix element $\langle \alpha, N + 2|b_k{}^+|0, N\rangle$ is large not only for the $N + 2$ particle ground state $|0, N + 2\rangle$ but that there is another state α giving a large matrix element. Specifically, the two-quasi-particle state $|k\uparrow, -k\downarrow, N + 2\rangle$ gives

$$|\langle k\uparrow, -k\downarrow, N + 2|b_k{}^+|0, N\rangle| = u_k{}^2 = \frac{1}{2}\left(1 + \frac{\epsilon_k}{E_k}\right) \tag{2-86}$$

compared to the matrix element

$$|\langle 0, N + 2|b_k{}^+|0, N\rangle| = u_k v_k = \frac{\Delta_k}{2E_k} \tag{2-87}$$

which we retained. For $\mathbf{k}$ on the Fermi surface, both of matrix elements are equal to $1/(2)^{1/2}$. However, the matrix element $\langle p\uparrow, N + 1|c_{-p\downarrow}|k\uparrow, -k\downarrow, N + 2\rangle$ which multiplies (2-86) in (2-60) is zero since $c_{-p\downarrow}$ does not affect the quasi-particles in $k\uparrow$ and $-k\downarrow$. On the other hand, the matrix element $\langle p\uparrow, N+1|c_{-p\downarrow}|0, N+2\rangle$ which multiplies (2-87) in (2-60) is equal to $-v_p$ which is $1/(2)^{1/2}$ for $\mathbf{p}$ on the Fermi surface. Therefore we are justified in retaining the single term $\alpha = 0$ in the intermediate state sum.

6. While the ground-state wave function $|\psi_0\rangle$ [see (2-25)] represents an ensemble average of ground-state wave functions $|\psi_{0N}\rangle$ for systems having an even number of electrons,

$$|\psi_0\rangle = \sum_{N(\text{even})} A_N|\psi_{0N}\rangle \tag{2-88}$$

we can obtain $|\psi_{0N}\rangle$ from $|\psi_0\rangle$ if A_N is arranged to be of the form $|A_N|e^{iN\varphi}$. Then

$$|\psi_0{}^{\varphi}\rangle = \sum_{N(\text{even})} |A_N|e^{iN\varphi}|\psi_{0N}\rangle \tag{2-89}$$

and

$$|A_{N'}|\,|\psi_{0N'}\rangle = \frac{1}{2\pi}\int_0^{2\pi} e^{-iN'\varphi}|\psi_0{}^{\varphi}\rangle\, d\varphi \tag{2-90}$$

By our choice of phases (2-32) is just $|\psi_0{}^0\rangle$ so that $|\psi_0{}^{\varphi}\rangle$ is given by

$$|\psi_0{}^{\varphi}\rangle = \prod_k (u_k + e^{2i\varphi}v_k b_k{}^+)|0\rangle \tag{2-91}$$

that is, a factor of $e^{i\varphi}$ is contributed by each creation operator c^+. Therefore the normalized N' particle ground state is given by

$$|\psi_{0N'}\rangle = \frac{1}{2\pi|A_{N'}|}\int_0^{2\pi} e^{-iN'\varphi}\prod_k (u_k + e^{2i\varphi}v_k b_k{}^+)|0\rangle\, d\varphi \tag{2-92}$$

where the amplitude of the N' particle state is given by

$$|A_{N'}|^2 = \frac{1}{2\pi}\int_0^{2\pi} e^{-iN'\varphi}\prod_k (u_k{}^2 + e^{2i\varphi}v_k{}^2)\, d\varphi \tag{2-93}$$

The probability $|A_{N'}|^2$ is sharply peaked about the average number N_0, having a width[60] of the order of $N_0{}^{1/2}$. The fact that the average energy $\langle\psi_0{}^{\varphi}|H_{\text{red}}|\psi_0{}^{\varphi}\rangle$ is independent of φ should not be interpreted as a degeneracy of the ground state of a physical N particle system. Since the $N-2$, N, $N+2, \ldots$ particle systems are completely independent, the average energy of these systems should be independent of the relative phase of their wave functions.

For a large system, whether the physical system has an even or an odd total number of electrons makes no difference in its macroscopic properties; thus the wave functions above apply for any N. The situation is distinctly different for pairing correlations in atomic nuclei, where these differences lead to the well-known even–odd effects.[61]

CHAPTER 3

APPLICATIONS OF THE PAIRING THEORY

Since the BCS theory was originally proposed, attempts to justify the pairing correlations basic to the theory have proceeded along two lines. The first approach has been to apply the BCS theory to a wide variety of phenomena in superconductors and check the theoretical predictions of the pairing approximation against experiment. The second approach has been to treat by various approximate methods the residual interactions neglected within the pairing scheme, hoping to show that these residual interactions introduce no major change in the predicted properties of the system. Both approaches have enjoyed considerable success. Owing to the remarkably good agreement between the pairing theory and a broad class of experimentally observed phenomena, it would appear that the first approach has successfully established the validity of the pairing concept upon which the theory is based.[9, 16]

3-1 JUSTIFICATION OF THE PAIRING HYPOTHESIS

In this chapter we shall follow the first approach and review the calculation of a number of system properties within the pairing approximation. In the few cases where a discrepancy between theory and experiment exists, one can often attribute the difference to limitations of our understanding of normal state effects (details of band structure, phonon spectra, electron–phonon interactions, etc.). In Chapters 7 and 8 we take the second approach and discuss several theoretical advances which take proper account of retardation and damping effects. We also treat certain classes of residual interactions, neglected within the simple pairing approximation. Within the framework of these more elaborate treatments the predicted system properties are essentially in agreement with those given by the pairing model; where differences appear, the agreement between theory and experiment is generally improved by the more elaborate treatment.

3-2 ACOUSTIC ATTENUATION RATE

While the majority of the electron–phonon interaction has been accounted for in forming the wave functions for the normal and superconducting states, there remains the part corresponding to resonant phonon absorption and emission processes. These resonant processes give rise to attenuation of acoustic waves (dressed phonons). To calculate the time rate of change of $\langle N_{q_0\lambda_0}\rangle$, the number of phonons of wave vector $\mathbf{q}_0$ and polarization λ_0, we consider a typical state $|I\rangle$ excited at temperature T.[62] Within the pairing approximation $|I\rangle$ is of the form

$$|I\rangle = \Big[\prod_{k,\, s(\mathrm{occ.})} \gamma_{ks}^{+}\Big]\Big[\prod_{q,\,\lambda} (a_{q\lambda}^{+})^{\bar{N}_{q\lambda}}\Big]|0, T\rangle \qquad (3\text{-}1)$$

where the quasi-particle operators γ_{ks}^{+} are defined as in Section 2-6 appropriate to the temperature T and act on states with a fixed number of particles. The state $|0, T\rangle$ of N_0 electrons is the vacuum state for these operators, that is,

$$\gamma_{ks}(T)|0, T\rangle = 0 \qquad \text{for all } \mathbf{k} \text{ and } s \qquad (3\text{-}2)$$

The average of the quasi-particle occupation numbers in a small region r of $\mathbf{k}$-space taken about a particular value $\mathbf{k}$ is given by the Fermi distribution

$$\sum_{\mathbf{k}\text{ in }r} \langle I|\gamma_{ks}^{+}\gamma_{ks}|I\rangle = \sum_{\mathbf{k}\text{ in }r} \frac{1}{e^{\beta E_k}+1} \equiv \sum_{\mathbf{k}\text{ in }r} f_k \qquad (3\text{-}3)$$

although the occupation number $\bar{\nu}_{ks} \equiv \langle I|\gamma_{ks}{}^{+}\gamma_{ks}|I\rangle$ for a given state $\mathbf{k}$, s is either one or zero for the state vector given by (3-1). We assume the dressed electron–phonon interaction is of the form

$$H_{\text{el-ph}} = \sum_{\mathbf{pp}'s\lambda} \bar{g}_{pp'\lambda}(a_{q\lambda} + a_{-q\lambda}{}^{+})c_{p's}{}^{+}c_{ps} \tag{3-4}$$

where $\mathbf{q} = \mathbf{p}' - \mathbf{p} + \mathbf{K}$ (see Chapter 4). We shall use the golden rule[63] to calculate the transition rate $\langle \dot{N}_{q_0\lambda_0}\rangle$ so that we need matrix elements of $H_{\text{el-ph}}$ between $|I\rangle$ and all states $|F\rangle$ which are degenerate with $|I\rangle$ and differ from it by having a single particle change its state and a phonon $\mathbf{q}_0\lambda_0$ either absorbed or emitted. If the phonon energy $\omega_{q_0\lambda_0}$ is smaller than the energy gap $2\Delta(T)$, which is the case in most acoustic attenuation experiments, additional quasi-particles cannot be created from the superfluid so that only quasi-particle scattering processes enter. The final states are of the form

$$|F\rangle = \begin{cases} \gamma_{p_2s}{}^{+}\gamma_{p_1s}a_{q_0\lambda_0}|I\rangle & \text{absorption} \\ \gamma_{p_1s}{}^{+}\gamma_{p_2s}a_{q_0\lambda_0}{}^{+}|I\rangle & \text{emission} \end{cases} \tag{3-5}$$

where $\mathbf{q}_0 = \mathbf{p}_2 - \mathbf{p}_1 + \mathbf{K}$. Since $H_{\text{el-ph}}$ conserves the total number of electrons (as does $\gamma_{p's'}{}^{+}\gamma_{ps}$), $|I\rangle$ and $|F\rangle$ both describe states of the N_0 particle system. The matrix elements are readily evaluated by transforming (3-4) to the γ-representation (i.e., make a particle conserving B–V transformation). From (2-67) we find

$$c_{p\uparrow}{}^{+} = u_p\gamma_{p\uparrow}{}^{+} + v_p\gamma_{-p\downarrow}R^{+} \tag{3-6a}$$

$$c_{p\uparrow} = u_p\gamma_{p\uparrow} + v_pR\gamma_{-p\downarrow}{}^{+} \tag{3-6b}$$

$$c_{-p\downarrow}{}^{+} = u_p\gamma_{-p\downarrow}{}^{+} - v_p\gamma_{p\uparrow}R^{+} \tag{3-6c}$$

$$c_{-p\downarrow} = u_p\gamma_{-p\downarrow} - v_pR\gamma_{p\uparrow}{}^{+} \tag{3-6d}$$

Since the γ's are linear combinations of c's and c^{+}'s it follows that there are *two* terms in $H_{\text{el-ph}}$ which connect $|I\rangle$ with a given final state $|F\rangle$; in particular, the factors $c_{p_2\uparrow}{}^{+}c_{p_1\uparrow}$ and $c_{-p_1\downarrow}{}^{+}c_{-p_2\downarrow}$ lead to the same quasi-particle transition. The matrix elements arising from these two factors must be added before squaring the total matrix element. Note that if one makes a single-particle approximation for the normal state, only a single term contributes in the normal state. This difference is characteristic of the superconducting state and appears in most dynamical properties of the system. The cross terms which enter in squaring the total matrix element are known as "coherence effects" and have important experimental consequences, as we shall see below. From (3-4) we see that the factors multiplying $c_{p'\uparrow}{}^+c_{p\uparrow}$ and $c_{-p\downarrow}{}^+c_{-p'\downarrow}$ are identical so that we are interested in the combination

$$c_{p'\uparrow}{}^+c_{p\uparrow} + c_{-p\downarrow}{}^+c_{-p'\downarrow} = n(\mathbf{p},\mathbf{p}')[\gamma_{p'\uparrow}{}^+\gamma_{p\uparrow} + \gamma_{-p\downarrow}{}^+\gamma_{-p'\downarrow}] + m(\mathbf{p},\mathbf{p}')[\gamma_{p'\uparrow}{}^+\gamma_{-p\downarrow}{}^+R - \gamma_{p\uparrow}\gamma_{-p'\downarrow}R^+] \quad (3\text{-}7)$$

where we have used the transformation (3-6). The so-called "coherence factors" $m(\mathbf{p},\mathbf{p}')$ and $n(\mathbf{p},\mathbf{p}')$ are defined by

$$m(\mathbf{p},\mathbf{p}') = u_pv_{p'} + v_pu_{p'} \quad (3\text{-}8a)$$

and

$$n(\mathbf{p},\mathbf{p}') = u_pu_{p'} - v_pv_{p'} \quad (3\text{-}8b)$$

(We shall meet two more coherence factors l and p below in discussing spin-flip processes and the electromagnetic response of the system.) Returning to (3-5), the matrix element for phonon absorption is given (for $s = \uparrow$) by

$$\langle F|H_{\text{el-ph}}|I\rangle = n(\mathbf{p}_1,\mathbf{p}_2)\bar{\nu}_{p_1\uparrow}(1 - \bar{\nu}_{p_2\uparrow})[\bar{N}_{q_0\lambda_0}]^{1/2}\,\bar{g}_{p_1p_2\lambda_0} \quad (3\text{-}9)$$

where $\bar{\nu}_{p_1\uparrow}$ and $\bar{\nu}_{p_2\uparrow}$ are the quasi-particle occupation numbers for the initial state and are either one or zero, as mentioned above. $\bar{N}_{q_0\lambda_0}$ is the phonon occupation number for this state. The rate of absorption of $q_0\lambda_0$ phonons is then

$$w_{\text{abs}} = 2\pi \times 2\sum_{\mathbf{p}_1\mathbf{p}'}|\bar{g}_{pp'\lambda_0}|^2n^2(\mathbf{p},\mathbf{p}')\bar{\nu}_{p\uparrow}(1 - \bar{\nu}_{p'\uparrow})\bar{N}_{q_0\lambda_0} \times \delta(E_{p'} - E_p - \omega_{q_0\lambda_0}) \quad (3\text{-}10)$$

where the sum is restricted by momentum conservation $\mathbf{p}' = \mathbf{p} + \mathbf{q}_0 + \mathbf{K}$. The factor of two accounts for the absorption by quasi-particles of both spin orientations. In a similar way one finds the emission rate is given by

$$w_{\text{emiss}} = 2\pi \times 2 \sum_{\mathbf{pp}'} |\bar{g}_{pp'\lambda_0}|^2 n^2(\mathbf{p}, \mathbf{p}')\bar{\nu}_{p'\uparrow}(1 - \bar{\nu}_{p\uparrow})(\bar{N}_{q_0\lambda_0} + 1) \times \delta(E_{p'} - E_p - \omega_{q_0\lambda_0}) \quad (3\text{-}11)$$

so that the net rate of absorption is

$$-\frac{d\bar{N}_{q_0\lambda_0}}{dt} = \alpha_{q_0\lambda_0}\bar{N}_{q_0\lambda_0} - S_{q_0\lambda_0} \quad (3\text{-}12)$$

where the acoustic attenuation rate is given by

$$\alpha_{q_0\lambda_0} = 4\pi \sum_{\mathbf{pp}'} |\bar{g}_{pp'\lambda_0}|^2 n^2(\mathbf{p}, \mathbf{p}')(\bar{\nu}_{p\uparrow} - \bar{\nu}_{p'\uparrow})\, \delta(E_{p'} - E_p - \omega_{q_0\lambda_0}) \quad (3\text{-}13)$$

and S is the spontaneous emission rate. When the sum is performed, the occupation numbers can be replaced by their local average values and one has

$$\alpha_{q_0\lambda_0} = 4\pi \sum_{\mathbf{pp}'} |\bar{g}_{pp'\lambda_0}|^2 n^2(\mathbf{p}, \mathbf{p}')(f_p - f_{p'})\, \delta(E_{p'} - E_p - \omega_{q_0\lambda_0}) \quad (3\text{-}14)$$

where $\mathbf{p}' = \mathbf{p} + \mathbf{q}_0 + \mathbf{K}$ and $\mathbf{K}$ is a reciprocal lattice vector. The spontaneous emission rate $S'_{q_0\lambda_0}$ is given by (3-11) with $\bar{N}_{q_0\lambda_0} \equiv 0$.

The expression for α simplifies if we assume that g depends only on the momentum transfer (i.e., $\mathbf{q}_0$ for normal processes). For $|\mathbf{q}_0|$ small compared to k_F only normal processes enter the sum with appreciable weight, so that α is given by

$$\alpha_{q_0\lambda_0} = 2\pi|\bar{g}_{q_0\lambda_0}|^2 \sum_p \left(1 + \frac{\epsilon_p\epsilon_{p'} - \Delta_p\Delta_{p'}}{E_pE_{p'}}\right)(f_p - f_{p'}) \times \delta(E_{p'} - E_p - \omega_{q_0\lambda_0}) \quad (3\text{-}15)$$

where $\mathbf{p}' = \mathbf{p} + \mathbf{q}_0$, and the relation

$$n^2(\mathbf{p}, \mathbf{p}') = (u_pu_{p'} - v_pv_{p'})^2 = \frac{1}{2}\left(1 + \frac{\epsilon_p\epsilon_{p'} - \Delta_p\Delta_{p'}}{E_pE_{p'}}\right) \quad (3\text{-}16)$$

has been used. Since the speed of sound is small compared to the Fermi velocity, $\sim 10^{-3}v_F$ in typical cases, energy and momentum conservation require that $\mathbf{q}_0$ be essentially tangent to the

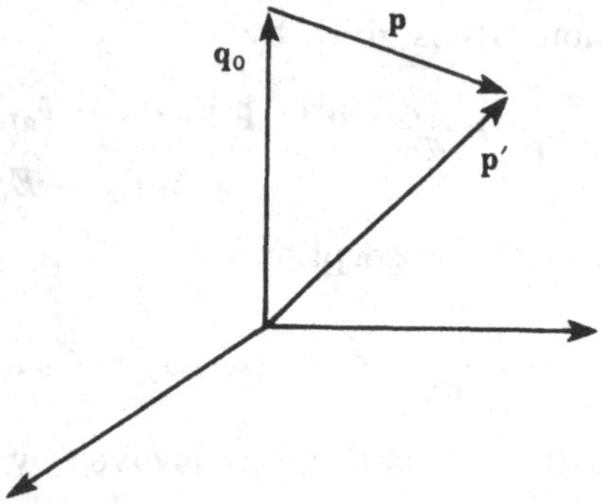

FIGURE 3-1 Coordinate system with $\mathbf{q}_0$ chosen as the polar axis and $\mathbf{p}' = \mathbf{p} + \mathbf{q}_0$.

Fermi surface for those states **p** which contribute to (3-15). For spherical energy surfaces, only those states near the Fermi surface lying near the equatorial plane (perpendicular to $\mathbf{q}_0$) enter the problem. If we approximate the energy contours of the Bloch states by spheres with the effective mass m^*, the sum in (3-15) can be expressed as

$$\sum_p \to \frac{1}{(2\pi)^3}\int p^2\, dp\, d\mu\, d\varphi \to \frac{m^{*2}}{(2\pi)^2|\mathbf{q}_0|}\int d\epsilon_p\, d\epsilon_{p'} \tag{3-17}$$

where $\mathbf{q}_0$ is the polar axis as shown in Figure 3-1. The $\epsilon_p\epsilon_{p'}$ term in the coherence factor vanishes upon integration since the remaining factors in the integral are even in ϵ_p and $\epsilon_{p'}$;[64] thus (3-15) reduces to

$$\begin{aligned}\alpha_{q_0\lambda_0} &= \frac{1}{2\pi}|\bar{g}_{q_0\lambda_0}|^2\frac{m^{*2}}{|\mathbf{q}_0|}\\ &\quad\times\int d\epsilon_p\, d\epsilon_{p'}\left(1-\frac{\Delta_p\Delta_{p'}}{E_pE_{p'}}\right)(f_p - f_{p'})\,\delta(E_{p'} - E_p - \omega_{q_0\lambda_0})\\ &\cong \frac{2}{\pi}|\bar{g}_{q_0\lambda_0}|^2\frac{m^{*2}}{|\mathbf{q}_0|}\\ &\quad\times\int_\Delta^\infty dE\,\frac{E}{(E^2-\Delta^2)^{1/2}}\,\frac{E'}{(E'^2-\Delta^2)^{1/2}}\left(1-\frac{\Delta^2}{EE'}\right)\\ &\quad\times[f(E) - f(E')]\end{aligned} \tag{3-18}$$

where $E' = E + \omega_{q_0\lambda_0}$ and we have taken Δ to be independent of **p**. For most experiments $\omega_{q_0\lambda_0} \ll \Delta$ so that we may set $E = E'$ in (3-18) except in the Fermi factor and obtain

$$\begin{aligned}\alpha_{q_0\lambda_0} &= \frac{2}{\pi}|\bar{g}_{q_0\lambda_0}|^2 \frac{m^{*2}}{|\mathbf{q}_0|}\int_\Delta^\infty dE\left(-\frac{\partial f}{\partial E}\right)\omega_{q_0\lambda_0} \\ &= \frac{2}{\pi}|\bar{g}_{q_0\lambda_0}|^2 \frac{m^{*2}\omega_{q_0\lambda_0}}{|\mathbf{q}_0|} f(\Delta)\end{aligned} \tag{3-19}$$

This expression applies to the normal state if we set $\Delta = 0$, so that the ratio of the acoustic attenuation rates in the S- and N-phases at the temperature T is

$$\frac{\alpha_S(T)}{\alpha_N(T)} = \frac{f(\Delta)}{f(0)} = \frac{2}{\exp\left[\dfrac{\Delta(T)}{k_B T}\right] + 1} \tag{3-20}$$

Therefore the temperature-dependent energy gap can be obtained

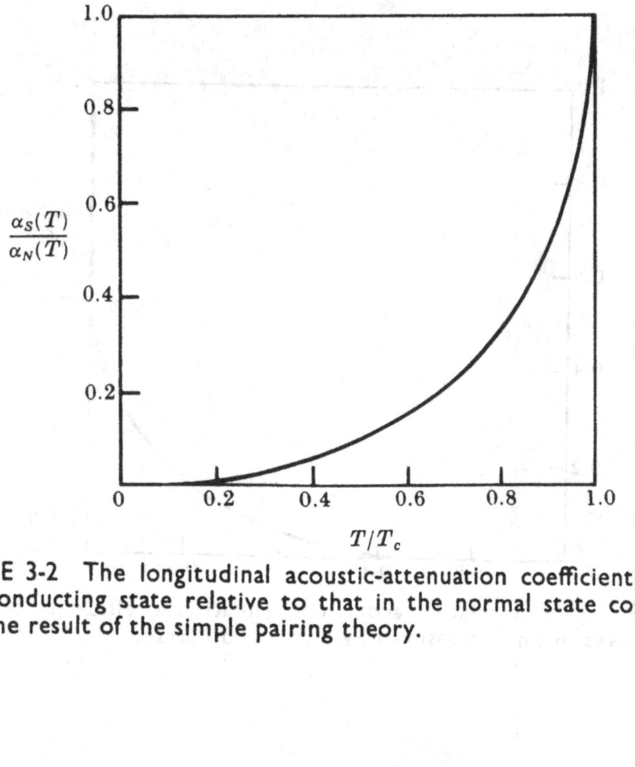

FIGURE 3-2 The longitudinal acoustic-attenuation coefficient in the superconducting state relative to that in the normal state compared with the result of the simple pairing theory.

from α_S/α_N. In Figure 3-2 the theoretical ratio is compared with experimental results for longitudinal phonons in tin and indium.[27a, 65] The rapid drop near T_c reflects the rapid decrease in the number of excitations as the energy gap opens up below T_c. We also note that the large density-of-states factors

$$\frac{EE'}{(E^2 - \Delta^2)^{1/2}(E'^2 - \Delta^2)^{1/2}}$$

are cancelled by the coherence factor $(1 - \Delta^2/E^2)$ in (3-18) for $\omega_{q_0\lambda_0} \ll \Delta$ so that only the Fermi factors enter the attenuation rate in this limit.

The energy bands and the gap parameter are anisotropic in real metals so that α_S/α_N measures a complicated average of Δ over the regions discussed above. Variations of the averaged gap of the order of 10 per cent have been observed in single crystals as $\mathbf{q}_0$ is rotated relative to the crystal axes.[27]

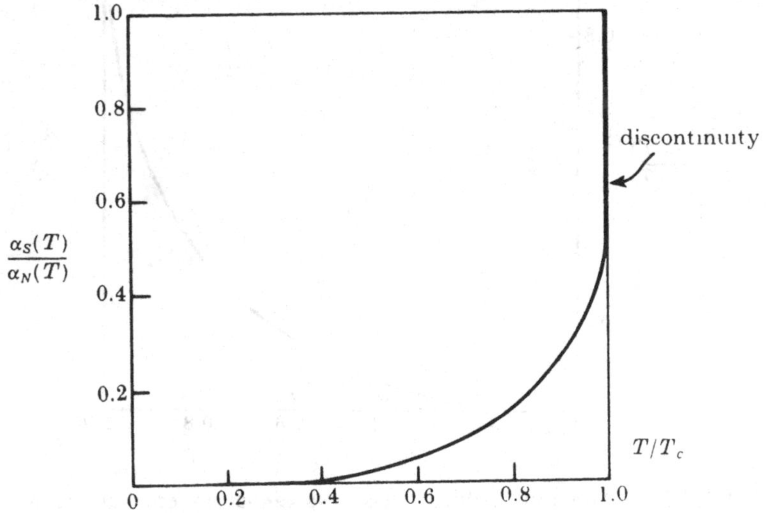

FIGURE 3-3 The relative acoustic-attenuation coefficient for transverse waves in tin as measured by Bohm and Morse.

In the above discussion it was tacitly assumed that the electron–phonon matrix element is the same in the N- and S-phases. Although this is most likely true for purely longitudinal phonons,[66] this assumption is incorrect for transverse phonons. A transverse phonon sets up transverse electromagnetic fields in addition to crystal potential effects.[27a, 67] While the screening of the crystal potential is essentially the same in the N- and S-states, the Meissner effect drastically reduces the electromagnetic coupling between the transverse phonons and the electrons. Thus, one expects an essentially discontinuous drop of the transverse acoustic attenuation rate upon entering the superconducting state at T_c due to the Meissner effect eliminating the electromagnetic coupling. The remaining coupling should be accurately treated by the above analysis and one indeed finds this to be the case, as shown in Figure 3-3.[67b]

3-3 NUCLEAR-SPIN RELAXATION RATE

The above calculation of the acoustic attenuation rate is easily modified to account for the relaxation rate of oriented nuclear spins due to their hyperfine coupling to the valence electrons. The interaction for a given nuclear spin **I** is of the form

$$H_{\mathbf{I}\cdot\mathbf{S}} = A \sum_{k,k'} a_{k'}^* a_k [I_z(c_{k'\uparrow}^+ c_{k\uparrow} - c_{k'\downarrow}^+ c_{k\downarrow}) + I_+ c_{k'\downarrow}^+ c_{k\uparrow} + I_- c_{k'\uparrow}^+ c_{k\downarrow}] \quad (3\text{-}21)$$

where a_k is proportional to the amplitude of the Bloch function $\chi_k(r)$ at the nuclear site in question, so that $a_{-k} = a_k^*$, and $I_\pm \equiv I_x \pm iI_y$.[68] To calculate the rate at which a given nucleus decreases its z-component of spin we observe that the Zeeman and hyperfine energies are in general small compared to the energy gap so that only quasi-particle spin-flip processes enter. We consider a typical initial state $|I\rangle$ excited at the temperature T, as in the preceding section, and notice that the final states are of the form

$$|F\rangle = \begin{cases} \gamma_{p_2\uparrow}^+ \gamma_{p_1\downarrow} |I\rangle & \text{nuclear spin flips down} \quad (3\text{-}22a) \\ \gamma_{p_1\downarrow}^+ \gamma_{p_2\uparrow} |I\rangle & \text{nuclear spin flips up} \quad (3\text{-}22b) \end{cases}$$

As before, $H_{I\cdot S}$ contains two terms which connect $|T\rangle$ and a given final state; for the final state (3-22a) one has the terms

$$AI_-(a_{p_2}{}^*a_{p_1}c_{p_2\uparrow}{}^+c_{p_1\downarrow} + a_{-p_1}{}^*a_{-p_2}c_{-p_1\uparrow}{}^+c_{-p_2\downarrow}) \\ = AI_-a_{p_2}{}^*a_{p_1}(c_{p_2\uparrow}{}^+c_{p_1\downarrow} + c_{-p_1\uparrow}{}^+c_{-p_2\downarrow}) \quad (3\text{-}23)$$

By transforming to the γ-representation these terms become

$$AI_-a_{p_2}{}^*a_{p_1}[l(\mathbf{p}_1, \mathbf{p}_2)(\gamma_{p_2\uparrow}{}^+\gamma_{p_1\downarrow} + \gamma_{-p_1\uparrow}{}^+\gamma_{-p_2\downarrow}) \\ + p(\mathbf{p}_1, \mathbf{p}_2)(\gamma_{p_2\uparrow}{}^+\gamma_{-p_1\uparrow}{}^+R - \gamma_{p_1\downarrow}\gamma_{-p_2\downarrow}R^+)] \quad (3\text{-}24)$$

where the coherence factors l and p are defined by

$$\begin{aligned} l(\mathbf{p}_1, \mathbf{p}_2) &= u_{p_1}u_{p_2} + v_{p_1}v_{p_2} \\ p(\mathbf{p}_1, \mathbf{p}_2) &= u_{p_1}v_{p_2} - v_{p_1}u_{p_2} \end{aligned} \quad (3\text{-}25)$$

The nuclear spin-flip-down matrix element is then

$$Aa_{p_2}{}^*a_{p_1}l(\mathbf{p}_1, \mathbf{p}_2)\bar{\nu}_{p_1\downarrow}(1 - \bar{\nu}_{p_2\uparrow})(I_-)_{fi} \quad (3\text{-}26)$$

where the last factor gives the nuclear matrix element. The transition rate for flipping down the nuclear spins is proportional to

$$w_{\text{down}} = 2\pi|A|^2 \sum_{\mathbf{p}_1,\mathbf{p}_2} |a_{p_1}|^2|a_{p_2}|^2 l^2(\mathbf{p}_1, \mathbf{p}_2) \\ \times f_{p_1}(1 - f_{p_2})\,\delta(E_{p_2} - E_{p_1} - \omega)N_\uparrow \quad (3\text{-}27)$$

Thus the rate of decay of the z-component of nuclear magnetization is proportional to

$$\alpha_S = 2\pi|A|^2 \sum_{\mathbf{p}_1,\mathbf{p}_2} |a_{p_1}|^4 \frac{1}{2}\left(1 + \frac{\Delta_{p_1}\Delta_{p_2}}{E_{p_1}E_{p_2}}\right) \\ \times f_{p_1}(1 - f_{p_2})\,\delta(E_{p_2} - E_{p_1} - \omega) \quad (3\text{-}28)$$

if we neglect crystalline anisotropy, since $l^2(\mathbf{p}_1, \mathbf{p}_2)$ is given by

$$l^2(\mathbf{p}_1, \mathbf{p}_2) = \frac{1}{2}\left(1 + \frac{\epsilon_{p_1}\epsilon_{p_2} + \Delta_{p_1}\Delta_{p_2}}{E_{p_1}E_{p_2}}\right) \quad (3\text{-}29)$$

and the term in $\epsilon\epsilon'$ vanishes on integration as above. If Δ is independent of $\mathbf{p}$ near the Fermi surface, (3-28) can be written for $\omega \ll \Delta$ as

$$\alpha_S = 4\pi|A|^2|a|^4N^2(0) \\ \times \int_\Delta^\infty \left[1 + \frac{\Delta^2}{E(E + \omega)}\right] \frac{E(E + \omega)k_BT(-\partial f/\partial E)\,dE}{(E^2 - \Delta^2)^{1/2}[(E + \omega)^2 - \Delta^2]^{1/2}} \quad (3\text{-}30)$$

so that the ratio of the relaxation rates in the S- and N-phases is

$$\frac{\alpha_S}{\alpha_N} = 2\int_{\Delta}^{\infty} \frac{[E(E+\omega)+\Delta^2](-\partial f/\partial E)\,dE}{[E^2-\Delta^2]^{1/2}[(E+\omega)^2-\Delta^2]^{1/2}} \qquad (3\text{-}31)$$

If ω is set equal to zero in the (3-31) the integral is logarithmically singular at the lower limit. If ω is calculated from the Zeeman energy, Hebel and Slichter[28] found the ratio increased to about 10 before falling to zero as $T \to 0$. Experimentally they found the ratio increases to about 2 in aluminum so that gap anisotropy and spacial inhomogeneity may well be the limiting feature. Perhaps the most interesting and important feature of

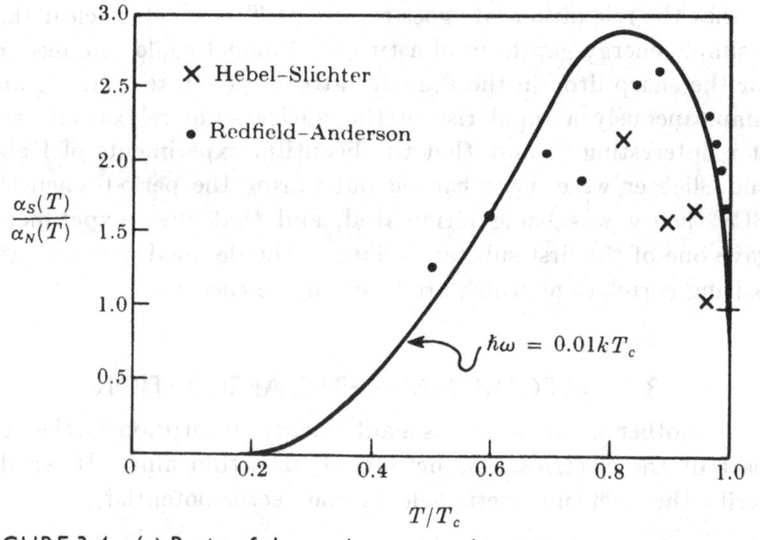

FIGURE 3-4 (a) Ratio of the nuclear-spin relaxation rates in the normal and superconducting states in aluminum. Solid curve calculated by L. C. Hebel for a smeared energy gap. (b) The nuclear relaxation time T_1 of superconducting aluminum. The solid curves are based on the pairing theory with the density of states smeared by folding the density of states with a square function of width $2d$ and height $(2d)^{-1}$. The dotted and solid curves were calculated with $2\Delta(0)/k_BT_c = 3.52$ and 3.25, respectively, the latter value being that found by Biondi and Garfunkel from microwave experiments.

the ratio α_S/α_N for spin relaxation rate is its predicted rise as T drops below T_c as shown in Figure 3-4. This result is distinctly different from the result for the acoustic attenuation rate. The only difference in the predicted rates for the two types of processes is that the coherence factor n^2 appears in the acoustic case while l^2 appears in the spin relaxation case. As we mentioned, the anomalously small matrix element for quasi-particles near the Fermi surface being scattered by phonons exactly cancels out the large density of quasi-particle states in this vicinity. On the other hand, the quasi-particles are coupled to the nuclear spins with essentially the same strength as single particles in the normal state so that the large density of quasi-particle states near the Fermi surface leads to an increased relaxation rate. Of course, at low enough temperatures few quasi-particles are excited so that the relaxation rate goes to zero as $T \to 0$. It is clear that a simple energy-gap form of a two-fluid model could not account for the sharp drop in the acoustic attenuation rate near T_c and simultaneously a rapid rise of the nuclear-spin relaxation rate. It is interesting to note that the beautiful experiments of Hebel and Slichter were being carried out during the period when the BCS theory was being formulated, and that their experiments gave one of the first substantiations of the detailed nature of the pairing correlations which are basic to the theory.

3-4 ELECTROMAGNETIC ABSORPTION

Another example of resonant energy absorption is the real part of the electrical conductivity σ_1 in a thin film. If we describe the electromagnetic field by the vector potential

$$\mathbf{A}(\mathbf{r}, t) = \mathbf{A}_0 e^{i(\mathbf{q}\cdot\mathbf{r} - \omega t)} + \text{c.c.} \tag{3-32}$$

the first-order coupling is of the form

$$\begin{aligned} H_A(t) &= -\frac{1}{c}\int \mathbf{j}(\mathbf{r}) \cdot \mathbf{A}(\mathbf{r}, t)\, d^3r \\ &= \frac{e}{2mc}\sum_{p,s} \mathbf{A}_0 \cdot (2\mathbf{p} + \mathbf{q}) c_{p+q,s}{}^{+} c_{ps} e^{-i\omega t} + \text{H.c.} \end{aligned} \tag{3-33}$$

As above, there are two terms in a given sum which lead to the same quasi-particle transition; for example, the combination

$$c_{p'\uparrow}{}^{+}c_{p\uparrow} - c_{-p\downarrow}{}^{+}c_{-p'\downarrow} = l(\mathbf{p}, \mathbf{p}')(\gamma_{p'\uparrow}{}^{+}\gamma_{p\uparrow} - \gamma_{-p\downarrow}{}^{+}\gamma_{-p'\downarrow}) - p(\mathbf{p}, \mathbf{p}')(\gamma_{p'\uparrow}{}^{+}\gamma_{-p\downarrow}{}^{+}R + \gamma_{p\uparrow}\gamma_{-p'\uparrow}R^{+}) \quad \text{(3-34)}$$

enters here due to $(2\mathbf{k} + \mathbf{q}) \to -(2\mathbf{k} + \mathbf{q})$ as $\mathbf{k} \to -(\mathbf{k} + \mathbf{q})$ and $(\mathbf{k} + \mathbf{q}) \to -\mathbf{k}$. The first terms on the right-hand side of (3-34) lead to quasi-particle scattering and contribute at $T \neq 0$ while the last terms lead to creation or destruction of two quasi-particles. They contribute only if $\omega \geqslant 2\Delta$.

For simplicity we consider only $T = 0$ so that absorption occurs only for $\omega \geqslant 2\Delta$. By calculating the rate of photon absorption just as we did for phonon absorption, one finds

$$\frac{\sigma_{1S}}{\sigma_{1N}} = \frac{1}{\omega}\int_{\Delta}^{\omega-\Delta} \frac{[E(\omega - E) - \Delta^2]\, dE}{(E^2 - \Delta^2)^{1/2}[(\omega - E)^2 - \Delta^2]^{1/2}} \quad \text{(3-35)}$$

Mattis and Bardeen[69] have carried out the integral in terms of the complete elliptic integrals E and K and find the ratio of the conductivities in the S- and N-phases is

$$\frac{\sigma_{1S}}{\sigma_{1N}} = \left(1 + \frac{1}{x}\right)E\left(\frac{1-x}{1+x}\right) - \frac{2}{x}K\left(\frac{1-x}{1+x}\right) \quad \text{(3-36)}$$

where $x = \omega/2\Delta \geqslant 1$. A plot of the theoretical ratio is shown in Figure 3-5 and is in quite good agreement with experiment. For general ω and temperature, the integrals must be done numerically. For $\omega \ll 2\Delta$, σ shows a rise as T decreases below T_c, as in the case of nuclear spin relaxation, followed by the low-temperature exponential drop. For $\omega \gtrsim kT_c/2$, the ratio no longer shows a peak. Several cases have been worked out by Miller.[70] Note: At the time of the first printing of this volume, it had been reported that a precursor absorption ($\omega/2\Delta \sim 0.85$) was observed at low temperature.[71,72] Subsequent measurements and improved processing of the data showed that the precursor was an artifact. Collective modes which were proposed to account for the precursor gave too weak an absorption to agree with experiment (see Ch. 8).

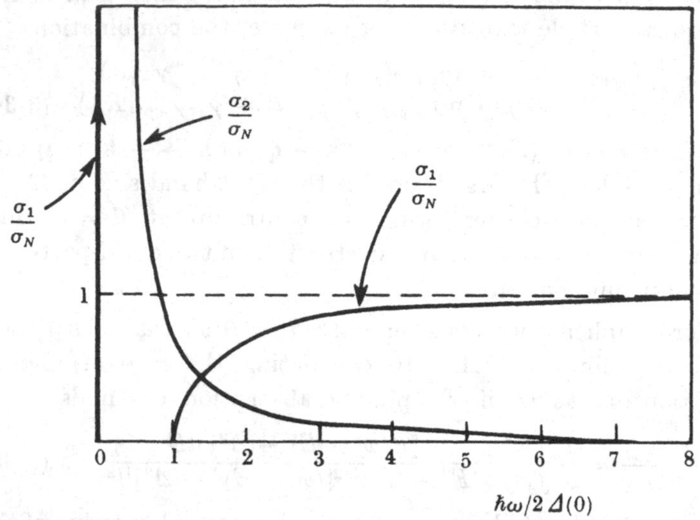

FIGURE 3-5 The frequency dependence of σ_1/σ_n and σ_2/σ_n at $T = 0$ as calculated by Tinkham from the work of Mattis and Bardeen.

3-5 PHYSICAL ORIGIN OF THE COHERENCE FACTORS

Aside from the coherence factors, one might have guessed the results of this chapter on the basis of a simple single-particle energy-gap model for the superconductor. The physical origin of the coherence factors is, however, fairly simple.

Suppose we are interested in a process in which a quasi-particle is scattered from an initial state, say $\mathbf{k}\uparrow$, to a final state, say $\mathbf{k}'\uparrow$, by absorbing a boson (a phonon or photon) of momentum $\mathbf{k}' - \mathbf{k}$. For simplicity we assume there are no quasi-particles in the states $-\mathbf{k}\downarrow$, $\mathbf{k}'\uparrow$, and $-\mathbf{k}'\downarrow$ initially. (The argument is easily generalized to include excitations in these states.)

As we saw in Chapter 2, a quasi-particle in $\mathbf{k}\uparrow$ (and none in $-\mathbf{k}\downarrow$) corresponds to an electron definitely occupying the Bloch state $\mathbf{k}\uparrow$ (i.e., with unit probability) and the mate state $-\mathbf{k}\downarrow$ being definitely empty. The pair state $(\mathbf{k}'\uparrow, -\mathbf{k}'\downarrow)$, with no quasi-particles in it, has a probability amplitude $u_{k'}$ of $\mathbf{k}'\uparrow$ *and* $-\mathbf{k}'\downarrow$ being empty, and an amplitude $v_{k'}$ of these states being simultaneously occupied. Since $u_{k'}^2 + v_{k'}^2 = 1$, there is zero amplitude for other possible occupancies of this pair state. Therefore, the initial state of the system can be viewed as having an

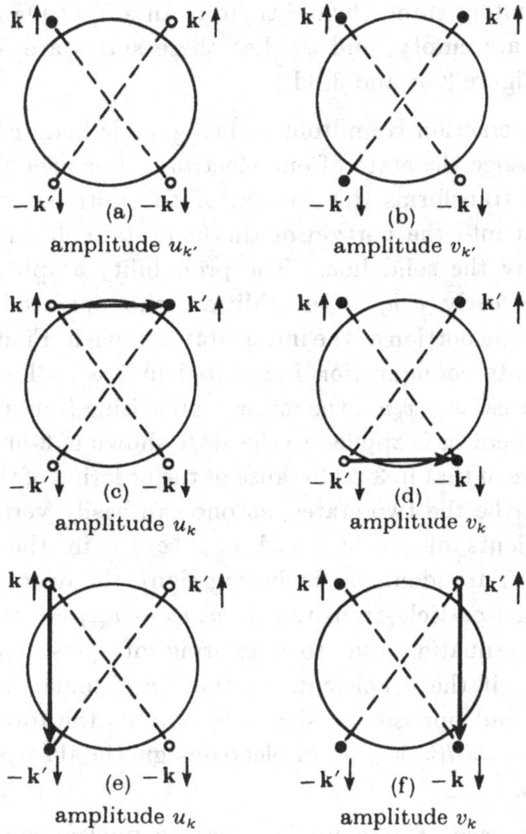

FIGURE 3-6 (a) and (b) The two configurations entering the wave function for a state with a quasi-particle in $k\uparrow$. (c) and (d) The two configurations entering the wave function for the state with a quasi-particle in $k\uparrow$, showing how $a) \to c)$ and $b) \to f)$ when the electrons couple to a field which does not flip the electronic spin (acoustic or electromagnetic fields are examples). (e) and (f) The two configurations entering for a state with a quasi-particle in $-k'\downarrow$, showing how $a) \to e)$ and $b) \to f)$when the electrons couple to a field which flips the electronic spin (the hyperfine coupling involved in nuclear spin relaxation is an example).

amplitude $u_{k'}$ for the single-particle occupancy shown in Figure 3-6a and an amplitude $v_{k'}$ for that shown in Figure 3-6b. For clarity we do not show how states other than the four mentioned above are occupied.

In analogy, the final state has $\mathbf{k}' \uparrow$ definitely occupied and $-\mathbf{k}' \downarrow$ definitely empty, while there is an amplitude u_k that $\mathbf{k} \uparrow$ *and* $-k \downarrow$ are empty, and v_k that these states are occupied, as shown in Figure 3-6c and 3-6d.

The interaction Hamiltonian, being a one-body operator, can at most change the state of one electron. Consider the operator $c_{k'\uparrow}{}^+ c_{k\uparrow}$: it transforms the portion of the initial state shown in Figure 3-6a into the portion of the final state shown in 3-6c, as indicated by the solid line. The probability amplitude for this process is clearly $u_k u_{k'}$. In addition, the operator $c_{-k\downarrow}{}^+ c_{-k'\downarrow}$ transforms the portion of the initial state shown in Figure 3-6b into the final-state configuration illustrated in 3-6d. The amplitude for this process is $-v_k v_{k'}$, the minus sign arising from the fact that when the operator is applied to the state shown in 3-6b it produces the negative of that in 3-6d because of the ordering of the operators which describe the two states, as one can easily verify. Now if the coefficients of $c_{k'\uparrow}{}^+ c_{k\uparrow}$ and $c_{-k\downarrow}{}^+ c_{-k'\downarrow}$ in the interaction Hamiltonian are identical (including sign), the overall amplitude for the quasi-particle transition is $u_k u_{k'} - v_k v_{k'} = m(\mathbf{k}, \mathbf{k}')$ as in acoustic attenuation due to scattering of quasi-particles. Alternatively, if the coefficients in the Hamiltonian are equal in magnitude but opposite in sign, one obtains the total amplitude $u_k u_{k'} + v_k v_{k'} = l(\mathbf{k}, \mathbf{k}')$, as in electromagnetic absorption by the excitations.

In the case of spin flip, as for the nuclear spin-relaxation problem, the corresponding final state would be a quasi-particle in $-k' \downarrow$ with the configurations shown in Figure 3-6e and f. In this case the operator $c_{-k'\downarrow}{}^+ c_{k\uparrow}$ transforms 3-6a into 3-6e with amplitude $u_k u_{k'}$ while $c_{-k\downarrow}{}^+ c_{k'\uparrow}$ transforms 3-6b into 3-6f with amplitude $v_k v_{k'}$. Since the coefficients of these operators are identical in the hyperfine interaction, the l-coherence factor enters here.

When two quasi-particles are created in $\mathbf{k}\uparrow$ and $-\mathbf{k}'\downarrow$ from the superfluid with no quasi-particles present initially, one has the unique final-state configuration shown in Figure 3-7a. The only configurations in the initial state which are connected to this final state by a one-body operator are shown in Figure 3-7b and c, which are transformed into 3-7a by $c_{k\uparrow}{}^{+}c_{k'\uparrow}$ and $c_{-k'\downarrow}{}^{+}c_{-k\downarrow}$ with amplitudes $u_k v_{k'}$ and $u_{k'} v_k$, respectively. For acoustic attenuation (i.e., coefficients of the *same* sign) one has the total amplitude $u_k v_{k'} + u_{k'} v_k \equiv m(\mathbf{k}, \mathbf{k}')$, while for electromagnetic absorption (opposite signs) one has $u_k v_{k'} - u_{k'} v_k \equiv p(\mathbf{k}, \mathbf{k}')$, in agreement with the results above. The spin-flip pair-creation process follows in an analogous manner.

From the above discussion it is clear that there are in general only two terms in the c, c^+ representation of a one-body operator which contribute to a given quasi-particle process and that the contribution due to each term can be understood by these simple pictures. Notice that the number of particles is explicitly

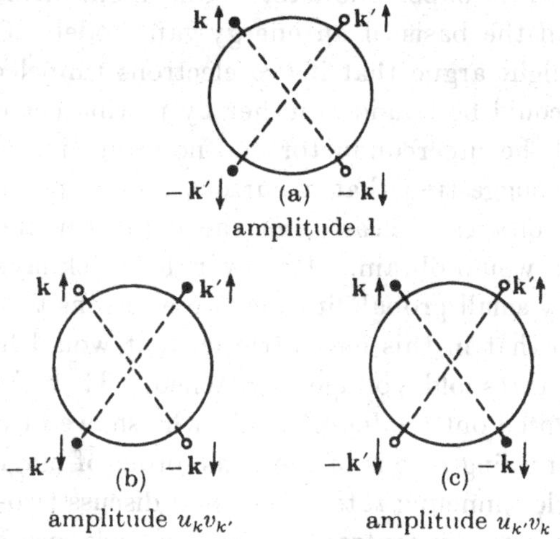

FIGURE 3-7 (a) The only configuration entering a state with quasi-particles in $k\uparrow$ and $-k'\downarrow$. (b) A configuration which is connected to that in (a) by a spin-independent one-body operator. (c) The only state other than that shown in (b) connected to that shown in (a).

conserved between the initial- and final-state configurations, although the number of particles in single-particle states other than $(\mathbf{k}\uparrow, -\mathbf{k}\downarrow)$ and $(\mathbf{k}'\uparrow, -\mathbf{k}'\downarrow)$ depend upon which configuration of these states one is considering. For example, the number of electrons not shown in Figure 3-6a is two larger than that shown in 3-6b since the total number of electrons in each configuration is exactly N_0.

3-6 ELECTRON TUNNELING

The three examples discussed above, acoustic and electromagnetic absorption and nuclear spin relaxation, all involve transitions between states of the N_0 particle system. The problem of electron tunneling between two metals which are separated by an insulating layer involves transitions between N_0 and $N_0 \pm n$ particle states of each metal. In his pioneering experiments, Giaever[26] observed that at sufficiently low temperature no current flowed between a normal metal and a superconducting metal separated by a thin oxide layer, unless the applied voltage V (multiplied by the electronic charge) exceeded the energy-gap parameter Δ of the superconductor. One might intuitively expect this result on the basis of an energy-gap model. On the other hand, one might argue that if *two* electrons tunneled simultaneously, they could be bound together by pairing correlations once they entered the superconductor and no excitation energy would be required, suggesting that a current flow is possible even for very small voltages. Clearly, as the oxide thickness vanishes, this situation would obtain. However, for thick layers one *might* expect a very small probability for two electrons to tunnel simultaneously so that in this case little current would flow until the one-particle threshold voltage is reached. That this is not the case was pointed out by Josephson,[83] who showed that the superfluid pair tunneling rate is of the same order of magnitude as the single-particle tunneling rate. We shall discuss two-particle processes below and concentrate on one-particle processes for the moment.

The foundation for a Hamiltonian formulation of the tunneling problem was laid by Bardeen[74] and refined by Cohen,

Falicov, and Phillips,[75] and more recently by Prange.[76] In this approach, one describes the system by an effective Hamiltonian

$$H = H_l + H_r + H_T \tag{3-37}$$

where H_l and H_r are the full many-body Hamiltonians for the left and right metals in the absence of tunneling, and H_T is a one-body operator which transfers electrons between the two metals,

$$H_T = \sum_{kk's} \{T_{kk'} c_{k's}^{r+} c_{ks}^{l} + \text{H.c.}\} \tag{3-38}$$

Bardeen has shown that $T_{kk'}$ is given by the matrix element of the current density at center of the oxide taken between single-particle states which decay exponentially as one moves into the oxide layer. Harrison[77] has evaluated $T_{kk'}$ within the WKB approximation and finds

$$|T_{kk'}|^2 = \frac{1}{4\pi^2} \frac{\delta_{k_\parallel, k_\parallel'}}{\rho_\perp^r \rho_\perp^l} \exp\left[-2\int_{x_l}^{x_r} k_\perp(x)\, dx\right] \tag{3-39}$$

where $\rho_\perp$ is the one-dimensional density of states for motion normal to the barrier which has boundaries of x_l and x_r.

To calculate the tunneling current,[78, 79] we begin with the zero-temperature case and treat H_T by first-order time-dependent perturbation theory. The rate of transferring electrons from l to r is

$$w_{l\to r} = 2\pi \sum_{\alpha,\beta} |\langle\alpha_l|\langle\beta_r| \sum_{kk's} T_{kk'} c_{k's}^{r+} c_{ks}^{l} |0_l\rangle|0_r\rangle|^2\, \delta(\epsilon_\alpha + \epsilon_\beta - V) \tag{3-40}$$

if the electrons decrease their potential energy by V in moving from l to r due to the applied bias. In (3-40) the state vectors are the exact many-body eigenstates of H_l and H_r, that is,

$$\begin{aligned} H_l|\alpha_l\rangle &= \epsilon_\alpha|\alpha_l\rangle \\ H_r|\beta_r\rangle &= \epsilon_\beta|\beta_r\rangle \end{aligned} \tag{3-41}$$

and the energies ϵ_α and ϵ_β are measured relative to the ground-state energies in l and r, respectively. At zero temperature electrons cannot tunnel in the reverse direction due to energy conservation. From (3-40) one can readily see that the current density is proportional to

$$I(V) \propto \int_0^V N_{T+}{}^r(E) N_{T-}{}^l(V - E)\, dE \tag{3-42}$$

where

$$\begin{aligned} N_{T+}{}^r(E) &= \sum_{k,\beta} |\langle \beta_r | c_{k'}{}^+ | 0_r \rangle|^2\, \delta(\epsilon_\beta - E) \\ &\cong N_r(0) \int_{-\infty}^{\infty} d\epsilon_k \rho_r{}^{(+)}(\mathbf{k}, E) \end{aligned} \tag{3-43a}$$

and

$$\begin{aligned} N_{T-}{}^l(E) &= \sum_{k,\alpha} |\langle \alpha_l | c_k | 0_l \rangle|^2\, \delta(\epsilon_\alpha - E) \\ &\cong N_l(0) \int_{-\infty}^{\infty} d\epsilon_k \rho_l{}^{(-)}(\mathbf{k}, E) \end{aligned} \tag{3-43b}$$

The spectral weight functions $\rho^{(+)}$ and $\rho^{(-)}$ are discussed in Section 5-7 and are related to the one-electron Green's function G by

$$\rho^{(+)}(\mathbf{k}, \omega) = -\frac{1}{\pi} \operatorname{Im} G(\mathbf{k}, \omega) \qquad \omega \geqslant 0 \tag{3-44a}$$

$$\rho^{(-)}(\mathbf{k}, \omega) = \frac{1}{\pi} \operatorname{Im} G(\mathbf{k}, -\omega) \qquad \omega \geqslant 0 \tag{3-44b}$$

(See Chapter 5.) Thus a knowledge of $G(\mathbf{k}, \omega)$ for each metal suffices to determine the tunneling current. We note that in deriving (3-42) we have assumed $T_{kk'}$ = const. in an energy region $V(\ll E_F)$ about the Fermi surface, which is a very good approximation for voltages of interest in investigating the superconductor aspects of the tunneling characteristic.

The effective tunneling density of states for a superconducting metal can be calculated in the pairing approximation with a nonretarded two-body potential directly from the definition (3-43). By using the particle conserving B–V transformation (3-6) we find

$$\begin{aligned} N_{T+}(E) &= \sum_{k,p} |\langle p | c_{k\uparrow}{}^+ | 0 \rangle|^2\, \delta(E_p - E) \\ &\cong N(0) \int_{-\infty}^{\infty} d\epsilon_k u_k{}^2\, \delta(E_k - E) \\ &= N(0) \left| \frac{d\epsilon_k}{dE_k} \right|_{E_k = E} \end{aligned} \tag{3-45}$$

where we have used the fact that if a given ϵ_k makes the argument of the delta function vanish, so does $\epsilon_{k'} \equiv -\epsilon_k$ so that $u_k^2 + u_{k'}^2 = 1$. This simple manner in which the coherence factors vanish in the expression for the tunneling current was first pointed out by Cohen, Falicov, and Phillips.[75] It is interesting to note that N_{T+} is just the density of quasi-particle states which we used earlier in this chapter, as one would have guessed on the basis of a simple energy-gap model without coherence effects.

In a similar manner one finds for a nonretarded pairing potential model

$$N_{T-}(E) = N(0) \int_{-\infty}^{\infty} v_k^2 \, \delta(E_k - E) \, d\epsilon_k$$
$$= N(0) \left| \frac{d\epsilon_k}{dE_k} \right|_{E_k = E} = N_{T+}(E) \tag{3-46}$$

where the coherence factors v_k^2 vanishes in the result just as u_k^2 did in N_{T+}.

As we shall see in Chapter 7, the expressions (3-45) and (3-46) are incorrect in real metals due to the strong retardation effects associated with the phonon interaction between electrons. There one finds the simple result

$$N_{T\pm}(E) = N(0) \, \mathrm{Re} \left\{ \frac{E}{[E^2 - \Delta^2(E)]^{1/2}} \right\} \tag{3-47}$$

as opposed to $N(0)(E - \frac{1}{2} \, d\Delta^2/dE)/[E^2 - \Delta^2(E)]^{1/2}$, which follows from (3-46).

Returning to expression (3-42) for $I(V)$ we find for tunneling between a normal and a superconducting metal that

$$\frac{dI_S/dV}{dI_N/dV} = \mathrm{Re} \left\{ \frac{V}{[V^2 - \Delta^2(V)]^{1/2}} \right\} \tag{3-48}$$

if we use (3-47), where I_S and I_N are the currents flowing when the superconductor is in the S- or N-state, respectively. Therefore, the tunneling experiment can give detailed information about the energy dependence of the gap parameter.

The finite-temperature tunneling current can be treated in an analogous manner if one includes a thermodynamic average over initial states rather than using $|0_l\rangle|0_r\rangle$, and further one includes currents from l to r and r to l.[78, 79] The tunneling current can then be expressed in terms of the spectral weight functions for the thermodynamic Green's functions. In the simple case of a non-retarded two-body pairing potential, the expression reduces to the golden rule result for a simple energy-gap model without coherence effects, as for $T = 0$. A typical I–V characteristic for tunneling between a normal and a superconducting metal is shown in Figure 3-8 for several temperatures for an energy independent Δ. Experimental curves are in general agreement with theory although small deviations exist, some of which we shall discuss below.

A pictorial view of the one-particle tunneling process between a normal and a superconducting metal is illustrated in Figure 3-9a and b. In 3-9a, an electron in $\mathbf{k}\uparrow$, beneath the Fermi surface in the normal metal tunnels through the oxide to state $\mathbf{k}'\uparrow$ above the Fermi surface in the superconductor. In the process a hole is left behind in l giving an excitation energy $\epsilon_\alpha = |\epsilon_k|$ for this metal. In addition, a quasi-particle is placed in $\mathbf{k}'\uparrow$ giving an excitation energy $\epsilon_\beta = E_{k'} = (\epsilon_{k'}{}^2 + \Delta_{k'}{}^2)^{1/2}$ for the

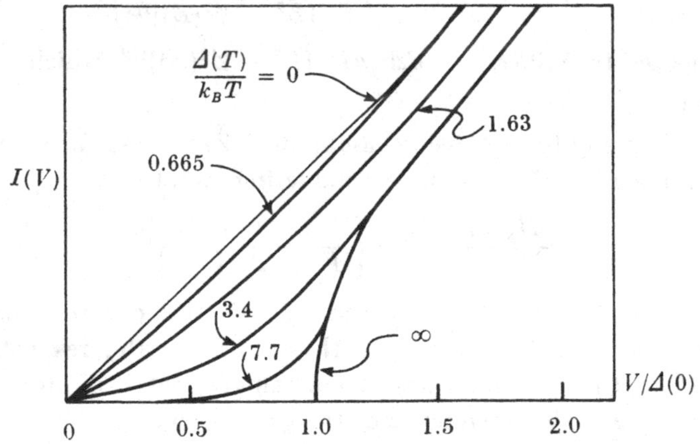

FIGURE 3-8 The tunneling current I between a normal and a superconducting metal as a function of the applied voltage V (multiplied by the electronic charge).

superconductor. This process can go only if the pair state $(\mathbf{k}'\uparrow \; -\mathbf{k}'\downarrow)$ is initially empty; this occurs with the probability $u_{k'}^2$. Energy is conserved if $|\epsilon_k| + E_{k'} = V$. Another energy-conserving process shown in Figure 3-9b is identical to that shown in 3-9a, except for the quasi-particle being placed in a state $\mathbf{k}''\uparrow$ beneath the Fermi surface. For Δ = constant, the two states are related by $\epsilon_{k'} = -\epsilon_{k''}$, as we saw above. The probability that $(\mathbf{k}''\uparrow, -\mathbf{k}''\downarrow)$ is initially empty is $u_{k''}^2 = v_{k'}^2$ so that the total probability for the process to go is $u_{k'}^2 + v_{k'}^2 = 1$, as far as the Pauli principle restrictions are concerned. The tunneling current is then given by summing only over states above (or below) the Fermi surface in the superconductor and replacing the coherence factor by unity, as we saw above.

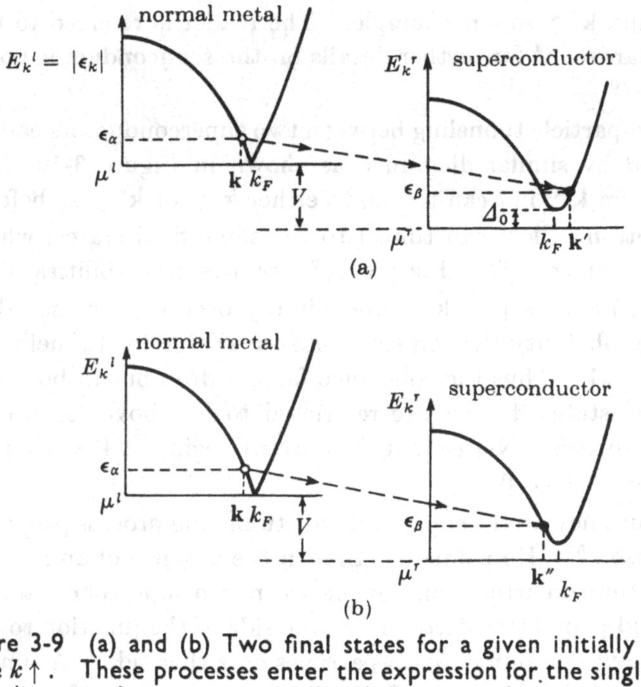

Figure 3-9 (a) and (b) Two final states for a given initially occupied state $k\uparrow$. These processes enter the expression for the single-particle tunneling rate between a normal and a superconducting metal.

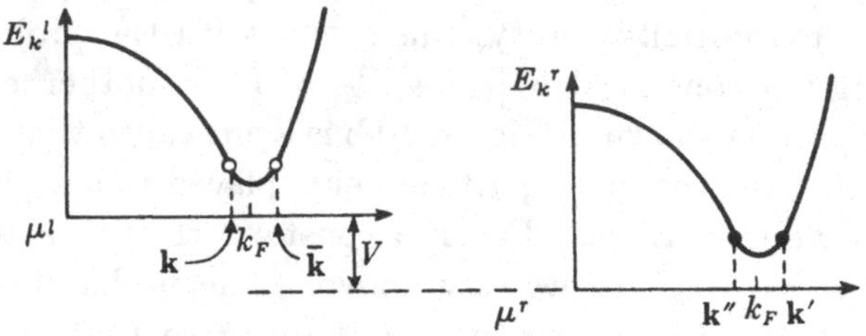

Figure 3-10 (a) and (b) Processes analogous to those in Figure 3-9, however the tunneling is between superconductors here.

A semiconductor model of the excitations in the superconductor is often helpful for discussing tunneling phenomena. However, the model must be used with caution since states "above the energy gap" in this model are really linear combinations of quasi-particle states above and below the Fermi surface, i.e., as are $\mathbf{k}' \uparrow$ and $\mathbf{k}'' \uparrow$ in our example. The reader is referred to the work of Bardeen[74] for further details on the semiconductor point of view.[79, 80]

Single-particle tunneling between two superconductors can be understood by similar diagrams, as shown in Figure 3-10. An electron from $\mathbf{k} \uparrow$ in l can tunnel to either $\mathbf{k}' \uparrow$ or $\mathbf{k}'' \uparrow$ as before, or an electron in $\bar{\mathbf{k}} \uparrow$ can tunnel to the same final states, where $\epsilon_k = -\epsilon_{\bar{k}}$. Since $v_k{}^2$ and $v_{\bar{k}}{}^2 = u_k{}^2$ are the probabilities that $(\mathbf{k}\uparrow, -\mathbf{k}\downarrow)$ and $(\bar{\mathbf{k}}\uparrow, -\bar{\mathbf{k}}\downarrow)$ are initially occupied, we see that the total probability that an electron is available for tunneling is $u_k{}^2 + v_k{}^2 = 1$. Thus the coherence factors drop out in both the initial final states if sums are restricted to be above (or below) the Fermi surface. Notice that the current begins at $V = \Delta_l + \Delta_r$ in this case at $T = 0$.

We turn now to the superfluid pair tunneling process proposed by Josephson.[83] He points out that in the absence of an applied voltage a tunnel current can flow between two superconductors if a superfluid pair is transferred from one side of the junction to the other without creation of quasi-particles on either side. A simple derivation of Josephson's effect, due to Josephson[83] and to

Anderson[84] makes the physics of the situation quite clear. In the absence of both an applied potential across the barrier and the tunneling Hamiltonian, no energy is required to transfer ν superfluid pairs from one side of the barrier to the other. If the state with ν pairs transferred from left to right (relative to a standard state) is denoted by Φ_ν, the "tight-binding approximation" for the exact eigenstates in the presence of the tunneling operator is

$$\Psi_\alpha = \sum_\nu e^{i\alpha\nu}\,\Phi_\nu \tag{3-49}$$

The canonical momentum α plays the role of the wave number k in band theory. Since H_T can only transfer one electron between the materials, the coupling between the Φ_ν's is second order in T and one finds for the energy shift due to H_T of the eigenstates

$$E_\alpha = \frac{\langle \Psi_\alpha | H_T^{(2)} | \Psi_\alpha \rangle}{\langle \Psi_\alpha | \Psi_\alpha \rangle} = -\frac{\hbar J_1}{2}\cos\alpha \tag{3-50}$$

where $H_T^{(2)}$ is the second-order tunneling Hamiltonian given by

$$H_T^{(2)} = H_T \frac{1}{E - H_0} H_T \tag{3-51}$$

and

$$\hbar J_1 = 4|\langle \Phi_{\nu+1} | H_T^{(2)} | \Phi_\nu \rangle| \tag{3-52}$$

To find the current, note that the rate of transfer of pairs is

$$\frac{d\langle\nu\rangle}{dt} = \left\langle \frac{dE_\alpha}{d\hbar_\alpha} \right\rangle = \frac{J_1}{2}\langle \sin\alpha \rangle \tag{3-53a}$$

where the average is taken in a wave-packet state formed from the Ψ_α's, in complete analogy with the tight-binding approach to the one-electron theory of metals. In the absence of an applied bias, the momentum $\hbar\alpha$ (canonically conjugate to the pair number ν) is a constant of motion; however, for $V \neq 0$ one has

$$\frac{d\langle\hbar\alpha\rangle}{dt} = 2V \tag{3-53b}$$

From (3-53a) and (3-53b) it follows that the rate at which electrons flow across the barrier is

$$J(t) = \frac{2\,d\langle\nu\rangle}{dt} = J_1 \sin\frac{2Vt}{\hbar} + \alpha_0 \tag{3–54}$$

so that an alternating current of frequency $2V/h = 483.6$ Mc/sec/μvolt is expected to flow for $V \neq 0$, while for $V = 0$, a steady current is expected, according to (3-53a). The d-c effect has been observed by Rowell and Anderson.[85] As Josephson pointed out,[83] the current is sharply reduced when a magnetic field is applied to the junction of such a strength that a multiple of the flux quantum occurs in the junction. This effect has been observed by Rowell.[86] The a–c effect has been observed by Shapiro.[86c]

Tunneling experiments by Burstein and Taylor[81] show that in many cases an excess current between two superconductors at low reduced temperature begins at an applied bias Δ_l or Δ_r, that

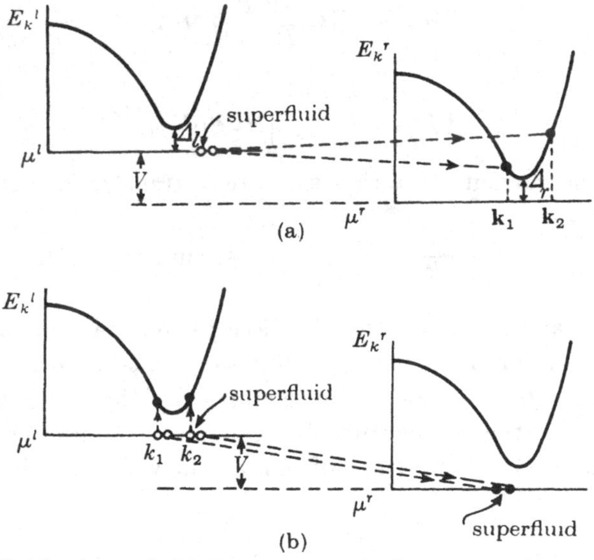

FIGURE 3-11 (a) and (b) Processes contributing to the two-particle tunneling mechanism which leads to current onsets at Δ_l and Δ_r.

is, below the one-particle threshold $\Delta_l + \Delta_r$. Wilkins and the author[82] have proposed a two-particle tunneling mechanism involving the superfluid electrons to explain the phenomenon. Their processes, illustrated in Figure 3-11a and b involve a second-order matrix element in which a pair of superfluid electrons in l is withdrawn without creating excitations in l, and the electrons tunnel to r, where they occupy quasi-particle states. This process is shown in 3-11a and gives a current onset at $\Delta_r + V$. A second process shown in 3-11b involves removing two superfluid electrons from l but leaving two extra quasi-particles behind. The two electrons tunnel to r, where they recombine to enter the superfluid, thereby creating no excitations in r. This process has an onset at $\Delta_l = V$. The processes give a polarity-independent, temperature-insensitive excess current, as observed. To fit the magnitude of the observed currents one must assume the oxide films are patchy, having a small fraction of thin regions so that the fourth-order matrix element $|T|^4$ which appears in the two-particle tunneling rate does not lead to a drastic reduction of the excess current relative to the single-particle rate which involves $|T|^2$. The oxide thicknesses and the ratio of the areas of thick and thin regions required to obtain agreement with experiment do not appear to be unreasonable, considering the imperfect nature of the oxide films.

3-7 OTHER APPLICATIONS OF THE PAIRING THEORY

In this chapter only a few of the simplest applications of the pairing theory have been discussed. In Chapter 8 we shall discuss the electromagnetic properties of superconductors, i.e., the Meissner effect, the persistence of supercurrents, magnetic flux quantization, etc., as well as the paramagnetic spin susceptibility and the resultant Knight shift. Further support for the pairing theory comes from the distinctly different effects magnetic and nonmagnetic impurities have on the energy gap. As Anderson showed, nonmagnetic impurities do not smear the gap edge as one might intuitively expect; on the contrary, nonmagnetic impurities remove the crystalline anisotrophy of the gap and thereby effectively sharpen the observed energy gap. In Anderson's discussion the time-reversal invariance of the impurity scattering potential plays an essential role. Abrikosov and Gor'kov have discussed the case of magnetic impurities. Owing to the lack of time-reversal invariance, they find a broadening of the gap edge. As the impurity concentration is increased to a critical value ($\sim 1\%$) the energy gap vanishes, although the density of states is smaller than that in the normal phase near the Fermi surface. For a small range of concentration above this critical value, the material exhibits "gapless" superconductivity. This effect was discovered experimentally by Reif and Woolf in tunneling experiments.

The reader is referred to the literature[9, 16] for a discussion of numerous other applications which give further empirical support for the theory, e.g., thermal conductivity, boundary effects and small specimens, type II superconductors, etc. In general the agreement between theory and experiment is remarkably good, as mentioned above, considering the simplicity of the models used in applying the basic ideas of the pairing theory.

CHAPTER 4

ELECTRON–ION SYSTEM

As we have seen above, many of the observed properties of superconductors can be explained on the basis of a model in which an attractive velocity-dependent potential acts between pairs of quasi-particles of the normal metal. However, from the isotope effect[11, 12] and from electron-tunneling current anomalies[87, 88] we know that the electron–phonon interaction plays an essential role in bringing about superconductivity in most (if not all) superconductors.

4-1 THE ELECTRON–ION HAMILTONIAN

To obtain a more complete understanding of superconductivity, we should study the full electron–ion system and show how the results of the simplified model discussed in Chapters 2 and 3 emerge from "first principles." In particular, one would hope to be able to (a) explain the isotope effect; (b) determine why certain metals are superconductors and others are not; (c) isolate those parameters which determine the transition temperature; (d) check the basic nature of the pairing interaction by explaining the observed tunneling density-of-states anomalies. In addition, one would like to account for the observed deviations from the empirical "law of corresponding states." This "law" states that all superconductors have identical properties when these properties are expressed in reduced units; e.g., the critical field curve $H_c(T)/H_0$ is a universal function of the reduced temperature T/T_c, etc. Although it is not possible to reach all of these goals at present, the basic theoretical framework which is required to treat many-body effects in real metals is developing rapidly. It is likely that the above questions will be resolved in the foreseeable future.

In our discussion we concentrate on a simplified model of a metal in which ions, with their core electrons rigidly attached, interact with a sea of conduction electrons. Of course, the ions interact among themselves, as do the conduction electrons. Thus we assume that the core electrons adiabatically follow the vibrating nuclei but are otherwise unexcited. While our approximation neglects core polarization, this effect presumably has a minor influence on the dynamics of the conduction electrons in most superconductors because of the large energy required to excite the cores. Core polarization may well play a more important role in determining the ion–ion interaction; however, this effect can be included approximately in the ion–ion potential.

For simplicity we shall also neglect spin–orbit interactions, although as we shall see later they have been invoked as a possible explanation for the nonvanishing Knight shift observed in the superconducting state.[89, 90] The hyperfine and electron spin–spin interactions will also be neglected for the moment; the former were treated as a perturbation when we considered nuclear-spin relaxation processes (see Chapter 3).

It is particularly convenient to omit for the moment the *magnetic* interaction between conduction electrons due to their orbital motion. This interaction is exceedingly important in understanding superconductivity because it leads to the Meissner effect. The reason that the omission is not damaging at the outset is that in the absence of external fields (and/or net currents in the material) the magnetic forces between electrons almost exactly cancel each other and therefore lead to weak effects.

When external fields and/or net currents exist there is a coherent addition of the fluctuating magnetic fields between electrons; the resultant field can be conveniently treated within a self-consistent field approximation and fluctuations about the average field again neglected. We shall adopt this view in treating the magnetic properties of superconductors in Chapter 8.

The Hamiltonian of the conduction electron–ion system is then

$$H = \sum_i \frac{\mathbf{p}_i^2}{2m} + \frac{1}{2}\sum_{i \neq j} \frac{e^2}{|\mathbf{r}_i - \mathbf{r}_j|} + \sum_\nu \frac{\mathbf{P}_\nu^2}{2M_\nu} + \frac{1}{2}\sum_{\nu \neq \nu'} W(\mathbf{R}_\nu, \mathbf{R}_{\nu'}) + \sum_{i,\nu} U(\mathbf{r}_i, \mathbf{R}_\nu) \quad (4\text{-}1)$$

where $\mathbf{r}_i$ is the position of the ith conduction electron, and $\mathbf{R}_\nu$ the position of the νth ion whose equilibrium position is at $\mathbf{R}_\nu^0$. The index ν labels both the cell n and the site α in this cell at which R_ν^0 is located. For a crystal with one atom per unit cell the label α is superfluous and the ions will be labeled by n. The first two terms in (4-1) represent the kinetic energy of the conduction electrons and the Coulomb interactions between them. The third and fourth terms give the kinetic energy of the ions and the ion–ion interaction. The last term in (4-1), representing the conduction electron–ion interaction, is in general nondiagonal in the electron coordinate representation because of exchange interactions between the conduction and core electrons.

We work with a system of unit volume and use periodic boundary conditions. Our basic approach is to cast this complicated many-body Hamiltonian into a form which will allow us to use well-known techniques of quantum field theory. There, one traditionally begins with a set of "bare" particles and through a consistent treatment of the coupling between these particles one derives a set of "dressed" particles in terms of which properties of the physical system can be described. The richness of various approximation procedures and the relative ease of calculation make this approach highly attractive. An introduction to field-theoretic techniques in the many-body problem[91] is given in Chapters 5 and 6.

The "bare" particles in our case will be of two types: single electrons occupying Bloch states, and quantized vibrations of the ionic lattice. The coupling between these excitations causes drastic changes in the system which cannot be treated by elementary perturbation theory. Nevertheless, the methods of quantum field theory are sufficiently powerful to allow us to understand the origin of the superconducting state and make detailed theoretical predictions regarding its properties.

4-2 BARE PHONONS

We introduce bare-phonon coordinates $Q_{q,\lambda}$ of wavevector $\mathbf{q}$ and polarization λ which describe the deviations $\delta\mathbf{R}_\nu$ of the ions from their equilibrium positions $\mathbf{R}_\nu{}^0$. This is done by performing the canonical transformation

$$\mathbf{R}_{n\alpha} = \mathbf{R}_{n\alpha}{}^0 + \frac{1}{(N_c M_c)^{1/2}} \sum_{q,\lambda} Q_{q\lambda} \boldsymbol{\epsilon}_{q\lambda}(\alpha) e^{i\mathbf{q}\cdot\mathbf{R}_{n\alpha}{}^0} \tag{4-2}$$

The Q's are normal coordinates of the vibrating ion system when the ion–ion interaction is treated within the harmonic approximation. This procedure is discussed in detail by Peierls.[92] In (4-2), N_c is the number of unit cells per unit volume and M_c the total ionic mass within a unit cell. The polarization vectors $\boldsymbol{\epsilon}_{q,\lambda}(\alpha)$ are determined by solving the above normal-mode problem. It is convenient to normalize these vectors by requiring

$$\sum_\alpha M_\alpha |\boldsymbol{\epsilon}_{q\lambda}(\alpha)|^2 = M_c \tag{4-3}$$

For one atom per unit cell, this reduces to $|\boldsymbol{\epsilon}_{q,\lambda}| = 1$. The normalization condition and an orthogonality condition for the $\boldsymbol{\epsilon}$'s can be written as

$$\sum_\alpha M_\alpha \boldsymbol{\epsilon}_{q\lambda}(\alpha) \cdot \boldsymbol{\epsilon}_{-q\lambda'}(\alpha) = M_c\, \delta_{\lambda\lambda'} \tag{4-4}$$

The wavevectors $\mathbf{q}$ are restricted to the first Brillouin zone (which contains N_c points). Since the density of states in $\mathbf{q}$-space of phonons with a given polarization is $1/(2\pi)^3$, we shall often replace sums over $\mathbf{q}$ by integrals through the relation

$$\sum_q \rightarrow \frac{1}{(2\pi)^3}\int d^3q \tag{4-5}$$

The number α_0 of independent polarizations (i.e., branches in the phonon spectrum) is three times the number of ions per unit cell, so that the total number of phonon modes is equal to the number of degrees of freedom of the ionic lattice, as expected. There also exists a completeness relation of the form

$$\sum_{q,\lambda} \boldsymbol{\epsilon}_{q\lambda}(\alpha)\cdot\boldsymbol{\epsilon}_{-q\lambda}(\alpha')e^{i\mathbf{q}\cdot(\mathbf{R}_{n\alpha}{}^0-\mathbf{R}_{n'\alpha'}{}^0)} = \delta_{nn'}\,\delta_{\alpha\alpha'} \tag{4-6}$$

It is conventional to call the three branches, which in the long wavelength limit have all ions in a unit cell moving in phase, acoustic branches. The remaining branches are called optical. If the wave vector $\mathbf{q}$ is along certain symmetry directions in the crystal, the polarization vectors $\boldsymbol{\epsilon}_{q,\lambda}$ are either parallel (for longitudinal phonons) or perpendicular (for transverse phonons) to $\mathbf{q}$. In general, the polarization vectors bear no simple relation to $\mathbf{q}$ although one continues to speak of longitudinal and transverse phonons, keeping the same designation of a given branch as $\mathbf{q}$ moves away from a symmetry direction.

While the transverse acoustic modes tend to zero frequency as $\mathbf{q}\rightarrow 0$, the longitudinal modes tend to the ionic-plasma frequency $\Omega_p = (4\pi N_c z_c{}^2 e^2/M_c)^{1/2}$ due to the long-range nature of the Coulomb force.[93] We know that in real solids the frequency of a longitudinal acoustic sound wave is proportional to q and vanishes as $q\rightarrow 0$; clearly, the difference comes from the conduction electrons responding to the electric field set up by the ionic oscillations and screening out the long-range force. In our scheme, the screening is due to the electron–phonon and electron–electron interactions. The main point is that, while the shift in the phonon frequency is large (from $\Omega_p \rightarrow \sim 0$), the problem is simply treated by standard field-theoretic techniques. Of course, simpler methods could be used for this problem (for example, the Thomas–Fermi approximation or a time-independent self-consistent field approach);

however, these methods are not general enough to treat superconductivity. By working out simple problems with the more elaborate scheme we shall be better able to understand how to approach the problem of superconductivity.

To complete the dynamics of the bare phonons, the momentum part of the canonical transformation (4-2) is given by

$$\mathbf{P}_{n\alpha} = \left(\frac{M_c}{N_c}\right)^{1/2} \sum_{q,\lambda} \Pi_{q\lambda} \boldsymbol{\epsilon}_{-q\lambda}(\alpha) e^{-i\mathbf{q}\cdot\mathbf{R}_{n\alpha}{}^0} \tag{4-7}$$

where $\Pi_{q,\lambda}$ is the phonon momentum. It follows from the canonical commutation relations for the ion variables,

$$[\mathbf{P}_{n\alpha}, \mathbf{R}_{n'\alpha'}] = \frac{\hbar}{i} \delta_{nn'} \delta_{\alpha\alpha'} \mathbf{1} \tag{4-8a}$$

$$[\mathbf{P}_{n\alpha}, \mathbf{P}_{n'\alpha'}] = [\mathbf{R}_{n\alpha}, \mathbf{R}_{n'\alpha'}] = 0 \tag{4-8b}$$

that the phonon variables also satisfy canonical commutation relations

$$[\Pi_{q\lambda}, Q_{q'\lambda'}] = \frac{\hbar}{i} \delta_{qq'} \delta_{\lambda\lambda'} \tag{4-9a}$$

$$[\Pi_{q\lambda}, \Pi_{q'\lambda'}] = [Q_{q\lambda}, Q_{q'\lambda'}] = 0 \tag{4-9b}$$

In (4-8a), $\mathbf{1}$ is the unit tensor. As we shall see, the fact that phonons are bosons has nothing to do with the spin of the individual ions. This is clear since we have treated the ions as being distinguishable by localizing them near lattice sites. The Bose character of a phonon simply reflects the quantum-mechanical commutation rules applied to the individual ions.

With the aid of above relations it is straightforward to transform the ionic kinetic energy plus the ion–ion interaction (treated to second order in $\delta\mathbf{R}$) into phonon coordinates. One finds

$$\sum_{\nu} \frac{\mathbf{P}_\nu{}^2}{2M_\nu} + \frac{1}{2} \sum_{\nu \neq \nu'} W_{\nu,\nu'} \simeq \frac{1}{2} \sum_{q,\lambda} \{\Pi_{q\lambda}{}^+ \Pi_{q\lambda} + \Omega_{q\lambda}{}^2 Q_{q\lambda}{}^+ Q_{q\lambda}\} + \text{const.} \tag{4-10}$$

where $\Omega_{q,\lambda}$ is the normal-mode frequency and the constant term is the energy of the ion system when the ions are located on their equilibrium sites. We introduce bare-phonon creation and destruction operators $a_{q,\lambda}^+$ and $a_{q,\lambda}$ by

$$Q_{q\lambda} = \left(\frac{\hbar}{2\Omega_{q\lambda}}\right)^{1/2} (a_{q\lambda} + a_{-q\lambda}^+) \qquad \text{(4-11a)}$$

and

$$\Pi_{q\lambda} = i\left(\frac{\hbar\Omega_{q\lambda}}{2}\right)^{1/2} (a_{q\lambda}^+ - a_{-q\lambda}) \qquad \text{(4-11b)}$$

The formalism of second quantization, which deals with these operators, is discussed in the Appendix. It follows from (4-9) that the a's satisfy Bose commutation relations:

$$[a_{q\lambda}, a_{q'\lambda'}^+] = \delta_{qq'}\,\delta_{\lambda\lambda'} \qquad \text{(4-12a)}$$

$$[a_{q\lambda}, a_{q'\lambda'}] = [a_{q\lambda}^+, a_{q'\lambda'}^+] = 0 \qquad \text{(4-12b)}$$

The bare-phonon Hamiltonian (4-10) then becomes

$$H_{\text{ph}} = \sum_{q,\lambda} \hbar\Omega_{q\lambda}(N_{q\lambda} + \tfrac{1}{2}) \qquad \text{(4-13)}$$

where $N_{q,\lambda} = a_{q,\lambda}^+ a_{q,\lambda}$ is the phonon number operator and we have dropped the constant term in (4-10). H_{ph} will be one of the two terms in our zeroth-order Hamiltonian for the system.

The anharmonic terms neglected in (4-10) presumably have little effect on superconductivity since the volume change between the N- and S-states is small.[4] Also, since the transition temperature T_c is small compared to the Debye temperature, the amplitude of the ionic vibration is expected to be small below T_c, as our approximation requires.

4-3 BARE ELECTRONS

We would like to introduce a set of one-electron eigenstates χ_k to describe the bare conduction electrons; however, a difficulty arises. If we introduce a one-body potential U_0 and require that χ_k satisfies the Schrödinger equation

$$\left[\frac{\mathbf{p}^2}{2m} + U_0\right]\chi_k = \epsilon_k\chi_k \qquad \text{(4-14)}$$

then the states χ_k are not in general orthogonal to the core states. Even if we arrange U_0 so that the "conduction band" solutions of

(4-14) are orthogonal to the core states when the ions are in their equilibrium positions, the orthogonality is not maintained when the ions vibrate. A partial solution to this difficulty has been given by Wilkins,[94] using the pseudo-potential method of Kleinman and Phillips. This work was generalized to the many-body problem in the work of Bassani, Robinson, Goodman, and the author,[95] who treat the case of rigid cores described within the one-electron approximation. We shall not discuss this treatment here because of the mathematical complications necessary to carry through the analysis. It suffices to say that an auxiliary wave field describing the conduction electrons can be introduced in such a manner that the conduction and core states are properly orthogonal even if the cores vibrate. The equations of motion of this auxiliary wave field are the same as for the original wave field except for a redefinition of the potentials involved. Since these potentials are difficult to estimate from first principles at present, we shall simply disregard the above complication and proceed using the one-electron states (4-14) as the bare conduction electron states.

To make the states χ_k precise, we must define U_0. In order that the electrons are not scattered by the lattice when no phonons are present, U_0 should include the electron–ion interaction with the ions fixed on their equilibrium positions. Since the ion system has a large positive charge, this leads to a very large negative potential acting on a conduction electron. Since the Coulomb interactions with the remaining conduction electrons cancel most of this interaction, we include in U_0 the potential due to the remaining conduction electrons occupying a standard configuration. This configuration could be a uniform distribution of electronic charge or the distribution given by treating the conduction electrons within the Hartree–Fock approximation. Of course the better one does in choosing U_0, the less there is to take into account as coupling between the bare particles. In any event, U_0 should be chosen to have the periodicity of the lattice (although it may not be diagonal in the coordinate representation) so that Bloch's theorem[96] holds

$$\chi_k(\mathbf{r} + \mathbf{a}) = e^{i\mathbf{k}\cdot\mathbf{a}}\chi_k(\mathbf{r}) \tag{4-15}$$

where $\mathbf{a}$ is any translation under which the crystal lattice is invariant. This relation defines the "crystal momentum" $\hbar\mathbf{k}$ of the state in question. While χ_k is an eigenfunction of crystal momentum, it is not in general an eigenfunction of physical momentum (i.e., a plane wave), since it can be represented as a linear combination of plane waves of wave vectors $\mathbf{k} + \mathbf{K}_n$. The vectors $\mathbf{K}_n$ are reciprocal lattice vectors defined by

$$\mathbf{K}_n \cdot \mathbf{a} = 2\pi \times \text{integer} \tag{4-16}$$

where $\mathbf{a}$ is an arbitrary allowed translation of the lattice. These reciprocal lattice vectors play an important role in the theory of solids since crystal momentum is conserved mod $\hbar\mathbf{K}_n$ in dynamical processes, and it is crystal rather than physical momentum we shall usually encounter.

While the curves of ϵ versus $\mathbf{k}$ are fairly simple for alkali metals, the situation becomes quite involved for polyvalent metals. The reader is referred to Ziman's book[96] for a discussion of theoretical and experimental results in this rapidly developing field. The over-all picture is greatly simplified if the energy states are represented in an extended zone scheme as one does for a free electron, rather than folding the curves back into the first Brillouin zone. In many cases one finds that a large fraction of the energy surfaces resemble the free-electron case except for discontinuities of the energy as one crosses zone boundaries. Since a major part of our calculations will emphasize states near the Fermi surface and since in general a small fraction of the Fermi surface which is effective lies near zone boundaries, one hopes that an effective mass approximation will adequately represent the gross features of the band structure. We shall often make this approximation to simplify the mathematics although the approximation is not essential.

We shall use the formalism of second quantization to treat the conduction electrons; this scheme is reviewed in the Appendix. The creation and destruction operators for an electron in state $\mathbf{k}$ with z-component of spin s are defined to be $c_{k,s}{}^{+}$ and $c_{k,s}$, respectively. These operators satisfy Fermi anticommutation relations:

$$\{c_{ks}, c_{k's'}^{+}\} = \delta_{kk'}\,\delta_{ss'} \tag{4-17a}$$

$$\{c_{ks}, c_{k's'}\} = \{c_{ks}^{+}, c_{k's'}^{+}\} = 0 \tag{4-17b}$$

The bare electron Hamiltonian is then

$$H_{\text{el}} = \sum_{k,s} \epsilon_k n_{ks} \tag{4-18}$$

where $N_{ks} = c_{ks}^{+}c_{ks}$ is the electron number operator for state **k**, s. We work in an extended zone scheme with the **k**-sum running over all but the core states.

The spin–orbit interaction in the periodic lattice could be included in (4-18) by working with spin-orbitals rather than the orbital functions χ_k. The electron–phonon interaction would then contain spin-flip terms. We shall neglect these complications for the moment.

Thus, the total zero-order Hamiltonian is the sum of the bare-particle energies,

$$H_0 = \sum_{k,s} \epsilon_k n_{ks} + \sum_{q,\lambda} \hbar\Omega_{q\lambda}(N_{q\lambda} + \tfrac{1}{2}) \tag{4-19}$$

4-4 BARE ELECTRON–PHONON INTERACTION

We included in H_0 the interaction of the electrons with the ions in their equilibrium positions. The difference between this potential and the full electron–ion potential remains as a perturbation (along with several other terms). It turns out that it is sufficient for most purposes to expand this difference in powers of the ionic displacements $\delta\mathbf{R}_\nu$, and retain only the leading term. Thus, the bare electron–phonon interaction is of the conventional form for a boson–fermion coupling, that is, linear in the boson field, bilinear in the fermion field.

A reliable first-principles calculation of the coupling is not possible at present for most superconductors, since one requires accurate one-electron wave functions as well as reliable ionic potentials. Unfortunately, the most reliable calculations of the interaction are for small-momentum transfers; the phase space for these processes is small and they contribute little in bringing about superconductivity. It is rather the large-momentum transfers (of order the Fermi momentum $\hbar k_F$) which are important, but their coupling is difficult to estimate accurately since, in this case, one is sampling short-range details of the core potential rather than the Coulomb tail. In addition, details of the Bloch functions near the cores become important for these processes and it is in this vicinity that a free-electron approximation for χ_k is poorest. In practice one can carry through calculations without specifying details of the electron–phonon matrix elements and replace certain averages of these matrix elements by parameters to be determined from electrical resistivity, thermal conductivity, superconducting transition temperature, etc.

We can make a few general statements about the coupling. The perturbing potential acting on the ith electron is

$$\begin{aligned}\sum_\nu [U_{i\nu} - U_{i\nu}{}^0] &= -\sum_\nu \delta\mathbf{R}_\nu \cdot \nabla_i U_{i\nu} \\ &= -\frac{1}{(N_c M_c)^{1/2}} \sum_{q,\lambda} Q_{q\lambda} \sum_\nu \boldsymbol{\epsilon}_{q\lambda}(\nu) \cdot \nabla_i U_{i\nu} e^{i\mathbf{q}\cdot\mathbf{R}_\nu{}^0}\end{aligned} \tag{4-20}$$

where we have used (4-2) and have denoted $U(\mathbf{r}_i, \mathbf{R}_\nu)$ by $U_{i,\nu}$ as well as $U(\mathbf{r}_i, \mathbf{R}_\nu{}^0)$ by $U_{i,\nu}{}^0$. For given values of $\mathbf{q}$ and λ the matrix element of this potential between bare electron states $\mathbf{k}$ and $\mathbf{k}'$ is

$$-\frac{Q_{q\lambda}}{(N_c M_c)^{1/2}} \sum_\nu \langle k' | \nabla_i U_{i\nu} | k \rangle \cdot \boldsymbol{\epsilon}_{q\lambda}(\nu) e^{i\mathbf{q}\cdot\mathbf{R}_\nu{}^0} \tag{4-21}$$

If we introduce a cell location $\mathbf{R}_n{}^0$ and the relative position $\boldsymbol{\rho}_\alpha{}^0$ such that

$$\mathbf{R}_{n\alpha}{}^0 = \mathbf{R}_n{}^0 + \boldsymbol{\rho}_\alpha{}^0 \tag{4-22}$$

we can reduce the matrix element to

$$-Q_{q\lambda}\left(\frac{N_c}{M_c}\right)^{1/2} \sum_\alpha \langle k' | \nabla_i U_{i\alpha} | k \rangle \cdot \boldsymbol{\epsilon}_{q\lambda}(\alpha) e^{i\mathbf{q}\cdot\boldsymbol{\rho}_\alpha{}^0} \sum_{K_n} \delta_{\mathbf{k}'-\mathbf{k},\mathbf{q}+\mathbf{K}_n} \tag{4-23}$$

In the reduction we used Bloch's theorem (4-15) plus the fact that $U_{i,\nu}$ depends only on the relative separation between the electron and ion. In (4-23) $U_{i,\alpha}$ refers to ions in the unit cell located at the origin, i.e., at $\mathbf{R}_n^0 = 0$, and $\mathbf{K}_n$ is a reciprocal lattice vector. We see from (4-23) that the electron–phonon interaction conserves momentum mod $\hbar\mathbf{K}$, as mentioned above. The number of scripts in (4-23) is somewhat confusing so that we introduce the abbreviation

$$g_{kk'\lambda} \equiv -\left(\frac{\hbar N_c}{2\Omega_{q\lambda} M_c}\right)^{1/2} \sum_\alpha \langle k'|\nabla_i U_{i\alpha}|k\rangle \cdot \boldsymbol{\epsilon}_{q\lambda}(\alpha) e^{i\mathbf{q}\cdot\boldsymbol{\rho}_\alpha^0} \tag{4-24}$$

Since $\mathbf{q}$ is restricted to the first zone we use the convention that if the momentum transfer $\mathbf{k}' - \mathbf{k}$ falls outside the first zone we shall use the corresponding reduced wave vector for $\mathbf{q}$ and suppress $\mathbf{K}$. We note that

$$g_{k'k\lambda} = g_{kk'\lambda}^* \tag{4-25}$$

Going over to the second-quantization language for electrons, the bare electron–phonon interaction becomes

$$H_{\text{el-ph}} = \sum_{k,k',s,\lambda} g_{kk'\lambda} \varphi_{\mathbf{k}'-\mathbf{k},\lambda} c_{k's}^+ c_{ks} \tag{4-26}$$

where the phonon field amplitude is defined by

$$\varphi_{q\lambda} = a_{q\lambda} + a_{-q\lambda}^+ \tag{4-27}$$

As we mentioned above, despite a large amount of good work, our first-principles understanding of the matrix elements $g_{k,k',\lambda}$ is in a rough state at present. Ziman[96] gives a detailed discussion of all but the most recent work.

An oversimplified but useful model of a solid consists of smearing the ions out into a continuous charged "jelly"; in the absence of vibrations the jelly is taken to be uniform so that the Bloch functions degenerate into simple plane waves. The bare phonons are quantized vibrations of the jelly. In this "jellium" model, the bare electron–phonon interaction is easily calculated and one finds for longitudinal phonons[93]

$$g_{kk'} \equiv g_q = -i\,\frac{4\pi e^2}{q}\left(\frac{\hbar Z_c^2 N_c}{2\Omega_p M_c}\right)^{1/2} \tag{4-28}$$

where $Z_c e$ is the total ionic charge per unit cell and $\Omega_p = (4\pi N_c Z_c^2 e^2/M_c)^{1/2}$ is the ionic plasma frequency. In the long wavelength limit ($\mathbf{q} \to 0$), where the model is presumably reasonable, the coupling of the electrons to bare longitudinal phonons is singular. The singularity is clearly due to the long-range Coulomb force. When screening is taken into account, the dressed interaction vanishes as $\mathbf{q} \to 0$, again showing the importance of screening in metals. We note the relation

$$\frac{2g_q^2}{\hbar\Omega_p} = \frac{4\pi e^2}{q^2} \tag{4-29}$$

holds for longitudinal phonons in jellium. This relation has significance in superconductivity since the left-hand side turns out to be related to the bare electron–electron interaction resulting from the exchange of virtual phonons. The equality (4-29) states that the phonon attraction is exactly cancelled by the Coulomb repulsion if the electrons scatter without changing their energy (i.e., the net interaction vanishes in the static limit). This result is peculiar to jellium; however, the relative scale of the phonon and Coulomb interactions in real metals is roughly set by (4-29).

The jellium model has the added simplicity that the electrons and transverse phonons are uncoupled since their interaction is proportional to $\mathbf{q} \cdot \boldsymbol{\epsilon}_{q\lambda}$, which vanishes in this case.

Actual metals are considerably more complicated than jellium since transverse phonons play a strong role in umklapp processes[96] (transitions in which the momentum transfer $\mathbf{k}' - \mathbf{k}$ lies outside of the first Brillouin zone). Transverse phonons can also enter normal (non-umklapp) processes if the electronic energy contours in $\mathbf{k}$-space are not spherical or if $\mathbf{q}$ is not in a symmetry direction. Also, there is an electromagnetic coupling between the transverse phonons and the electrons.[67] In addition, the bare longitudinal matrix elements are certainly more complicated than (4-28) since they will reflect crystalline anisotropy as well as details of the core potential and the behavior of the Bloch functions near the cores. The reader is referred to Ziman's book for further details about the electron–phonon interaction in real metals. It appears that the orthogonalized plane wave method will prove to be very useful in gaining insight into this difficult problem.[94, 97, 98]

4-5 THE ELECTRON–PHONON HAMILTONIAN

To complete the program of expressing the electron–ion Hamiltonian (4-1) in terms of bare electron and phonon operators we must include the Coulomb interaction between conduction electrons. As shown in the Appendix, this can be expressed in second-quantization language as

$$H_{\text{el-el}} = \tfrac{1}{2} \sum_{k_1 \cdots k_4,\, s,\, s'} \langle \mathbf{k}_3, \mathbf{k}_4 | V | \mathbf{k}_1, \mathbf{k}_2 \rangle c_{k_3 s}{}^+ c_{k_4 s'}{}^+ c_{k_2 s'} c_{k_1 s} \tag{4-30}$$

where the Coulomb matrix element is given by

$$\langle \mathbf{k}_3, \mathbf{k}_4 | V | \mathbf{k}_1, \mathbf{k}_2 \rangle = \int \chi_{k_3}{}^*(\mathbf{r}) \chi_{k_4}{}^*(\mathbf{r}') \frac{e^2}{|\mathbf{r} - \mathbf{r}'|} \chi_{k_1}(\mathbf{r}) \chi_{k_2}(\mathbf{r}')\, d^3r\, d^3r' \tag{4-31}$$

It follows from Bloch's theorem (4-15) that the Coulomb interaction conserves crystal-momentum mod reciprocal-lattice vector **K**; thus the matrix element vanishes unless

$$\mathbf{k}_1 + \mathbf{k}_2 = \mathbf{k}_3 + \mathbf{k}_4 + \mathbf{K} \tag{4-32}$$

The final term in H is the difference between the one-body potential U_0 introduced in defining the Bloch functions [see (4-14)], and the interaction $U_{\iota,\nu}{}^0$ of the electrons with the ions fixed on their equilibrium positions. This contribution is

$$H_{\tilde{U}} = \sum_{k,K,s} \langle \mathbf{k} + \mathbf{K} | \tilde{U} | \mathbf{k} \rangle c_{\mathbf{k}+\mathbf{K},s}{}^+ c_{ks} \tag{4-33}$$

where

$$\tilde{U} = \sum_{\nu} U_{\iota\nu}{}^0 - U_0 \tag{4-34}$$

The full electron–phonon Hamiltonian is then

$$H = H_{\text{el}} + H_{\text{ph}} + H_{\text{el-ph}} + H_{\text{el-el}} + H_{\tilde{U}} \tag{4-35}$$

The system is a complicated one from a field-theoretic point of view since it involves the interaction of a Bose field with a self-coupled Fermi field, and, as we shall see, the coupling constants are *not* small. In the next chapter we shall discuss how field-theoretic methods can be applied to this system.

CHAPTER 5

FIELD-THEORETIC METHODS IN THE MANY-BODY PROBLEM

We shall introduce field-theoretic methods in the many-body problem by discussing three well-known "pictures" or representations used in discussing quantum mechanical problems.

5-1 THE SCHRÖDINGER, HEISENBERG, AND INTERACTION PICTURES

In elementary discussions of quantum mechanics one usually works in the "Schrödinger picture" in which the dynamical variables are taken to be time independent so that the wave function contains the time dependence of the problem. In this picture, the wave function $\Psi_S(t)$ satisfies

$$i\hbar \frac{\partial \Psi_S(t)}{\partial t} = H(t)\Psi_S(t) \tag{5-1}$$

While the dynamical variables are independent of time, the Hamiltonian may contain an explicit time dependence because of an external field acting on the system; (5-1) includes this possibility. If we consider an isolated system, H is independent of time and an exact solution of (5-1) is given by

$$\Psi_S(t) = e^{-iH(t-t_0)/\hbar}\Psi_S(t_0) \tag{5-2}$$

It is convenient for many purposes to make a unitary transformation to the "Heisenberg picture" in which the wave function Ψ_H is time independent and the time dependence of the problem is transferred to the operators. If we choose phases so that the wave functions Ψ_S and Ψ_H are identical at a time t_0, these functions are related by

$$\Psi_H(t) = \Psi_H = e^{iH(t-t_0)/\hbar}\Psi_S(t) \tag{5-3}$$

Thus the unitary operator factors out the time dependence of Ψ_S. In order that all observable quantities (i.e., matrix elements) be unaltered by the transformation, the operators Θ in the two pictures must be related by

$$\Theta_H(t) = e^{iH(t-t_0)/\hbar}\Theta_S(t)e^{-iH(t-t_0)/\hbar} \tag{5-4}$$

The Hamiltonian $H(p, q)$ has the same form in each picture although the time dependence of the p's and q's which express H differ according to (5-4). From (5-4) it follows that the time dependence of a Heisenberg operator is given by

$$i\hbar \frac{d\Theta_H(t)}{dt} = [\Theta_H(t), H] + i\hbar \frac{\partial\Theta_H(t)}{\partial t} \tag{5-5}$$

where the partial derivative accounts for any time dependence of the operator $\Theta_S(t)$.

For the purpose of perturbation expansions it is often convenient to write the Hamiltonian as

$$H = H_0 + H' \tag{5-6}$$

and define an interaction picture through the relations

$$\Psi_I(t) = e^{iH_0(t-t_0)/\hbar}\Psi_S(t) \tag{5-7a}$$

and

$$\Theta_I(t) = e^{iH_0(t-t_0)/\hbar}\Theta_S(t)e^{-iH_0(t-t_0)/\hbar} \tag{5-7b}$$

If H' vanishes, the Heisenberg and interaction pictures are identical. In the presence of H', the time dependence of the operator Θ_I is given by the zero-order Hamiltonian H_0, and the wave function Ψ_I is time-dependent solely due to the perturbation H'. We shall choose H_0 so that the time dependence of the operators Θ_I is very simple; this allows one to construct a simple set of rules for treating H' in a perturbation series corresponding to Feynman diagrams.

By inserting (5-7a) into Schrödinger's equation (5-1) we find

$$i\hbar \frac{\partial \Psi_I(t)}{\partial t} = H_I'(t)\Psi_I(t) \tag{5-8}$$

where H_I' is the perturbing Hamiltonian expressed in the interaction picture,

$$H_I'(t) = e^{iH_0(t-t_0)/\hbar} H_S' e^{-iH_0(t-t_0)/\hbar} \tag{5-9}$$

From now on we shall work with units such that $\hbar = 1$.

5-2 THE GREEN'S FUNCTION APPROACH

In the many-body problem we are ultimately interested in predicting such quantities as the thermodynamic and mechanical properties of the system as well as nonequilibrium properties such as electrical and thermal conductivities, and absorption of quanta of external fields. It is clear that we cannot determine these quantities by solving for the exact many-body eigenfunctions except for extremely simple systems; even if these functions were available they would be hopelessly complicated unless expressed in a form suitable for calculating a particular property of the system. One would prefer to work with dynamical quantities which are more closely related to experiment and contain less information than the full wave functions. One would then approximate these dynamical quantities directly rather than working with the Ψ's. Quantities which satisfy these conditions are the Green's functions of quantum field theory.[91, 99] The one-electron Green's function gives information about the spin and charge densities and the momentum distribution of the electrons as well as information about the excitation spectrum of the system. When the system consists only of fermions interacting via an instantaneous two-body potential, the one-particle Green's function also suffices to determine the ground-state energy (or more generally the free energy) of the system.[100] The electrical conductivity, magnetic susceptibility, and many other nonequilibrium properties can be obtained from the two-particle Green's function.

Perhaps the greatest advantage of the Green's function approach is that approximation schemes are readily developed which allow one to use physical insight in solving many-body problems. By studying simple problems we shall learn how to transcribe our physical ideas into field-theoretic language. Alternatively, mathematical approximations can often be better understood by reversing this process. The interplay of these two possibilities will allow us to go a long way in understanding complex systems.

We consider for the moment a system of fermions of spin $\frac{1}{2}$ interacting via a two-body spin-independent potential. The one-particle Green's function G is defined by

$$G_s(\mathbf{r}_1, t_1; \mathbf{r}_2, t_2) = -i\langle 0|T\{\psi_s(\mathbf{r}_1, t_1)\psi_s^+(\mathbf{r}_2, t_2)\}|0\rangle \tag{5-10}$$

where $|0\rangle$ represents the exact ground state of the interacting system expressed in the Heisenberg picture and the field operators ψ_s and ψ_s^+ (see the Appendix) are also expressed in the Heisenberg picture. The time-ordering symbol T is defined to order chronologically the operators inside the braces so that operators with earlier times are placed to the right with minus signs arising from anticommutation of the operators included. Thus

$$G_s(\mathbf{r}_1, t_1; \mathbf{r}_2, t_2) = \begin{cases} -i\langle 0|\psi_s(\mathbf{r}_1, t_1)\psi_s^+(\mathbf{r}_2, t_2)|0\rangle & t_1 > t_2 \\ i\langle 0|\psi_s^+(\mathbf{r}_2, t_2)\psi_s(\mathbf{r}_1, t_1)|0\rangle & t_1 \leqslant t_2 \end{cases} \tag{5-11}$$

We note that G depends on time only through $\tau \equiv t_1 - t_2$. If the system is translationally invariant, G depends only on the relative coordinates $\mathbf{r} \equiv \mathbf{r}_1 - \mathbf{r}_2$ and $\tau \equiv t_1 - t_2$:

$$G_s(\mathbf{r}_1, t_1; \mathbf{r}_2, t_2) \equiv G_s(\mathbf{r}, \tau) \tag{5-12}$$

(The latter simplification does not hold for electrons in solids because of the crystal lattice.) The Fourier transform of $G(\mathbf{r}, \tau)$, defined by

$$G_s(\mathbf{p}, p_0) = \int e^{-i(\mathbf{p}\cdot\mathbf{r} - p_0\tau)} G_s(\mathbf{r}, \tau)\, d^3r\, d\tau \tag{5-13}$$

is particularly useful since as we shall see later its poles as a function of p_0 are related to the elementary excitation spectrum of the system. The Fourier inverse of (5–13) is

$$G_s(\mathbf{r}, \tau) = \int e^{i(\mathbf{p}\cdot\mathbf{r} - p_0\tau)} G_s(\mathbf{p}, p_0) \frac{d^3p\, dp_0}{(2\pi)^4} \tag{5-14}$$

We are again using periodic boundary conditions and a box of unit volume so that we may use the relation

$$\sum_{\mathbf{p}} \leftrightarrow \int \frac{d^3p}{(2\pi)^3}$$

To save writing we use the abbreviations

$$\begin{aligned} x &\equiv (\mathbf{r}, \tau) \\ p &\equiv (\mathbf{p}, p_0) \\ px &\equiv \mathbf{p}\cdot\mathbf{r} - p_0\tau \\ d^4x &\equiv d^3r\, d\tau \\ d^4p &\equiv d^3p\, dp_0 \end{aligned}$$

The Fourier transforms (5-13) and (5-14) then become

$$G_s(p) = \int e^{-ipx} G_s(x)\, d^4x \tag{5-13$'$}$$

and

$$G_s(x) = \int e^{ipx} G_s(p) \frac{d^4p}{(2\pi)^4} \tag{5-14$'$}$$

For a translationally invariant system one often works with the spatial Fourier transform of $G(\mathbf{r}, \tau)$,

$$G_s(\mathbf{p}, \tau) = \int e^{-i\mathbf{p}\cdot\mathbf{r}} G_s(\mathbf{r}, \tau)\, d^3r \tag{5-15}$$

By using the relation (see the Appendix)

$$\psi_s(\mathbf{r}, t) = \sum_{p} c_{ps} e^{i\mathbf{p}\cdot\mathbf{r}} \tag{5-16}$$

where $c_{p,s}$ destroys an electron of spin s in plane wave state $\mathbf{p}$, one readily establishes

$$G_s(\mathbf{p}, \tau) = -i\langle 0|T\{c_{ps}(\tau)c_{ps}^+(0)\}|0\rangle \tag{5-17}$$

5-3 THE FREE FERMI GAS

To illustrate the structure of G we study a system of noninteracting fermions of spin $\frac{1}{2}$, contained in our box of unit volume. In the ground state $|0\rangle$, all plane wave states with momentum less than the Fermi momentum p_F are occupied with an up and a down spin electron, while all other states are empty. The Hamiltonian of the system is

$$H = \sum_{p,s} \epsilon_p c_{ps}^+ c_{ps} \tag{5-18}$$

where $\epsilon_p = p^2/2m$. Since the system is symmetric in the spin variable, we shall concentrate on the up-spin electrons and suppress s. From the anticommutation relations (see the Appendix)

$$\begin{aligned} \{c_{ps}, c_{p's'}^+\} &= \delta_{pp'}\,\delta_{ss'} \\ \{c_{ps}, c_{p's'}\} &= \{c_{ps}^+, c_{p's'}^+\} = 0 \end{aligned} \tag{5-19}$$

it follows that in the Heisenberg picture one has

$$c_p(\tau) = c_p(0)e^{-i\epsilon_p\tau} \tag{5-20}$$

With the definition (3-17) for $G(\mathbf{p}, \tau)$ we find

$$G(\mathbf{p}, \tau) = \begin{cases} -i\langle 0|c_p c_p^+|0\rangle e^{-i\epsilon_p\tau} & \tau > 0 \\ i\langle 0|c_p^+ c_p|0\rangle e^{-i\epsilon_p\tau} & \tau \leqslant 0 \end{cases} \tag{5-21}$$

With the aid of (5-19) this can be written as

$$G(\mathbf{p}, \tau) = \begin{cases} -i(1 - f_p)e^{-i\epsilon_p\tau} & \tau > 0 \\ if_p e^{-i\epsilon_p\tau} & \tau \leqslant 0 \end{cases} \tag{5-22}$$

where the Fermi function f_p is one for $p < p_F$ and zero for $p > p_F$.

In a nontrivial many-body problem, the time Fourier transform of G is of great interest since as we mentioned its singularities determine the elementary excitation spectrum of the system. For the free Fermi gas one finds

$$G(\mathbf{p}, p_0) \equiv G(p) = \frac{1}{p_0 - \epsilon_p + i\eta_p}\, e^{ip_0\delta} \tag{5-23}$$

where η_p is a positive (negative) infinitesimal for p greater (less) than p_F. δ is a positive infinitesimal. Note that $G(p)$ has a pole at $p_0 = \epsilon_p$, the energy required to add an electron of momentum $\mathbf{p}$ to the system. To establish (5-23) we consider the inverse Fourier transform,

$$G(\mathbf{p}, \tau) = \int_{-\infty}^{\infty} e^{-ip_0\delta} \frac{e^{ip_0\delta}}{p_0 - \epsilon_p + i\eta_p} \frac{dp_0}{2\pi} \tag{5-24}$$

The integral is evaluated by using Cauchy's theorem. For $\tau \leqslant 0$ the contour can be closed around an infinite semicircle in the upper half of the complex p_0-plane, as shown in Figure 5-1, since $e^{-ip_0(\tau-\delta)}$ vanishes on this added piece of contour. For $\tau > 0$ the contour may be closed in the lower half-plane for similar reasons. Because of the factor $i\eta_p$, the pole is in the upper half-plane for $p < p_F$ and in the lower half-plane for $p > p_F$. Thus, for $\tau > 0$ we obtain

$$G(\mathbf{p}, \tau) = \begin{cases} 0 & p < p_F \\ -ie^{-i\epsilon_p\tau} & p > p_F \end{cases}$$

while for $\tau \leqslant 0$ we find

$$G(\mathbf{p}, \tau) = \begin{cases} ie^{-i\epsilon_p\tau} & p < p_F \\ 0 & p > p_F \end{cases}$$

as required by (5-22). While the factor $e^{ip_0\delta}$ in (5-23) can be replaced by unity for most purposes and hence we suppress it, there are cases where we shall need this factor (an example is in calculating the exchange energy within the Hartree–Fock approximation).

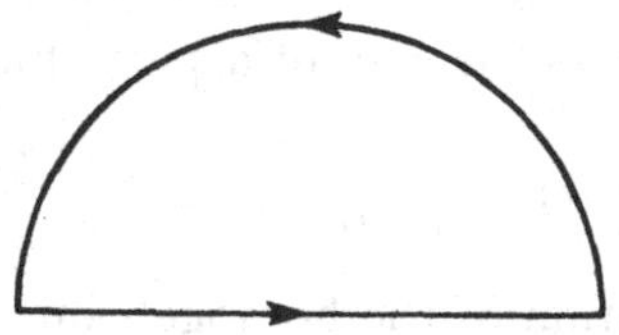

FIGURE 5-1 Integration contour for evaluating (5-24) when $\tau \leqslant 0$.

Another method to obtain $G(\mathbf{p}, p_0)$ for the free Fermi gas is to construct an equation of motion for $G(\mathbf{p}, \tau)$. This is easily done by writing (5-17) as

$$G(\mathbf{p}, \tau) = -i\langle 0|c_p(\tau)c_p^+(0)|0\rangle\theta(\tau) + i\langle 0|c_p^+(0)c_p(\tau)|0\rangle\theta(-\tau) \tag{5-25}$$

where the θ-function is defined as

$$\theta(\tau) = \begin{cases} 1 & \tau > 0 \\ 0 & \tau < 0 \end{cases}$$

By differentiating (5-25) with respect to τ we find

$$i\frac{\partial G(\mathbf{p}, \tau)}{\partial \tau} = \langle 0|T\left\{\frac{\partial c_p(\tau)}{\partial \tau} c_p^+(0)\right\}|0\rangle + \langle 0|\{c_p(0), c_p^+(0)\}|0\rangle\, \delta(\tau) \tag{5-26}$$

the delta-function terms coming from differentiating the θ-functions. With the aid of the anticommutation relations (5-19) and the expression (5-20), it follows that $G(\mathbf{p}, \tau)$ satisfies the differential equation

$$\left(i\frac{\partial}{\partial \tau} - \epsilon_p\right)G(\mathbf{p}, \tau) = \delta(\tau) \tag{5-27}$$

It is clear from this equation why G is called a "Green's function": it is in fact the Green's function for the operator $(i\partial/\partial\tau - \epsilon_p)$. We can solve (5-27) by taking its Fourier transform with respect to τ; then

$$G(\mathbf{p}, p_0) = \frac{1}{p_0 - \epsilon_p} \tag{5-28}$$

The essential difference between this expression and the correct result (5-23) is that (5-28) does not specify how one is to treat the singularity at $p_0 = \epsilon_p$. The difficulty is that (5-27) is a first-order differential equation in τ and we must specify a boundary condition in solving the equation. The most general solution of (5-27) can be written as the sum of a particular solution

$$P(\mathbf{p}, \tau) = \begin{cases} -ie^{-i\epsilon_p\tau} & \tau > 0 \\ 0 & \tau < 0 \end{cases} \tag{5-29}$$

plus the general solution of the homogeneous equation

$$h(\mathbf{p}, \tau) = i\bar{f}_p e^{-i\epsilon_p\tau} \qquad \text{(for all } \tau)$$

The arbitrary constant $\tilde{f}_p$ is determined, for example, from the boundary condition $G(\mathbf{p}, 0^-) = if_p$. With this choice of f_p, the Fourier transform of the full solution $P + h$ agrees with (5-23), as required. Thus the boundary conditions uniquely specify how we are to treat the singularity of $G(p)$.

At this point the reader may wonder how G is to be determined for the interacting system, since in the derivation (5-20) to (5-23) for the free system, we explicitly used the ground-state wave function; yet we argued that the Green's functions free us from considering the many-body wave function. There are two principal methods for determining G (and higher-order Green's functions). The first is to construct an equation of motion for G analogous to (5-27); in the presence of a two-body interaction an extra term appears on the right-hand side of this equation involving the "two-particle" Green's function, defined in the x-representation by

$$G_2(x_1, x_2, x_3, x_4) = (-i)^2\langle 0|T\{\psi(x_1)\psi(x_2)\psi^+(x_4)\psi^+(x_3)\}|0\rangle \quad (5\text{-}30)$$

We have suppressed spin variables. Since G_2 is also unknown, we write an equation of motion for it and find the three-particle Green's function appearing. In this way one can generate a hierarchy of equations which unfortunately does not terminate. The chain is often truncated by neglecting correlations between more than two or three particles at a time. In principle, one can then solve the coupled equations for the functions that are retained. While the scheme sounds straightforward, the mathematical problem is exceedingly complicated, even if one neglects all but two-body correlations. One is usually satisfied by retaining only pieces of the two-body correlations. This scheme has been extensively discussed by Schwinger, Martin, Kadanoff, etc.[99, 101]

The second method is the perturbation expansion of Feynman, Tomonaga, and Dyson,[102] familiar in quantum electrodynamics. While the most naive version of this scheme is formally limited to problems where the expansion converges, this approach is so closely related to one's physical intuition that partial summations of the diverging series can often be found which lead to physically meaningful results. The discussion below follows the second line of approach.

5-4 SPECTRAL REPRESENTATION OF $G(\mathbf{p}, \tau)$

The one-particle Green's function G for the interacting Fermi gas is closely related to G for the noninteracting system. To see this, consider the Green's function for spin-up particles:

$$G(\mathbf{p}, \tau) = -i\langle 0|T\{c_p(\tau)c_p^+(0)\}|0\rangle$$

$$= \begin{cases} -i\langle 0|c_p(0)e^{-iH\tau}c_p^+(0)|0\rangle e^{iE_0^n\tau} & \tau > 0 \\ i\langle 0|c_p^+(0)e^{iH\tau}c_p(0)|0\rangle e^{-iE_0^n\tau} & \tau \leqslant 0 \end{cases} \qquad (5\text{-}31)$$

where E_0^n is the ground-state energy of the interacting n-particle system. By inserting between c and c^+ a complete set of eigenstates $|\Psi_m^{n\pm 1}\rangle$ of H for the $n \pm 1$ particle system, we find

$$G(\mathbf{p}, \tau) = \begin{cases} -i\sum_m |(c_p^+)_{m,0}|^2 e^{-i(E_m^{n+1}-E_0^n)\tau} & \tau > 0 \\ i\sum_m |(c_p)_{m,0}|^2 e^{i(E_m^{n-1}-E_0^n)\tau} & \tau \leqslant 0 \end{cases} \qquad (5\text{-}32)$$

The matrix elements are defined by

$$(c_p^+)_{m,0} = \langle\Psi_m^{n+1}|c_p^+|\Psi_0^n\rangle \qquad (5\text{-}33a)$$

$$(c_p)_{m,0} = \langle\Psi_m^{n-1}|c_p|\Psi_0^n\rangle \qquad (5\text{-}33b)$$

that is, matrix elements of the operators for bare electrons taken between exact eigenstates of the full Hamiltonian. For convenience we introduce the energies $\omega_m^{n\pm 1}$ defined by

$$E_m^{n+1} - E_0^n = \omega_m^{n+1} + \mu_n \qquad (5\text{-}34a)$$

and

$$E_m^{n-1} - E_0^n = -\omega_m^{n-1} - \mu_{n-1} \qquad (5\text{-}34b)$$

where $\mu_n = E_0^{n+1} - E_0^n$ is the chemical potential of the n-particle system. Since we are concerned with large systems we have $\mu_n \cong \mu_{n-1} \equiv \mu$. The energy ω_m^{n+1} is necessarily nonnegative while the extra minus signs introduced for convenience in (5-34b) require that ω_m^{n-1} be *nonpositive*. By inserting these expressions into (5-32) we obtain

$$G(\mathbf{p}, \tau) = \begin{cases} -i \sum_m |(c_p^+)_{m,0}|^2 e^{-i(\omega_m^{n+1}+\mu)\tau} & \tau > 0 \\ i \sum_m |(c_p)_{m,0}|^2 e^{-i(\omega_m^{n-1}+\mu)\tau} & \tau \leqslant 0 \end{cases} \tag{5-35}$$

An important quantity is the spectral weight function $A(\mathbf{p}, \omega)$, defined by

$$A(\mathbf{p}, \omega) = \sum_m |(c_p^+)_{m,0}|^2 \, \delta(\omega - \omega_m^{n+1}) + \sum_m |(c_p)_{m,0}|^2 \, \delta(\omega - \omega_m^{n-1}) \tag{5-36}$$

The one-particle Green's function for the interacting system can then be expressed as

$$G(\mathbf{p}, p_0) = \int_{-\infty}^{\infty} \frac{A(\mathbf{p}, \omega)\, d\omega}{p_0 - \omega - \mu + i\omega\,\delta} \tag{5-37}$$

where δ is a positive infinitesimal. To ensure the correct definition of G for $\tau = 0$, the factor $e^{ip_0\delta}$ should multiply the right-hand side of (5-37) as in (5-23). The spectral representation (5-37) for $G(p, p_0)$ can be checked by calculating its Fourier inverse $G(\mathbf{p}, \tau)$ and noticing that the result is identical with (5-32). We see from (5-37) that $G(\mathbf{p}, p_0)$ is simply a weighted sum of Green's functions (5-23) for the noninteracting system.

For the free Fermi gas G is given by taking A to be a delta function,

$$\begin{aligned} A(\mathbf{p}, \omega) &= (1 - f_p)\, \delta(\omega - [\epsilon_p - \mu]) + f_p\, \delta(\omega - [\epsilon_p - \mu]) \\ &= \delta(\omega - [\epsilon_p - \mu]) \end{aligned} \tag{5-38}$$

since for $p > p_F$ the first term when inserted into (5-37) gives $1/(p_0 - \epsilon_p + i\delta)$ and the second term vanishes; for $p < p_F$ the second term gives $1/(p_0 - \epsilon_p - i\delta)$ while the first term vanishes, in agreement with (5-23).

According to (5-36) the spectral weight function is a positive real quantity

$$A(\mathbf{p}, \omega) = A^*(\mathbf{p}, \omega) \tag{5-39}$$

and satisfies the sum rule

$$\int_{-\infty}^{\infty} A(\mathbf{p}, \omega)\, d\omega = 1 \tag{5-40}$$

The latter follows by inserting the expression (5-36) under the integral

$$\int_{-\infty}^{\infty} A(\mathbf{p}, \omega)\, d\omega = \sum_m |(c_p{}^+)_{m,0}|^2 + \sum_m |(c_p)_{m,0}|^2$$

and using the completeness of the states $|\Psi_m{}^{n\pm 1}\rangle$ to write this as

$$\int_{-\infty}^{\infty} A(\mathbf{p}, \omega)\, d\omega = \langle 0|c_p c_p{}^+ + c_p{}^+ c_p|0\rangle = 1$$

where we have used the anticommutation relation (5-19) in the last step. We note that the free gas expression (5-38) trivially satisfies the sum rule.

From the definition of A it is clear that its positive-frequency part ($\omega > 0$) contains information about processes involving the addition of an electron to the system while its negative frequency part is related to hole-injection (electron-extraction) processes. Thus the sum rule connects these two types of processes.

As we mentioned above, one usually sets up equations to determine G directly rather than first finding A. A simple relation between the two functions is given by taking the imaginary part of the spectral representation (5-37):

$$\operatorname{Im} G(\mathbf{p}, \omega + \mu) = \begin{cases} -\pi A(\mathbf{p}, \omega) & \omega > 0 \\ \pi A(\mathbf{p}, \omega) & \omega < 0 \end{cases} \tag{5-41}$$

where we have used the well-known relation for factors occurring under an integral

$$\frac{1}{x \pm i\eta} = \frac{P}{x} \mp i\pi\, \delta(x) \tag{5-42}$$

P denotes the principal part of the singularity is to be taken in performing the integral. A dispersion relation connecting the real and imaginary parts of G follows by inserting (5-41) into the spectral representation,

$$\text{Re}\, G(\mathbf{p}, p_0) = -\frac{1}{\pi} P \int_{\mu}^{\infty} \frac{\text{Im}\, G(\mathbf{p}, p_0')\, dp_0'}{p_0 - p_0'} + \frac{1}{\pi} P \int_{-\infty}^{\mu} \frac{\text{Im}\, G(\mathbf{p}, p_0')\, dp_0'}{p_0 - p_0'} \quad (5\text{-}43)$$

5-5 ANALYTIC PROPERTIES OF G

We can neglect the infinitesimal $i\omega\,\delta$ in the spectral representation (5-37) if we allow the ω-integration to be along the contour C, shown in Figure 5-2; hence

$$G(\mathbf{p}, p_0) = \int_C \frac{A(\mathbf{p}, \omega)\, d\omega}{(p_0 - \mu) - \omega} \quad (5\text{-}44)$$

Thus far, $G(\mathbf{p}, p_0)$ has been defined only for real values of $p_0 - \mu$. We now extend the definition by writing

$$\hat{G}(\mathbf{p}, p_0) = \int_C \frac{A(\mathbf{p}, \omega)\, d\omega}{(p_0 - \mu) - \omega} \quad (5\text{-}45)$$

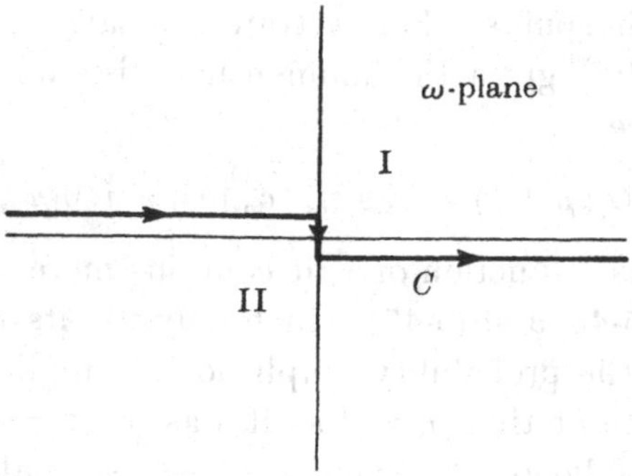

FIGURE 5-2 Integration contour for the spectral representation of $G(\mathbf{p}, p_0)$.

as a function of the complex variable p_0.[100] Integrals of this type define functions of p_0 which are analytic over the entire plane except possibly for values of $(p_0 - \mu)$ lying on the contour C. In our case the contour C cuts the plane into two regions so that two different functions f_{I} and f_{II} can be defined. As shown in Figure 5-2, $f_{\mathrm{I}}(p_0 - \mu)$ is analytic in region I and $f_{\mathrm{II}}(p_0 - \mu)$ is analytic in region II. For real p_0, $G(\mathbf{p}, p_0)$ coincides with $f_{\mathrm{I}}(p_0 - \mu)$ for $p_0 - \mu > 0$ and with $f_{\mathrm{II}}(p_0 - \mu)$ for $p_0 - \mu < 0$. The function f_{I} can be defined in region II by its analytic continuation across the cut. This continued function is not in general analytic in region II. Similarly, the analytic continuation of f_{II} into region I is not analytic in region I. The functions f_{I} and f_{II} will play an important role in later calculations.

Since $A(\mathbf{p}, \omega)$ is real, it is seen from (5-45) that at points lying infinitely close to the contour, f_{I} in region I and f_{II} in region II are complex conjugates. This property will also be useful later in our discussion.

5-6 PHYSICAL INTERPRETATION OF $G(\mathbf{p}, p_0)$

There are many ways of interpreting G. If we work in the x-representation, and let $x_1 = x_2$, we have

$$G_s(\mathbf{r}, t; \mathbf{r}, t) = i\langle 0|\psi_s^+(\mathbf{r}, t)\psi_s(\mathbf{r}, t)|0\rangle = i\langle 0|\rho_s(\mathbf{r}, t)|0\rangle \qquad (5\text{-}46)$$

In this limit, $-iG$ gives the expected value of the density of particles with spin s. For a translationally invariant system, $-iG(p, \tau = 0^-)$ gives the momentum distribution of the bare particles, since

$$G_s(p, 0^-) = i\langle 0|c_{ps}^+ c_{ps}|0\rangle = i\langle 0|n_{ps}|0\rangle \qquad (5\text{-}47)$$

Since G is a function of τ, it contains more information than is given by (5-46) and (5-47), which only use its value at $\tau = 0^-$. G also gives the probability amplitude for finding the system in the same state at time $\tau > 0$ as it was in at $\tau = 0^+$. This can be shown as follows. We choose phases so that the Heisenberg and Schrödinger pictures are identical at $t = 0$. For the moment we speak in the Schrödinger language. Let us create at $t = 0$ a particle in the bare state **p**. The wave function for the system is then

$$c_p{}^{+}|\Psi_{S,0}(0)\rangle$$

if $\Psi_{S,0}$ represents the ground state. At a later time τ the wave function has evolved into

$$e^{-iH\tau}c_p{}^{+}|\Psi_{S,0}(0)\rangle$$

If the particle were created in state **p** at $t = \tau$ rather than $t = 0$, the wave function would be

$$c_p{}^{+}e^{-iH\tau}|\Psi_{S,0}(0)\rangle$$

Thus, the probability amplitude for finding the system at $t = \tau > 0$ in the same state as it was in at $t = 0$ is the scalar product of the two Schrödinger states at time τ:

$$\langle\Psi_{S,0}(0)|e^{iH\tau}c_p e^{-iH\tau}c_p{}^{+}|\Psi_{S,0}(0)\rangle$$

which in the Heisenberg picture becomes

$$\langle 0|c_p(\tau)c_p{}^{+}(0)|0\rangle = -iG(\mathbf{p}, \tau) \qquad (\tau > 0)$$

On physical grounds we might expect this probability amplitude to oscillate with some frequency $(E_p + \mu)$ and decrease in magnitude with a decay rate $|\Gamma_p|$:

$$G(\mathbf{p}, \tau) \propto e^{-i(E_p+\mu)\tau}e^{-|\Gamma_p|\tau}$$

As is apparent from (5-22), this form is exact for the free Fermi gas with $E_p + \mu = \epsilon_p$ and $\Gamma_p = 0$.

Let us now investigate the behavior of $G(\mathbf{p}, \tau)$ for large positive times. By taking the Fourier inverse of the spectral representation (5-37) one finds for $\tau > 0$

$$G(\mathbf{p}, \tau) = -i\int_0^\infty A(\mathbf{p}, \omega)e^{-i(\omega+\mu)\tau}\,d\omega \tag{5-48}$$

While $A(\mathbf{p}, \omega)$ is a sum of delta functions for a finite system, it is a continuous function in the limit of a large system. We carry out this integral by pushing the contour into the lower half-plane as shown in Figure 5-3 and we use the analytic continuation of A in this region. If the continuation of A is analytic in the lower

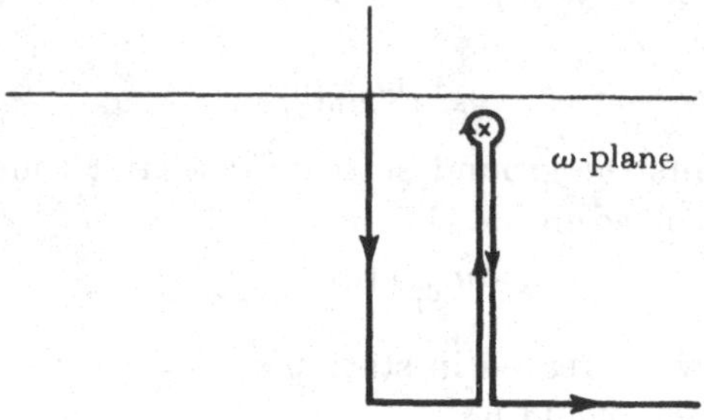

FIGURE 5-3 Integration contour for determining $G(\mathbf{p}, t)$ for large positive t.

half-plane except for a simple pole at $\omega = E_p - i|\Gamma_p|$ with residue r_p the integral can be written as the sum of the pole contribution

$$-2\pi r_p e^{-i(E_p+\mu)\tau} e^{-|\Gamma_p|\tau}$$

plus the line integral

$$\int_0^{-i\infty} A(\mathbf{p}, \omega) e^{-i\omega\tau}\, d\omega$$

where the path of integration is along the negative imaginary axis. For times τ such that $E_p\tau \gg 1$ but $e^{-|\Gamma_p|\tau} \gtrsim 1/E_p\tau$, the pole term dominates and we have the simple result guessed above. If we interpret the pole contribution in terms of the contribution of a quasi-stationary state called a quasi-particle, we see that the poles of the analytic continuation of $A(\mathbf{p}, \omega)$ into the lower half-plane gives the energy and damping rate of these quasi-particles.[100, 101]

It is often convenient to express this result in terms of the Green's function itself, rather than in terms of $A(\mathbf{p}, \omega)$. Consider $G(\mathbf{p}, p_0)$ for real values of p_0 which are greater than μ. The analytic continuation of this function into the complex p_0-plane $\tilde{G}(\mathbf{p}, p_0)$ is given by

$$\begin{aligned} \tilde{G}(\mathbf{p}, p_0) &= \hat{G}(\mathbf{p}, p_0) & \operatorname{Im} p_0 > 0 \\ &= \hat{G}(\mathbf{p}, p_0) - 2\pi i A(\mathbf{p}, p_0 - \mu) & \operatorname{Im} p_0 < 0 \end{aligned} \qquad (5\text{-}49a)$$

Since $\hat{G}(\mathbf{p}, p_0)$ has singularities only along the real axis [see (5-45)], the poles of $A(\mathbf{p}, p_0 - \mu)$ in the lower half-plane coincide with those of the analytic continuation of $G(\mathbf{p}, p_0)$ for $p_0 > \mu$. Thus, if there exists only a simple pole of the analytic continuation of $G(\mathbf{p}, p_0)$ into the lower half-plane for $p_0 - \mu > 0$, this pole may be interpreted as giving the energy and decay rate of a quasi-particle. If we assume there are several poles in the lower half-plane, the pole nearest the axis will presumably give the dominant contribution for large τ. The pole contribution may also be interpreted as stating that if a bare particle is created at time $t = 0$, it has a probability amplitude $2\pi i r_p$ of being in the quasi-particle state of momentum $\mathbf{p}$.

In a completely analogous way one can show that the poles of the analytic continuation of $A(\mathbf{p}, \omega)$ into the upper half-plane for $\omega < 0$ may be interpreted as giving the energy and damping rate of quasi-holes. Again, it is convenient to express this result in terms of G rather than A. Consider $G(\mathbf{p}, p_0)$ for real values of p_0 which are less than μ. The analytic continuation of this function into the complex p_0-plane $\tilde{\tilde{G}}(\mathbf{p}, p_0)$ is given by

$$\begin{aligned} \tilde{\tilde{G}}(\mathbf{p}, p_0) &= \hat{G}(\mathbf{p}, p_0) + 2\pi i A(\mathbf{p}, p_0 - \mu) \qquad & \operatorname{Im} p_0 > 0 \\ &= \hat{G}(\mathbf{p}, p_0) & \operatorname{Im} p_0 < 0 \end{aligned} \qquad (5\text{-}49\text{b})$$

Since $\hat{G}(\mathbf{p}, p_0)$ has poles only along the real axis, the poles of A and $\tilde{\tilde{G}}$ in the upper half-plane coincide. Therefore, the poles of the analytic continuation of $G(\mathbf{p}, p_0)$ (for real $p_0 < \mu$) into the upper half-plane give the energies and damping rates of the quasi-holes.

5-7 INTERPRETATION OF $A(\mathbf{p}, \omega)$

The above discussion is made clearer if we try to understand the spectral weight function

$$A(\mathbf{p}, \omega) = \sum_m |(c_p{}^+)_{m,0}|^2 \, \delta(\omega - \omega_m{}^{n+1}) + \sum_m |(c_p)_{m,0}|^2 \, \delta(\omega - \omega_m{}^{n-1}) \qquad (5\text{-}37\text{a})$$

in greater detail. Consider first the positive frequency part of A $(\omega > 0)$. If the system is initially in its ground state $|\psi_0{}^n\rangle$

and a bare electron is placed in the state $\mathbf{p}$ at $t = 0$, the state vector just after the electron has been added is given by

$$|\Phi_p\rangle = c_p^+|\Psi_0^n\rangle \tag{5-50}$$

We should keep in mind that unlike the states $|\Psi_m^\nu\rangle$, $|\Phi_p\rangle$ is not in general normalized to unity. This is easily understood if $|\Psi_0^n\rangle$ is expanded in occupation number states; according to the Pauli principle any component of $|\Psi_0^n\rangle$ in which the state $\mathbf{p}$ is occupied gives a vanishing contribution to (5-50), thereby changing the normalization of $|\Phi_p\rangle$.

Suppose we want to know the relative probability that the system described by $|\Phi_p\rangle$ (not an eigenfunction of H in general) is in the eigenstate $|\Psi_m^{n+1}\rangle$. Aside from a normalization factor, the probability is

$$P_m(\mathbf{p}) = |\langle\Psi_m^{n+1}|\Phi_p\rangle|^2 = |(c_p^+)_{m,0}|^2$$

Thus the strength of the delta function in the first term of (5-37) is just this relative probability. [For a translationally invariant system, $P_m(\mathbf{p})$ vanishes unless the momentum of state m is $\mathbf{p}$.] Suppose we want the relative probability that just after the electron is injected, the system is in the group of eigenstates whose energy is in the interval $\omega \to \omega + \delta\omega$. Again, aside from the normalization factor, the probability is

$$\begin{aligned} P_\omega(\mathbf{p})\,\delta\omega &= \int_\omega^{\omega+\delta\omega} \sum_m P_m(\mathbf{p})\,\delta(\omega' - \omega_m^{n+1})\,d\omega' \\ &= \int_\omega^{\omega+\delta\omega} A(\mathbf{p}, \omega')\,d\omega' \end{aligned}$$

Therefore, $A(\mathbf{p}, \omega)$ gives the relative probability per unit energy that the system described by $|\Phi_p\rangle$ has an energy $\omega + \mu$ greater than that of $|\Psi_0^n\rangle$.

If $|\Phi_p\rangle$ were normalized, the completeness of the states $|\Psi_m^{n+1}\rangle$ for the $n + 1$ particle system would require the total probability

$$\sum_m P_m(\mathbf{p}) = \int_0^\infty A(\mathbf{p}, \omega)\,d\omega$$

to be unity. Instead, one finds the integral is

$$\int_0^\infty A(\mathbf{p}, \omega)\, d\omega = \langle \Phi_p | \Phi_p \rangle = \langle \Psi_0{}^n | (1 - n_p) | \Psi_0{}^n \rangle = 1 - \langle n_p \rangle \tag{5-51}$$

The free-gas weight function (5-38) again trivially satisfies this condition, since for $p > p_F$ the excitation energy $\epsilon_p - \mu$ is greater than zero, so that the integral on the left-hand side of (5-51) is unity and $\langle n_p \rangle = 0$ in this case, while the integral vanishes for $p < p_F$ in agreement with $1 - \langle n_p \rangle = 0$, in this second case.

Turning now to the negative frequency part of $A(\mathbf{p}, \omega)$ we consider the state

$$|\tilde{\Phi}_p\rangle = c_p | \Psi_0{}^n \rangle$$

which describes the system just after a bare electron of momentum **p** has been withdrawn. Again, $|\tilde{\Phi}_p\rangle$ is in general not normalized to unity. If we ask for the relative probability that the system described by $|\tilde{\Phi}_p\rangle$ is in the eigenstate $|\Psi_m{}^{n-1}\rangle$ of the $n - 1$ particle system we find

$$\tilde{P}_m(\mathbf{p}) = |\langle \Psi_m{}^{n-1} | \Phi_p \rangle|^2 = |(c_p)_{m,0}|^2$$

which gives the strength of the delta function in the second term of (5-37). Owing to our choice of signs in the definition of $\omega_m{}^{n-1}$ [see (5-34b)], the excitation energy $E_m{}^{n-1} - E_0{}^{n-1}$ is defined to be the negative of $\omega_m{}^{n-1}$. Then, if $A(\mathbf{p}, \omega)$ is plotted as a function of ω, one should think of the excitation energy of states involved in electron extraction as being measured along the *negative* ω-axis. Our choice of signs unfolds the electron and hole parts of A; otherwise we would have to introduce two functions, say $\rho^{(+)}$ and $\rho^{(-)}$, defined on the interval $0 \to \infty$ by

$$\rho^{(+)}(\mathbf{p}, \omega) = \sum_m |(c_p{}^+)_{m,0}|^2\, \delta(\omega - \omega_m{}^{n+1}) \tag{5-36a}$$

and

$$\rho^{(-)}(\mathbf{p}, \omega) = \sum_m |(c_p)_{m,0}|^2\, \delta(\omega - |\omega_m{}^{n-1}|) \tag{5-36b}$$

so that

$$G(\mathbf{p}, p_0) = \int_0^\infty \frac{\rho^{(+)}(\mathbf{p}, \omega)\, d\omega}{p_0 - \omega - \mu + i\delta} + \int_0^\infty \frac{\rho^{(-)}(\mathbf{p}, \omega)\, d\omega}{p_0 + \omega - \mu - i\delta} \tag{5-37b}$$

The ρ-functions are somewhat more convenient than A for finite-temperature problems where the excitation energies are not necessarily positive.

To connect these spectral weight functions with those discussed by Kadanoff and Baym, we note that our A, $\rho^{(+)}$, and $\rho^{(-)}$ are related to their A_{BK}, $G^{>}$, and $G^{<}$ by

$$
\begin{aligned}
A_{BK}(\mathbf{p}, \omega) &= 2\pi A(\mathbf{p}, \omega) \\
G^{>}(\mathbf{p}, \omega) &= 2\pi\rho^{(+)}(\mathbf{p}, \omega) \\
G^{<}(\mathbf{p}, \omega) &= 2\pi\rho^{(-)}(\mathbf{p}, -\omega)
\end{aligned}
$$

The quasi-particle approximation is now simply understood: the probability distribution for the energy of the system just after a bare electron that has either been added to or subtracted from the ground state is approximated by one or more Lorentzian functions:

$$
A(\mathbf{p}, \omega) = \frac{u_p{}^2 \dfrac{|\Gamma_p{}^{(+)}|}{\pi}}{[\omega - E_p{}^{(+)}]^2 + \Gamma_p{}^{(+)2}} + \frac{v_p{}^2 \dfrac{|\Gamma_p{}^{(-)}|}{\pi}}{[\omega + E_p{}^{(-)}]^2 + \Gamma_p{}^{(-)2}} \tag{5-52}
$$

This expression satisfies the sum rule (5-40) if $u_p{}^2 + v_p{}^2 = 1$. This weight function is more complicated than that for the free Fermi gas, $\delta(\omega - [\epsilon_p - \mu])$, for three reasons.

1. The quasi-particle energy $E_p{}^{(\pm)}$ is not in general equal to $\pm|\epsilon_p - \mu|$ but includes "self-energy" effects arising from interactions of the injected particle (or hole) with the medium.

2. The delta function has been smeared out into a finite-width Lorentzian function, owing to $|\Phi_p\rangle$ not being an eigenstate of the interacting system in general.

3. There are two peaks in the weight function (5-52), one for positive and one for negative ω, both of which have finite weights $u_p{}^2$ and $v_p{}^2$, respectively. In accordance with our discussion above, the two peaks reflect the possibility of either adding an electron to momentum state $\mathbf{p}$ (leading to the positive energy peak) or removing an electron from this state (giving the negative energy peak). Both processes are possible in the interacting system for a given state $\mathbf{p}$ since the probability that this state is occupied is neither one nor zero.

It is interesting to note that the presence of large self-energy effects does not necessarily imply that the level widths Γ are large. In fact, for a system described by the BCS reduced Hamiltonian (see Chapter 2), the spectral weight function is exactly of the form (5-52) with $\Gamma_p^{(+)} = \Gamma_p^{(-)} \to 0$:

$$A_{\text{BCS}}(\mathbf{p}, \omega) = u_p^2\, \delta(\omega - E_p) + v_p^2\, \delta(\omega + E_p) \tag{5-53}$$

Thus, the reduced Hamiltonian has the peculiar property in common with the free-electron gas that $c_p|0\rangle$ and $c_p^+|0\rangle$ are eigenstates of the system. The quasi-particle energy E_p is given by

$$E_p = [(\epsilon_p - \mu)^2 + \Delta_p^2]^{1/2} \tag{5-54}$$

where Δ_p is the energy-gap parameter. For $\epsilon_p \cong \mu$, E_p is drastically different from $\epsilon_p - \mu$. The functions u_p^2 and v_p^2 are

$$u_p^2 = \frac{1}{2}\left(1 + \frac{\epsilon_p - \mu}{E_p}\right) \tag{5-55a}$$

$$v_p^2 = \frac{1}{2}\left(1 - \frac{\epsilon_p - \mu}{E_p}\right) \tag{5-55b}$$

and give the probability that the bare particle state $\mathbf{p}$ is unoccupied or occupied, respectively. For $\Delta_p \to 0$, the BCS spectral function goes over to that of the free Fermi gas. If one goes beyond the BCS approximation, finite level widths appear (as well as more complicated "continuum" contributions, which will be discussed later). While $A(\mathbf{p}, \omega)$ is in general an extremely complicated function, its form for *small* $\epsilon_p - \mu$ *and small* ω is expected to be simply the sum of two Lorentzian functions as in (5-52); however, the sum of the residues may be less than unity. This simplicity is exploited in Landau's theory of the Fermi liquid, discussed in Chapter 2, although a correspondingly simple picture holds for the superconducting state. Since most transport properties emphasize this low-energy part of the spectral function, a quasi-particle approximation is often sufficient for calculating the electrical and thermal conductivities, etc.

5-8 THE ONE-PHONON GREEN'S FUNCTION

In a completely analogous manner we introduce the one-phonon Green's function D defined by

$$D_\lambda(\mathbf{r}_1, t_1; \mathbf{r}_2, t_2) = -i\langle 0|T\{\varphi_\lambda(\mathbf{r}_1, t_1)\varphi_\lambda{}^+(\mathbf{r}_2, t_2)\}|0\rangle \quad (5\text{-}56)$$

where the T-product is given by

$$T\{\varphi_\lambda(\mathbf{r}_1, t_1)\varphi_\lambda{}^+(\mathbf{r}_2, t_2)\} = \begin{cases} \varphi_\lambda(\mathbf{r}_1, t_1)\varphi_\lambda{}^+(\mathbf{r}_2, t_2) & t_1 > t_2 \\ \varphi_\lambda{}^+(\mathbf{r}_2, t_2)\varphi_\lambda(\mathbf{r}_1, t_1) & t_1 < t_2 \end{cases}$$

D depends on time only through the difference $\tau = t_1 - t_2$, and for a translationally invariant system depends on the spatial variables only through $\mathbf{r} = \mathbf{r}_1 - \mathbf{r}_2$. With the aid of the expansion

$$\varphi_\lambda(\mathbf{r}, t) = \sum_q \varphi_{q\lambda}(t)e^{i\mathbf{q}\cdot\mathbf{r}}$$

we see that the propagator of phonons of wave vector $\mathbf{q}$ and polarization λ is

$$D_\lambda(\mathbf{q}, \tau) = -i\langle 0|T\{\varphi_{q\lambda}(\tau)\varphi_{q\lambda}{}^+(0)\}|0\rangle \quad (5\text{-}57)$$

for a translationally invariant system. We recall that the phonon-field amplitude is related to the creation and destruction operators by

$$\varphi_{q\lambda} = a_{q\lambda} + a_{-q\lambda}{}^+$$

As for G, we can write a spectral representation of D by introducing the weight function

$$B_\lambda(\mathbf{q}, \omega) = \sum_m |\langle m|\varphi_{q\lambda}{}^+|0\rangle|^2\, \delta(\omega - \omega_m) - \sum_m |\langle m|\varphi_{q\lambda}|0\rangle|^2\, \delta(\omega + \omega_m) \quad (5\text{-}58)$$

where ω_m is the excitation energy $E_m{}^n - E_0{}^n$ of the n-particle system. Then the time Fourier transform of $D_\lambda(\mathbf{q}, \tau)$ is given by

$$D_\lambda(\mathbf{q}, q_0) = \int_{-\infty}^{\infty} \frac{B_\lambda(\mathbf{q}, \omega)\, d\omega}{q_0 - \omega + i\omega\,\delta} \quad (5\text{-}59)$$

where $\delta = 0^+$. Since by definition B is a real quantity, the imaginary part of (5-59) gives the relation

$$\operatorname{Im} D_\lambda(\mathbf{q}, q_0) = -\pi B_\lambda(\mathbf{q}, q_0) \operatorname{sgn} q_0 \tag{5-60}$$

so that a dispersion relation holds for D:

$$D_\lambda(\mathbf{q}, q_0) = -\frac{1}{\pi}\int_{-\infty}^{\infty} \frac{\operatorname{Im} D_\lambda(\mathbf{q}, \omega) \operatorname{sgn} \omega}{q_0 - \omega + i\omega\,\delta}\, d\omega \tag{5-61}$$

Just as for $G(\mathbf{p}, p_0)$, we can extend the definition of $D(\mathbf{q}, q_0)$ into the complex q_0-plane with the aid of the spectral representation (5-59) and show that the poles of the analytic continuation of D into the lower half-plane for $q_0 > 0$ may be interpreted as giving the energy and decay rate of phonons (or phonon "holes") of wave vector $\mathbf{q}$ (or $-\mathbf{q}$) and polarization λ. The situation is somewhat more complicated here than in the electron case since it is the field amplitude $\varphi_{q\lambda}$ which is used in defining D rather than the destruction operator $a_{q,\lambda}$, as was used for the electron Green's function. We recall that the electron injection and extraction processes were separated by our definition of A. We see from the definition of B that both phonon creation and destruction processes contribute to the positive (and to the negative) frequency parts of B. In fact, B is an antisymmetric function of ω for a system with inversion symmetry since $\varphi_{q\lambda}{}^+ = \varphi_{-q,\lambda}$. In this case we can write D as

$$D_\lambda(\mathbf{q}, q_0) = \int_0^{\infty} B_\lambda(\mathbf{q}, \omega) \frac{2\omega}{q_0{}^2 - \omega^2 + i\,\delta}\, d\omega \tag{5-62}$$

For a system consisting only of bare phonons (i.e., uncoupled to the electrons), the spectral function is

$$B_\lambda(\mathbf{q}, \omega) = \delta(|\omega| - \Omega_{q,\lambda}) \operatorname{sgn} \omega \tag{5-63}$$

so that the bare phonon propagator is

$$D_{0\lambda}(\mathbf{q}, q_0) = \frac{2\Omega_{q\lambda}}{q_0{}^2 - \Omega_{q\lambda}{}^2 + i\delta} \tag{5-64}$$

In the next section we shall see how to build up the true one-electron and one-phonon Green's functions as a perturbation series involving the corresponding functions for the noninteracting system.

5-9 PERTURBATION SERIES

We shall not go into the details of deriving the perturbation-series expansions for the electron and phonon Green's functions for two reasons: there exist several good discussions of this derivation in print[103]; also one need not keep the derivation in mind when using the simple rules that follow from this derivation.

The perturbation expansion is based on two main assumptions. If the zero-order Hamiltonian which describes the bare particles (i.e., our Bloch electrons and bare phonons) is designated by H_0, then the ground state of H_0 is assumed to go over adiabatically to the ground state of the interacting system as the interaction is turned on adiabatically in time. In addition, it is assumed that the resulting time dependence of the ground state and the τ-dependence of G and D because of the presence of the interactions can be expanded as a power series in the strength of the interaction. These assumptions are less restrictive than they appear, since selected terms of these divergent expansions can often be summed to give physically meaningful results, as mentioned earlier. Unfortunately, one is often open to the criticism that neglected terms could drastically modify the results.

For simplicity we begin by assuming the total crystal momentum (i.e., the momentum $\mathbf{k}$ of a Bloch function, $\mathbf{q}$ of a phonon) is conserved by the interactions; that is, we neglect umklapp processes and all but the diagonal part of the one-body potential U. We can easily extend the treatment to include these effects. The rules for calculating the one-electron Green's function $G_s(\mathbf{p}, p_0)$ for the interacting system described by the Hamiltonian (4-35) are listed below.

1. Draw all diagrams in which an electron of spin s and four-momentum p (represented by a directed solid line) enters from the right and leaves from the left, undergoing all topologically distinct interactions in the process. Only "connected diagrams," those which cannot be separated into two or more unconnected pieces without breaking any lines, are to be included. The Coulomb interaction between electrons is represented by a dashed line and the one-body potential $\tilde{U}$ is represented by a dotted line

connecting the electron line to a cross (i.e., the source of the one-body potential). A phonon is represented by a wavy line. Label all the lines so that four-momentum and spin are conserved at each vertex. The "spin" of the phonon is taken to be zero in our definitions.

2. Associate with each bare electron line of four-momentum p the factor $iG_0(p)$, where $G_0(p)$ is the bare electron Green's function

$$G_0(p) = \frac{1}{p_0 - \epsilon_p + i\eta_p} \tag{5-23a}$$

3. Associate with each bare phonon line of four-momentum q and polarization λ the factor $iD_{0\lambda}(q)$, where $D_{0\lambda}(q)$ is the bare phonon Green's function

$$D_{0\lambda}(q) = \frac{2\Omega_{q\lambda}}{q_0{}^2 - \Omega_{q\lambda}{}^2 + i\delta} \tag{5-64a}$$

4. Associate with each vertex in which a phonon of polarization λ and momentum $\mathbf{q}$ is emitted (or $-\mathbf{q}$ is absorbed), scattering an electron from $\mathbf{p}$ to $\mathbf{p}'$, the factor $g_{p,\,p',\,\lambda}$.

5. Associate with each Coulomb line which involves one-electron scattering from $\mathbf{p}$ to $\mathbf{p}'$ and the other scattering from $\mathbf{k}$ to $\mathbf{k}'$ the factor $\langle \mathbf{p}', \mathbf{k}' | V | \mathbf{p}, \mathbf{k} \rangle$. If the bare electron states are approximated by plane waves the factor is

$$V(\mathbf{p} - \mathbf{p}') = \frac{4\pi e^2}{(\mathbf{p} - \mathbf{p}')^2}$$

6. Associate with each one-body line connected to an electron of momentum $\mathbf{p}$ the factor $\langle \mathbf{p} | \tilde{U} | \mathbf{p} \rangle$. The four-momentum is conserved in the process.

7. Include a factor $(-i)^n$, where n is the number of interactions (4), (5), and (6) taking place in the graph, and a factor $(-1)^l$, where l is the number of closed electron loops in the graph.

8. Multiply the above factors together and integrate over all free internal four-momenta $k_1 \cdots k_n$ according to

$$\int \frac{d^4k_1}{(2\pi)^4} \frac{d^4k_2}{(2\pi)^4} \cdots \frac{d^4k_n}{(2\pi)^4} [F_2F_3F_4F_5F_6F_7]$$

where $F_2, F_3, \ldots$ are the factors arising from rules 2, 3, ...

9. Sum the resulting contribution over all phonon polarizations and spin orientations of the internal electron lines, and sum the contributions from all topologically distinct graphs.

We illustrate the rules by calculating a few low-order contributions to $iG(p)$, shown graphically in Figure 5-4a, b, c, d, and e. These are

$$\text{(a)} \qquad iG_0(p) = \frac{i}{p_0 - \epsilon_p + i\eta_p} \tag{5-65a}$$

$$\text{(b)} \qquad [iG_0(p)]\left[(-i)(-1)\sum_s \int \frac{d^4p'}{(2\pi)^4} \langle \mathbf{p}, \mathbf{p}'|V|\mathbf{p}, \mathbf{p}'\rangle \times iG_0(p')\right][iG_0(p)] \tag{5-65b}$$

$$\text{(c)} \qquad [iG_0(p)][-i\langle \mathbf{p}|\tilde{U}|\mathbf{p}\rangle][iG_0(p)] \tag{5-65c}$$

$$\text{(d)} \qquad [iG_0(p)]\left[(-i)\int \frac{d^4p'}{(2\pi)^4} \langle \mathbf{p}', \mathbf{p}|V|\mathbf{p}, \mathbf{p}'\rangle iG_0(p')\right][iG_0(p)] \tag{5-65d}$$

$$\text{(e)} \qquad [iG_0(p)]\left[(-i)^2\sum_\lambda \int \frac{d^4p'}{(2\pi)^4} |g_{pp'\lambda}|^2 iD_{0\lambda}(p-p') \times iG_0(p')\right][iG_0(p)] \tag{5-65e}$$

The term (a) neglects all interactions and approximates G for the interacting system by its value G_0 for the noninteracting system. Term (b) describes the interaction of the added electron (or hole) with the average charge distribution of the bare conduction electrons of both spin orientations. While this term is formally infinite in the limit of a system with infinite volume, this infinity is cancelled by an infinite contribution of opposite sign arising from the one-body potential $\tilde{U}$ as shown in Figure 5-4c (see Sections 4-4 and 4-5 for the definition of $\tilde{U}$). By thinking through the definition of $\tilde{U}$, it becomes clear that the cancellation simply reflects the over-all electrical neutrality of the electron-ion system. A similar cancellation occurs whenever a diagonal matrix element of the Coulomb interaction appears, so that we need not worry

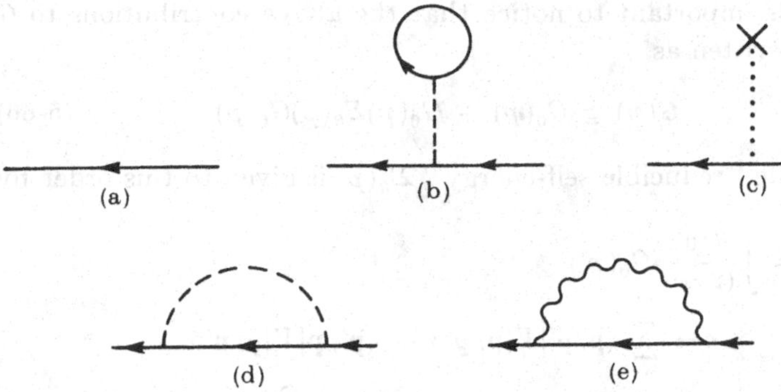

FIGURE 5-4 Terms entering the perturbation series for $iG(\mathbf{p}, p_0)$. (a) The zeroth approximation is $iG_0(\mathbf{p}, p_0)$. (b) The lowest-order "direct" contribution from the Coloumb interactions. (c) The lowest-order contribution from the one-body potential $\tilde{U}$. This cancels the divergent Coulomb contribution from (b). (d) The lowest-order "exchange" contribution from the Coulomb interaction. (e) The lowest-order contribution from the phonon field.

about infinities *of this nature*. In passing we note that there is no term corresponding to (b) in which the Coulomb line is replaced by a phonon line, since by definition there is no $q \equiv 0$ phonon mode. Term (d) represents the lowest-order exchange contribution due to the Coulomb potential. No spin sum appears in (d), as we expect. As we shall see, when the perturbation series is suitably rearranged, graphs which look like (b) and (c) will include the interactions within the Hartree approximation. By including the counterpart of (d), one gets the Hartree–Fock approximation.

Term (e), the lowest-order phonon process entering G, will lead to an effective mass correction and a finite decay rate of the added electron (or hole) due to phonon emission.

After inserting the expressions for G_0 and D_0, one can easily carry out the p_0'-integrals by the method of residues; however, the three-momentum integrals cannot be performed unless we have explicit expressions for the matrix elements involved. The result of the p_0'-integrations will be given below.

It is important to notice that the above contributions to G can be written as

$$G(p) \cong G_0(p) + G_0(p)\Sigma_R(p)G_0(p) \tag{5-66}$$

where the "reducible self-energy" $\Sigma_R(p)$ is given to this order by

$$\begin{aligned}\Sigma_R(p) = \int \frac{d^4p'}{(2\pi)^4}\, iG_0(p') \\ \times \Big\{ -\sum_s \langle \mathbf{p}, \mathbf{p}'|V|\mathbf{p}, \mathbf{p}'\rangle + \langle \mathbf{p}', \mathbf{p}|V|\mathbf{p}, \mathbf{p}'\rangle \\ + \sum_\lambda |g_{\mathbf{pp}'\lambda}|^2 D_{0\lambda}(p - p')\Big\} + \langle \mathbf{p}|\tilde{U}|\mathbf{p}\rangle\end{aligned} \tag{5-67}$$

Since every graph for iG starts and ends with a bare electron line, it follows that $G(p)$ can be exactly expressed in the form (5-66), where $\Sigma_R(p)$ contains no external lines entering or leaving the graph. An important observation is that $G(p)$ can also be expressed in terms of the "irreducible self-energy" $\Sigma(p)$:

$$G(p) = G_0(p) + G_0(p)\Sigma(p)G(p) \tag{5-68}$$

where $\Sigma(p)$ is given by the sum of all graphs giving Σ_R except those which can be separated into two unconnected pieces by cutting a single bare electron line. Thus, while the graphs shown in Figure 5-4b, c, d, and e contribute to Σ (and Σ_R), the graphs shown in Figure 5-5a and b contribute only to Σ_R and must not be included in computing Σ since these terms are already accounted for in (5-68) as one can see by solving this equation by iteration. On the other hand Figure 5-5c shows a valid contribution to Σ. Equation (5-68) is known as Dyson's equation (for G).[102] We note that

$$\frac{1}{G(p)} = \frac{1}{G_0(p)} - \Sigma(p)$$

or using the expression for $G_0(p)$ we find the simple relation

$$G(p) = \frac{1}{p_0 - \epsilon_p - \Sigma(p) + i\eta_p} \tag{5-69}$$

The power of the graphical methods is in part due to the fact that one can calculate $\Sigma(p)$ to low order in perturbation theory and often obtain a reasonable approximation for G, even though a straightforward series expansion of G in powers of the interaction strength is meaningless. As we shall see, the effect of electron–phonon interactions on G in normal metals comes almost entirely from the lowest-order graph for Σ, as was first pointed out by Migdal.[15]

In Section 5-5 we noted that the energy E_p and damping rate $|\Gamma_p|$ of a quasi-particle are given by the pole of the analytic continuation of G into the lower right (upper left) half-plane.[100] From (5-69) we have

$$E_p = \epsilon_p + \operatorname{Re} \tilde{\Sigma}(\mathbf{p}, E_p + i\Gamma_p) \tag{5-70a}$$

$$\Gamma_p = \operatorname{Im} \tilde{\Sigma}(\mathbf{p}, E_p + i\Gamma_p) \tag{5-70b}$$

where Σ is the appropriately continued function. While such poles usually exist, one must be careful to check that they give

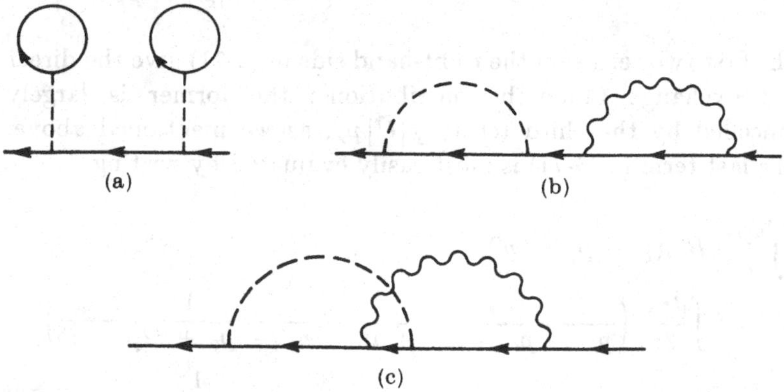

FIGURE 5-5 Higher-order corrections to $iG(\mathbf{p}, p_0)$. (a) This graph is automatically included if one includes the contribution shown in Figure 3-4b in the irreducible self-energy Σ. (b) This graph is also included automatically if Figures 3-4d and 3-4e are included in Σ. (c) This graph is not automatically included since it has an irreducible self-energy part. Its contribution must be included explicitly in Σ.

a good representation of the spectral weight function $A(\mathbf{p}, \omega)$ for real ω, and hence a good approximation of $G(\mathbf{p}, p_0)$. In general, the analytic behavior of G is more complicated than the idealized quasi-particle picture, as we shall see in Chapter 6.

Returning to the low-order contribution to Σ considered above, we have from (5-67):

$$\Sigma(p) \simeq \int \frac{d^3p'}{(2\pi)^3} \{2\langle \mathbf{p}, \mathbf{p}'|V|\mathbf{p}, \mathbf{p}'\rangle - \langle \mathbf{p}', \mathbf{p}|V|\mathbf{p}, \mathbf{p}'\rangle\} f_{\mathbf{p}'} + \langle \mathbf{p}|\tilde{U}|\mathbf{p}\rangle + \sum_\lambda i \int \frac{d^4p'}{(2\pi)^4} |g_{pp'\lambda}|^2 \times \frac{2\Omega_{q\lambda}}{(p_0 - p_0')^2 - \Omega_{q\lambda}{}^2 + i\delta} \frac{1}{p_0' - \epsilon_{p'} + i\eta_{p'}} \tag{5-71}$$

where $q \equiv p - p'$, and we have used the relation

$$i \int \frac{dp_0'}{2\pi} G_0(p') = i \int \frac{dp_0'}{2\pi} \frac{e^{ip_0'\delta}}{p_0' - \epsilon_{p'} + i\eta_{p'}} = -f_{p'} = \begin{cases} -1 & |\mathbf{p}'| < p_F \\ 0 & |\mathbf{p}'| > p_F \end{cases} \tag{5-72}$$

The first two terms on the right-hand side of (5-71) give the direct and exchange Coulomb contributions; the former is largely cancelled by the third term $\langle p|\tilde{U}|p\rangle$, as we mentioned above. The last term in (5-71) is most easily evaluated by writing

$$i \int \frac{dp_0'}{2\pi} D_{0\lambda}(p - p')G_0(p') = i \int \frac{dp_0'}{2\pi} \left(\frac{1}{p_0' - p_0 - \Omega_{q\lambda} + i\delta} - \frac{1}{p_0' - p_0 + \Omega_{q\lambda} - i\delta} \right) \times \frac{1}{p_0' - \epsilon_{p'} + i\eta_{p'}} \tag{5-73}$$

and noticing that the first factor in D_0 does not contribute for $|p'| > p_F$ since neither this factor nor $G_0(p')$ has poles in the upper half-plane; therefore, by closing the p_0'-contour in the upper half-plane we obtain zero in this case. Similarly, the second term in D_0 does not contribute for $|p'| < p_F$, since neither this factor nor $G_0(p')$ has poles in the lower half-plane, and we obtain zero by closing in the lower half-plane in this case. The remaining contributions are

$$\begin{cases} i \displaystyle\int \frac{dp_0'}{2\pi} \frac{1}{p_0' - p_0 - \Omega_{q\lambda} + i\delta} \frac{1}{p_0' - \epsilon_{p'} - i\delta} \\ \qquad = \dfrac{1}{p_0 - \epsilon_{p'} + \Omega_{q\lambda} - i\delta} \qquad |\mathbf{p}'| < p_F \\ -i \displaystyle\int \frac{dp_0'}{2\pi} \frac{1}{p_0' - p_0 + \Omega_{q\lambda} - i\delta} \frac{1}{p_0' - \epsilon_{p'} + i\delta} \\ \qquad = \dfrac{1}{p_0 - \epsilon_{p'} - \Omega_{q\lambda} + i\delta} \qquad |\mathbf{p}'| > p_F \end{cases} \tag{5-74}$$

Thus, the last term in (5-71), which arises from the lowest-order phonon process, is

$$\Sigma^{\text{ph}}(p) = \sum_{\lambda} \int \frac{d^3p'}{(2\pi)^3} |g_{pp'\lambda}|^2 \times \left\{ \frac{1 - f_{p'}}{p_0 - \epsilon_{p'} - \Omega_{q\lambda} + i\delta} + \frac{f_{p'}}{p_0 - \epsilon_{p'} + \Omega_{q\lambda} - i\delta} \right\} \tag{5-75}$$

The two terms in this expression can be easily understood within the framework of time-independent (Brillouin–Wigner) perturbation theory. Consider adding to the system an electron in the state $\mathbf{p}$ above the Fermi surface. The familiar second-order process shown in Figure 5-6, in which a phonon of energy $\Omega_{q,\lambda}$ is virtually emitted and reabsorbed by the added electron, leads to the energy shift

$$\Delta E = \sum_{\lambda} P \int \frac{d^3p'}{(2\pi)^3} \frac{|g_{pp'\lambda}|^2 (1 - f_{p'})}{p_0 - \epsilon_{p'} - \Omega_{q\lambda}} \tag{5-76}$$

Here P indicates the principal part of the integral is to be taken; the factor $(1 - f_{p'})$ reflects the Pauli principle requirement that the intermediate state $\mathbf{p}'$ is initially unoccupied. Equation (5-76) agrees with the real part of the first term of (5-75).

The second term in (5-75) is somewhat more subtle but illustrates a general feature of many-body perturbation theory. Consider for the moment the system in the absence of the added electron. Owing to the electron–phonon interaction there will be processes contributing to the ground-state energy of the system in which an electron, initially in a state $\mathbf{p}'$ within the Fermi sea, virtually emits a phonon of energy $\Omega_{q,\lambda}$ and jumps into an unoccupied state $\mathbf{p}$ above the Fermi surface. The excited electron then returns to its initial state by reabsorbing the phonon, as shown in Figure 5-7. For a given intermediate state $\mathbf{p}$ (above the Fermi surface) these processes contribute the energy shift

$$\Delta E' = \sum_\lambda P \int \frac{d^3p'}{(2\pi)^3} \frac{|g_{pp'\lambda}|^2 f_{p'}}{\epsilon_{p'} - \epsilon_p - \Omega_{q\lambda}} \tag{5-77}$$

to the ground state. Note the symbol P is not important here since the integrand is never singular. If an electron is now added to the system in state $\mathbf{p}$, the virtual excitation of the Fermi sea leading to (5-77) is forbidden by the Pauli principle. Therefore, this term must be subtracted in calculating the excitation energy. Thus, the total excitation energy of the system, when an electron is added in state $\mathbf{p}$, is given by

$$\begin{aligned} \epsilon_p + \Delta E - \Delta E' = \epsilon_p &+ \sum_\lambda P \int \frac{d^3p'}{(2\pi)^3} |g_{pp'\lambda}|^2 \\ &\times \left\{ \frac{(1 - f_{p'})}{p_0 - \epsilon_{p'} - \Omega_{q\lambda}} + \frac{f_{p'}}{\epsilon_p - \epsilon_{p'} + \Omega_{q\lambda}} \right\} \end{aligned} \tag{5-78}$$

We know from the discussion above that the excitation energy of the system is also given by the pole of $G(p, p_0)$; thus to order g^2

$$E_p = \epsilon_p + \operatorname{Re} \tilde{\Sigma}(\mathbf{p}, E_p) \cong \epsilon_p + \operatorname{Re} \Sigma(p, \epsilon_p) \tag{5-79}$$

If one uses (5-75) for Σ, the two expressions (5-78) and (5-79) are identical, as desired.

Throughout our application of many-body perturbation theory we shall see analogous examples of the Pauli principle bringing about modifications of the virtual fluctuations of the system when a particle (or hole) is added to the system. In the

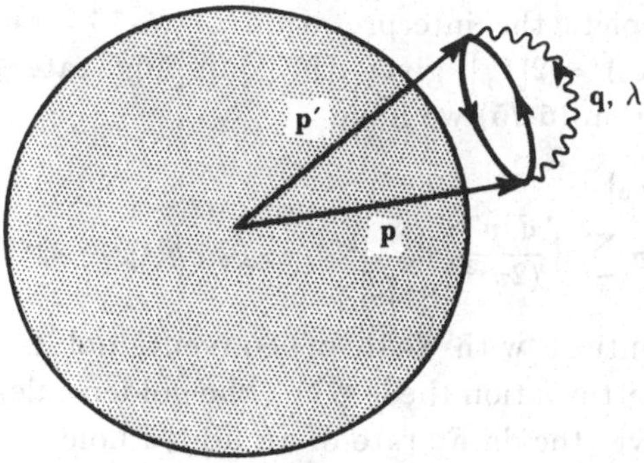

FIGURE 5-6 A typical contribution to the second-order time-independent perturbation series for the ground-state energy in the presence of an added particle of momentum **p**. The Pauli principle restricts the intermediate state to $|\mathbf{p}'| > p_F$.

superconducting state a major fraction of the energy gap arises from this type of effect. In a real sense the Pauli principle is ultimately responsible for absence of low-lying elementary excitations in superconducting metals.

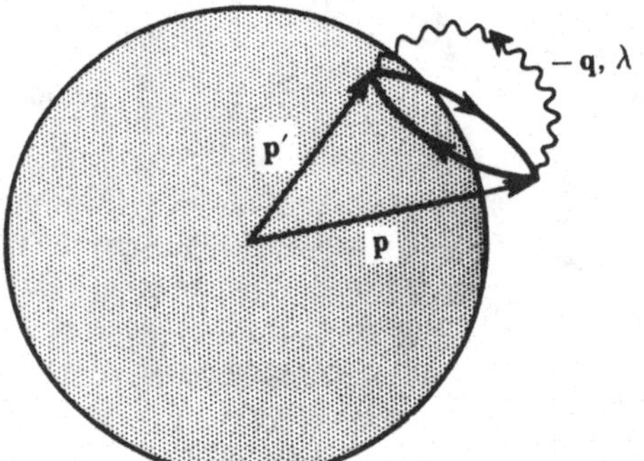

FIGURE 5-7 A typical contribution to the ground-state energy in the absence of the extra particle. The Pauli principle restricts the intermediate states to $|\mathbf{p}| > p_F$. This "vacuum-fluctuation" process is suppressed if a particle is added to **p**, thereby leading to a shift of the quasi-particle energy.

To complete the interpretation of (5-75), we observe that $-2\,\mathrm{Im}\,\Sigma(\mathbf{p}, \epsilon_p) = 2|\Gamma_p|$ gives the damping rate of the added electron. From (5-75) we have

$$-2\,\mathrm{Im}\,\Sigma(\mathbf{p}, \epsilon_p) = 2\pi \sum_{\lambda} \int \frac{d^3p'}{(2\pi)^3} |g_{pp'\lambda}|^2(1 - f_{p'})\,\delta(\epsilon_p - \epsilon_{p'} - \Omega_{q\lambda}) \quad (5\text{-}80)$$

a result identical with that of conventional first-order time-dependent perturbation theory, i.e., the golden rule.[104] A similar argument gives the decay rate of an added hole.

CHAPTER 6

ELEMENTARY EXCITATIONS IN NORMAL METALS

If one attempts to calculate the electronic self-energy in normal metals by the perturbation series described above, one quickly discovers that the series diverges, owing to the singular nature of the Coulomb matrix elements evaluated for small momentum transfer q.

6-1 THE ELECTRON GAS WITH COULOMB INTERACTIONS

This singularity, because of the long-range nature of the Coulomb potential, can be circumvented by summing a set of graphs which physically represents the screening of this potential by the valence electrons. We neglect the phonons for the moment. From the pioneering work of Bohm and Pines[105] and from the more recent work of Gell-Mann and Brueckner,[106] one knows that the most important screening effects in the limit $q \to 0$ are given

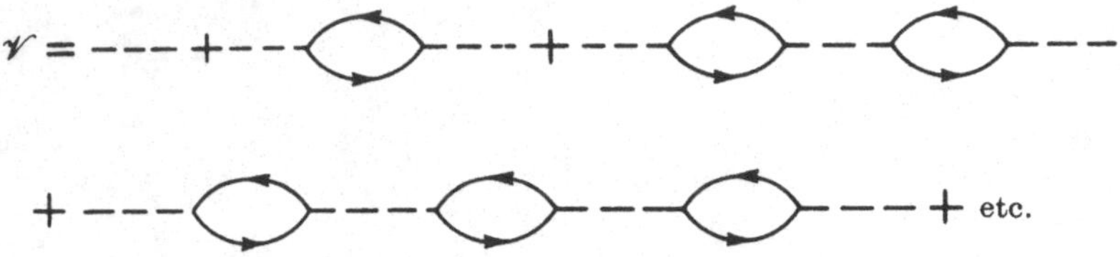

FIGURE 6-1 The random phase approximation (RPA) for the effective interaction in the electron gas.

by summing the so-called bubble graphs. In effect, one can replace the original bare Coulomb interaction V by an effective potential $\mathscr{V}_c$ given by the series shown in Figure 6-1. This approximation for the screening is known as the random phase approximation (RPA). To simplify the algebra we neglect momentum nonconserving processes and approximate the Bloch states by plane wave states so that matrix elements of the Coulomb interaction are given by $V(q) = 4\pi e^2/q^2$. By using the rules given in Chapter 3 for evaluating Feynman graphs, we have

$$\mathscr{V}_c^{\mathrm{RPA}}(q) = V(q) - V(q)P^{\mathrm{RPA}}(q)\mathscr{V}_c^{\mathrm{RPA}}(q) \tag{6-1}$$

where the "irreducible polarizability" $P(q)$, evaluated within the RPA, is given by

$$P^{\mathrm{RPA}} = 2i\int G_0(p+q)G_0(p)\frac{d^4p}{(2\pi)^4} \tag{6-2}$$

the factor of two coming from the spin sum. By rearranging (6-1) we have

$$\mathscr{V}_c^{\mathrm{RPA}}(q) = \frac{V(q)}{1 + V(q)P^{\mathrm{RPA}}(q)} = \frac{V(q)}{\kappa_0(q)} \tag{6-3}$$

where $q = (\mathbf{q}, q_0)$ and $V(q)$ is a function of $\mathbf{q}^2$ alone. From the form of (6-3), it is clear that the denominator is the wave vector and frequency-dependent dielectric function $\kappa_0(\mathbf{q}, q_0)$ of the valence electrons, evaluated within the random phase approximation.[103a, 107] By inserting the expression for G_0 into (6-3), it immediately follows that

$$\begin{aligned}\operatorname{Re} \kappa_0(\mathbf{q}, q_0) &= 1 + V(q) \operatorname{Re} P^{\mathrm{RPA}}(\mathbf{q}, q_0) \\ &= 1 - 2V(q)P \int \frac{d^3p}{(2\pi)^3} f_p(1 - f_{p+q}) \frac{2(\epsilon_{p+q} - \epsilon_p)}{q_0{}^2 - (\epsilon_{p+q} - \epsilon_p)^2}\end{aligned} \tag{6-4}$$

where P indicates the principal part of the integral is to be taken. In addition we have

$$\begin{aligned}\operatorname{Im} \kappa_0(\mathbf{q}, q_0) &= 2\pi V(q) \int \frac{d^3p}{(2\pi)^3} f_p(1 - f_{p+q}) \\ &\quad \times [\delta(q_0 - \epsilon_{p+q} + \epsilon_p) + \delta(q_0 + \epsilon_{p+q} - \epsilon_p)]\end{aligned} \tag{6-5}$$

The integrations (6-4) and (6-5) are straightforward and lead to the results first given by Lindhard[108]:

$$\begin{aligned}\operatorname{Re} \kappa_0(\mathbf{q}, q_0) = 1 + \frac{2e^2mk_F}{\pi q^2} \Bigg\{ 1 &+ \frac{k_F}{2q}\left[1 - \left(\frac{mq_0}{qk_F} + \frac{q}{2k_F}\right)^2\right] \\ &\times \ln \left| \frac{1 + \left(\frac{mq_0}{qk_F} + \frac{q}{2k_F}\right)}{1 - \left(\frac{mq_0}{qk_F} + \frac{q}{2k_F}\right)} \right| \\ &- \frac{k_F}{2q}\left[1 - \left(\frac{mq_0}{qk_F} - \frac{q}{2k_F}\right)^2\right] \ln \left| \frac{1 + \left(\frac{mq_0}{qk_F} - \frac{q}{2k_F}\right)}{1 - \left(\frac{mq_0}{qk_F} - \frac{q}{2k_F}\right)} \right| \Bigg\}\end{aligned} \tag{6-6a}$$

and

$$\operatorname{Im} \kappa_0(\mathbf{q}, q_0) = \begin{cases} 0 & \text{for } 2m|q_0| > q^2 + 2qk_F \\ 0 & \text{for } q > 2k_F \text{ and } 2m|q_0| < q^2 - 2qk_F \\ 2e^2m^2 \dfrac{q_0}{q^3} & \text{for } q < 2k_F \text{ and } 2m|q_0| < |q^2 - 2qk_F| \\ \dfrac{e^2mk_F{}^2}{q^3}\left\{1 - \left(\dfrac{mq_0}{qk_F} - \dfrac{q}{2k_F}\right)^2\right\} & \text{for } |q^2 - 2qk_F| < 2m|q_0| < |q^2 + 2qk_F| \end{cases} \tag{6-6b}$$

For $q_0 = 0$, (6-6) gives Im $\kappa = 0$ so that the static dielectric constant is given by

$$\kappa_0(\mathbf{q}, 0) = 1 + 0.66 r_s \left(\frac{k_F}{q}\right)^2 u\left(\frac{q}{2k_F}\right) \tag{6-7a}$$

where

$$u(x) = \frac{1}{2}\left[1 + \frac{(1 - x^2)}{2x} \ln\left|\frac{1 + x}{1 - x}\right|\right] \tag{6-7b}$$

and

$$\frac{4\pi r_s^3 a_0^3}{3} = \frac{1}{n} \tag{6-7c}$$

a_0 being the Bohr radius. The quantity r_s is a dimensionless measure of the strength of the Coulomb interaction. A plot of $1/\kappa_0(\mathbf{q}, 0)$ as obtained from (6-7) is given in Figure 6-2, along with a plot of the Thomas–Fermi result. The main features of the plot are given below.

1. κ_0 approaches the Thomas–Fermi result

$$\kappa_0(\mathbf{q}, 0) = 1 + \frac{k_s^2}{q^2} \tag{6-8}$$

as $q \to 0$, where the screening wave number is given by

$$k_s^2 = \frac{6\pi n e^2}{E_F}$$

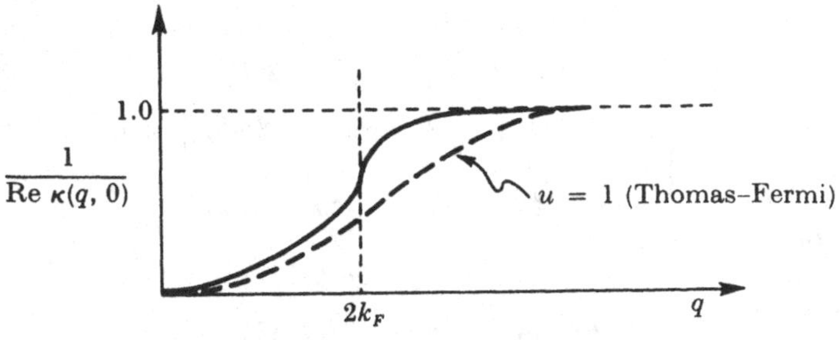

FIGURE 6-2 A plot of the zero-frequency RPA dielectric function κ as a function of the wave number q. Also shown is the Thomas–Fermi approximation for κ.

2. $d\kappa_0/dq \to \infty$ as $q \to 2k_F$. This fact leads to the result, first discussed by Kohn, Langer, and Vosko,[109a, b] that the asymptotic form of the screened Coulomb potential for large distance is not a Yukawa potential but rather the oscillatory function

$$\mathscr{V}(r) \propto \frac{\cos(2k_F r + \phi)}{r^3} \tag{6-9}$$

There is good experimental evidence to support this result.[110]

3. $\kappa_0 \to 1$ as $q \to \infty$. Thus, screening is ineffective for very large momentum transfers.

For $q_0 \neq 0$, the imaginary part of κ is nonzero only when q_0 and q are related so that the argument of the delta function in (6-5) can vanish for some $|\mathbf{p}| < p_F$ and $|\mathbf{p} + \mathbf{q}| > p_F$, that is to say,

$$q^2 - 2qk_F \leqslant 2m|q_0| \leqslant q^2 + 2qk_F \tag{6-10}$$

The situation is illustrated in Figure 6-3. For large frequency, i.e., $|q_0| \gg q^2 + (2qk_F/2m)$, we have the familiar limiting form

$$\text{Re}\,\kappa_0(\mathbf{q}, q_0) = 1 - \frac{\omega_p{}^2}{q_0{}^2} \tag{6-11}$$

where $\omega_p{}^2 = 4\pi ne^2/m$. The two expressions (6-8) and (6-11) are useful limiting forms for making physical arguments.

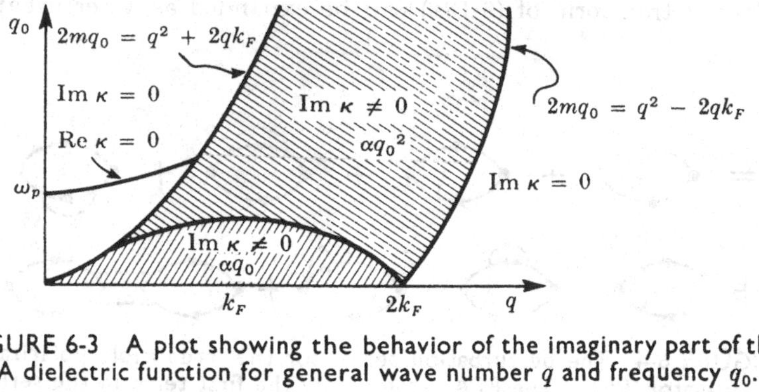

FIGURE 6-3 A plot showing the behavior of the imaginary part of the RPA dielectric function for general wave number q and frequency q_0.

While the RPA is valid for small momentum transfers, it is desirable to introduce an effective potential $\mathscr{V}_c(q)$ which includes *all* vacuum polarization processes.[103a] A partial summation of the series can be performed if we introduce the irreducible polarizability $P(q)$, defined as the sum of all graphs in which a single interaction line enters and leaves the graph and which cannot be divided into two disconnected graphs by cutting a single Coulomb line. $P(q)$ is shown in Figure 6-4. Then,

$$\mathscr{V}_c(q) = \frac{V(q)}{1 + V(q)P(q)} \equiv \frac{V(q)}{\kappa(q)} \tag{6-12}$$

The RPA result is given by retaining only the first term in the series for $P(q)$.

To obtain a compact expression for $\kappa(q)$, we consider the quantity (evaluated in the Heisenberg picture)

$$\langle 0 | T\{\rho_{-q}(-\tau)\rho_q(0)\} | 0\rangle \tag{6-13a}$$

Here ρ_q is the qth Fourier component of the electron density operator,

$$\rho_q(t) = \sum_s \int e^{-i\mathbf{q}\cdot\mathbf{r}}\psi_s^+(\mathbf{r}, t)\psi_s(\mathbf{r}, t)\, d^3r = \sum_{p,s} c_{p,s}^+(t)c_{p+q,s}(t) \tag{6-13b}$$

Again, the phonons are neglected for the moment. The time Fourier transform of (6-13a) can be expanded as a perturbation

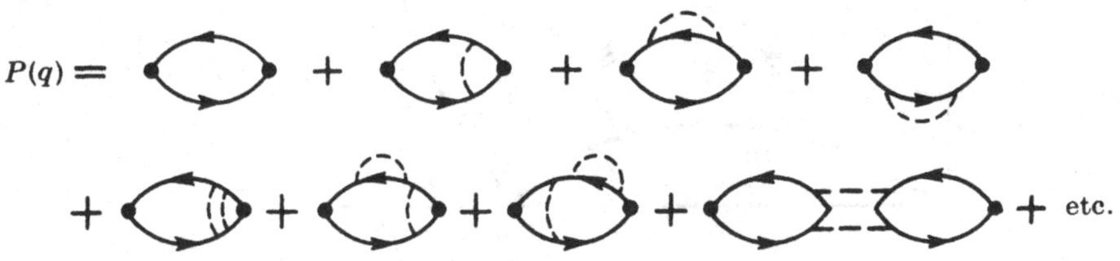

FIGURE 6-4 The perturbation series for the irreducible polarization propagator $P(q)$. The RPA retains only the first term in the series.

series in the Coulomb interaction strength and is easily seen to be related to the series for $\mathscr{V}_c(q)$:

$$\frac{\mathscr{V}_c(q) - V(q)}{V(q)} = \frac{1}{\kappa(q)} - 1$$
$$= -iV(q) \int_{-\infty}^{\infty} e^{iq_0\tau} \langle 0 | T\{\rho_{-q}(-\tau)\rho_q(0)\} | 0 \rangle \, d\tau \tag{6-13c}$$

This expression allows us to study the analytic structure of $\kappa(\mathbf{q}, q_0)$ as a function of q_0 for fixed $\mathbf{q}$. As for $G(\mathbf{p}, p_0)$, we can insert a complete set of eigenstates of H between ρ_q and ρ_{-q} in (6-13) and obtain the spectral representation

$$\frac{1}{\kappa(\mathbf{q}, q_0)} - 1 = \int_{-\infty}^{\infty} \frac{F(\mathbf{q}, \omega) \, d\omega}{q_0 - \omega + iq_0\delta} \tag{6-14a}$$

where

$$F(\mathbf{q}, \omega) = V(q) \sum_n |\langle n | \rho_{-q} | 0 \rangle|^2 \, \delta(\omega - \omega_{n0}) \quad \text{for } \omega > 0 \tag{6-14b}$$

and

$$F(\mathbf{q}, \omega) = -F(-\mathbf{q}, |\omega|) \quad \text{for } \omega < 0 \tag{6-14c}$$

For a system with inversion symmetry (6-14c) is equivalent to $F(\mathbf{q}, \omega) = -F(\mathbf{q}, -\omega)$ so that (6-14a) becomes

$$\frac{1}{\kappa(\mathbf{q}, q_0)} - 1 = \int_0^{\infty} F(\mathbf{q}, \omega) \left(\frac{1}{q_0 - \omega + iq_0\delta} - \frac{1}{q_0 + \omega + iq_0\delta} \right) d\omega \tag{6-15}$$

The right-hand side of (6-15) is of the same form as the spectral representation of the phonon (boson) Green's function. The essential difference between the two is that the weight function $F(\mathbf{q}, \omega)$ involves matrix elements of the *electronic* density fluctuation operator, while the longitudinal phonon weight function $\Phi(\mathbf{q}, \omega)$ involves matrix elements of the *ionic* density fluctuation operator. Thus, in analogy with the phonon Green's function, the poles of $1/\kappa(\mathbf{q}, q_0)$ give the *boson-like* excitations of the *electron gas* which are excited by a longitudinal field.

The nature of these Bose excitations is easily understood within the RPA. By taking the imaginary part of (6-14a) one finds

$$F(\mathbf{q}, q_0) = \begin{cases} -\dfrac{1}{\pi} \operatorname{Im} \dfrac{1}{\kappa(\mathbf{q}, q_0)} & \text{for } q_0 > 0 \\ \dfrac{1}{\pi} \operatorname{Im} \dfrac{1}{\kappa(\mathbf{q}, q_0)} & \text{for } q_0 < 0 \end{cases} \tag{6-16a}$$

or

$$F(q) = -\frac{1}{\pi} \frac{\kappa_2(q) \operatorname{sgn} q_0}{\kappa_1^2(q) + \kappa_2^2(q)} \tag{6-16b}$$

where κ_1 and κ_2 are the real and imaginary parts of κ, respectively.

If the RPA expressions (6-6a) and (6-6b) are used for κ_1 and κ_2, we see that, for a given momentum $\mathbf{q}$, the boson excitations form a continuum over the interval

$$q^2 - 2qk_F \leqslant q_0 \leqslant q^2 + 2qk_F$$

since κ_2 is finite on this interval but vanishes elsewhere, as stated by (6-10). These excitations are in one-to-one correspondence with the excitations of the noninteracting system, in which an electron initially in state $\mathbf{p}$ within the Fermi sea is raised above the Fermi surface to a state $\mathbf{p} + \mathbf{q}$.

While the RPA predicts the energies of these single-particle-like excitations to be identical with corresponding excitation energies in the noninteracting system, the "wave functions" of the excitations are markedly different because of correlations existing between the excited electron (or hole) and the background electrons. Crudely speaking, the excited electron pushes neighboring electrons away, making a "correlation hole" which follows the excited particle as it travels through the system. This local depletion of background electrons allows the fixed positive ions to terminate the electric field lines of the excited electron so that the net electric field associated with the excitation, i.e., the excited electron plus its correlation hole, vanishes far outside of the correlation hole. Since there is no long-range electric field associated with the excitation, Gauss' law states that the net effective charge of the excitation is zero. Therefore, if the excited electron moves through the system, it follows that a backflow of background electrons must accompany the excitation in such a way that the net charge transported is zero.

This backflow is the analog of the backflow discussed by Feynman and Cohen[111] in the case of excitations in superfluid He^4. In our case, the backflow is essential in obtaining consistent expressions for the response of the system to external fields. In particular, we shall see that the Meissner effect can be treated in a gauge-invariant manner once backflow is properly taken into account in the superconducting state.[112-114] Without backflow one violates the continuity equation for electronic charge and therefore violates gauge invariance. This problem is discussed in detail in Chapter 8.

As we mentioned above, κ_2 vanishes outside of the interval given by (6-10); however, $1/\kappa$ can still be singular if κ_1 and κ_2 both vanish. In the long wavelength, high-frequency limit, (6-11) shows that κ_1 vanishes when

$$q_0{}^2 = \omega_p{}^2 = \frac{4\pi ne^2}{m} \tag{6-17}$$

Thus, there is an excited state of the system of excitation energy equal to the electronic plasma frequency. Bohm and Pines[105] were the first to discover these excitations in metals and named them plasmons. Physically, they are simply density-fluctuation waves of the electron system. For sufficiently large momentum $\mathbf{q}$, the plasmons pass into the single particle-like continuum, as shown in Figure 6-5 and become so heavily damped that they are no longer useful entities. For small $\mathbf{q}$ they play a very important role in determining the response of the system to longitudinal perturbations.

Having discussed the problem of screening, we now return to the problem of determining the one-electron Green's function for the interacting electron gas. Since the system is electrically neutral and (we assume) translationally invariant, the Hartree term in the self-energy Σ vanishes. From our above discussion

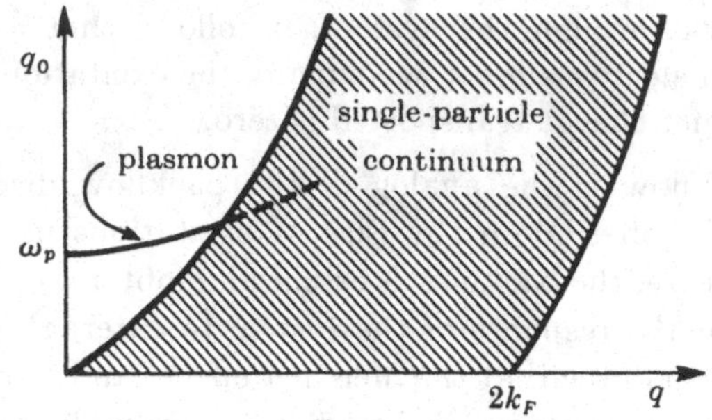

FIGURE 6-5 A plot of the excitation spectrum of the electron gas treated within the RPA.

of screening, it would appear natural to next include the exchange self-energy, not with the bare Coulomb matrix element $V(q)$ but rather with the screened potential $\mathscr{V}_c(q)$ evaluated within the RPA, as shown in Figure 6-6; thus,

$$\Sigma(p) = i \int G_0(p + q)\mathscr{V}_c(-q) \frac{d^4q}{(2\pi)^4} \tag{6-18}$$

While G_0 and $\mathscr{V}_c$ are known explicitly, the integration is difficult to carry out for general p. Calculations were first carried out by Quinn and Ferrell,[115] who found the energy of a quasi-particle near the Fermi surface is given by[91b]

$$E_p = E_F\left\{\frac{p^2}{k_F^2} - 0.166r_s\left[\frac{p}{k_F}(\ln r_s + 0.203) + \ln r_s - 1.80\right]\right\} \tag{6-19a}$$

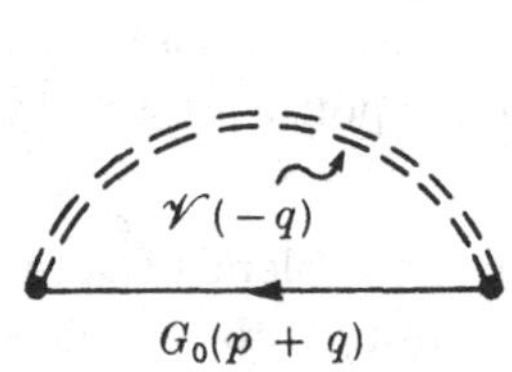

FIGURE 6-6 The electron self-energy evaluated within the screened exchange approximation. This graph presumably gives the most important contribution to $\Sigma(p)$ from small momentum transfer processes ($|\mathbf{q}| \ll k_F$).

and the damping rate due to electron–hole pair production is

$$|2\Gamma^{\text{pair}}(p)| = 2E_F(0.252r_s^{1/2})\left|\left(\frac{p}{k_F}\right)^2 - 1\right| \qquad \text{(6-19b)}$$

The effective mass of the quasi-particle at the Fermi surface is given by differentiating (6-19a):

$$\frac{1}{m^*} = \frac{1}{k_F}\frac{\partial E_p}{\partial p}\bigg|_{p=k_F} = \frac{1}{m}[1 - 0.083r_s(\ln r_s + 0.203)] \qquad \text{(6-19c)}$$

Since the electronic specific heat C is proportional to m^*, one obtains

$$\frac{C}{C_{\text{free}}} = 1 + 0.083r_s(\ln r_s + 0.203) \qquad \text{(6-20)}$$

a result first derived by Gell-Mann.[116] From (6-19b) it follows that the quasi-particles are well defined, in the sense that $|E(p) - E_F| \gg |\Gamma(p)|$, so long as $|\epsilon_p - E_F| \gtrsim \frac{1}{5}E_F$. As we shall see, this is no longer the case once the phonons are included.

The results (6-19a, b, c) are strictly valid only in the high-density limit ($r_s < 1$). Since r_s is typically $2 < r_s < 5$ for real metals, one must use these results with care. The long wavelength part of the Coulomb potential is probably well accounted for in Σ by the exchange graph screened within the RPA. For very short wavelengths (i.e., short-range part of the potential), parallel spin electrons do not interact since they are kept apart by the Pauli principle. One can presumably treat the interaction between antiparallel spin electrons by the second-order graph shown in Figure 6-7.[103, 117] Silverstein[118] and Pines have

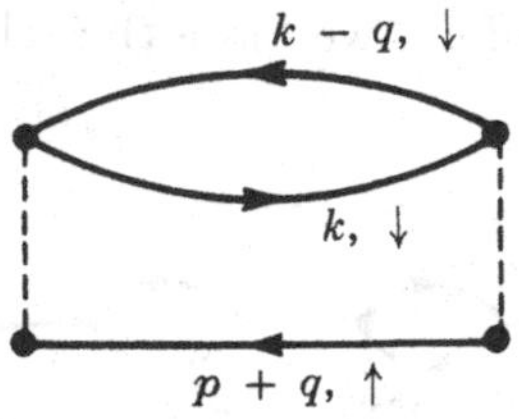

FIGURE 6-7 The leading contribution to Σ for large momentum transfers ($|q| \gg k_F$), after Figure 5-4d has been included.

exploited these limiting forms (i.e., large and small q-limits) by interpolating between these limits to obtain an estimate of Σ. The reader is referred to the literature for more details on this complicated problem.[91]

6-2 THE COUPLED ELECTRON–PHONON SYSTEM

Screening also plays an important role in determining the dressed phonon frequencies and the dressed electron–phonon interaction, since the long-range Coulomb forces between ions and between ions and electrons are screened out by the conduction electrons. For long wavelength effects we can again use the random phase approximation since it gives the leading corrections in this limit. For shorter wavelengths, processes neglected by the RPA begin to play a role. However, the RPA gives roughly correct answers for all wavelengths, since in the region where it is least accurate it gives small corrections to the bare quantities so that errors in these small corrections can often be neglected.

We begin by treating the entire electron–phonon system within the RPA and, for simplicity, neglect umklapp processes in both the Coulomb interaction and the electron–phonon interaction. The (irreducible) longitudinal phonon self-energy $\Pi_l(q)$ is given by the series of graphs shown in Figure 6-8. The string of bubbles, characteristic of the RPA, simply represents the screening of the ion–ion potential. From the series for $\mathscr{V}_c(q)$ shown in Figure 6-1 we see that $\mathscr{V}_c(q)$ and $\Pi_l(q)$ differ in two respects: (a) the incoming and outgoing bare Coulomb lines in $\mathscr{V}_c(q)$ have been replaced by electron–phonon matrix elements in $\Pi_l(q)$ and (b) the leading term $V(q)$ in the series for $\mathscr{V}_c(q)$ is missing in $\Pi_l(q)$. Therefore, we have the relation

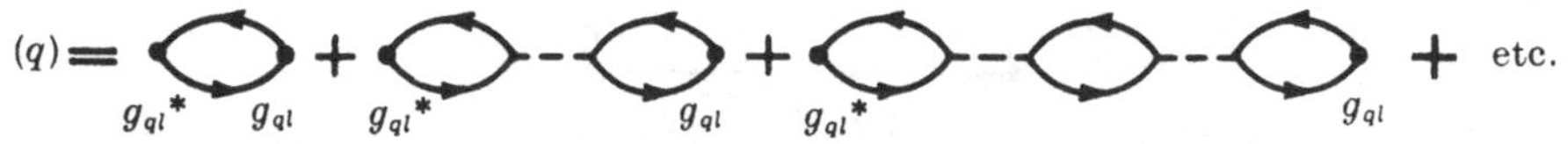

Figure 6-8 The random phase approximation for the self-energy of longitudinal phonons.

$$\frac{\mathscr{V}_c(q) - V(q)}{V^2(q)} = \frac{1}{V(q)}\left[\frac{1}{\kappa(q)} - 1\right] = |g_{ql}|^{-2}\Pi_l(q) \quad \text{(6-21a)}$$

where we have used (6-3) in the left equality and have assumed $g_{p,p',\lambda}$ is a function only of the momentum transfer for longitudinal phonons. By rearranging (6-21) we have

$$\Pi_l(q) = \frac{|g_{ql}|^2}{V(q)}\left[\frac{1}{\kappa(q)} - 1\right] \quad \text{(6-21b)}$$

From Dyson's equation,

$$D_l(q)^{-1} = D_{0l}(q)^{-1} - \Pi_l(q)$$

we find

$$D_l(q) = \frac{2\Omega_{ql}}{q_0{}^2 - \dfrac{2|g_{ql}|^2\Omega_{ql}}{V(q)\kappa(q)} - \left[\Omega_{ql}{}^2 + \dfrac{2|g_{ql}|^2\Omega_{ql}}{V(q)}\right] + i\delta} \quad \text{(6-22)}$$

As we mentioned in Chapter 4, the relation

$$\Omega_{ql}{}^2 = \frac{2|g_{ql}|^2\Omega_{ql}}{V(q)} \quad \text{(4-29a)}$$

holds for jellium, so that in this case $D_l(q)$ simplifies to[91a]

$$D_l(q) = \frac{2\Omega_{ql}}{q_0{}^2 - \dfrac{\Omega_{ql}{}^2}{\kappa(q)} + i\delta} \quad \text{(6-23)}$$

Since the poles of D give the dressed phonon frequencies $\omega_{q,l}$ we find

$$\omega_{ql}{}^2 = \frac{\Omega_{ql}{}^2}{\tilde{\kappa}(\mathbf{q}, \omega_{ql})} \quad \text{(6-24)}$$

where $\tilde{\kappa}$ is the analytic continuation of κ across the single-particle-like cut along the q_0-axis. Equation (6-24) agrees with the intuitive result that the dressed frequencies should be given by reducing the effective force between ions (or the square of the ionic charge) by the electronic dielectric constant. Since $\Omega_{q,l}{}^2$ is proportional to the ionic charge squared, (6-24) is the expected result. In calculating the real part of $\omega_{q,l}$ it is sufficient to use the static dielectric constant $\kappa(q, 0)$ since typical phonon and electron frequencies differ by $\sim (m/M)^{1/2} \sim 10^{-3}$. In the long wavelength limit we have, using (6-8),

$$\omega_{ql}^2 = \frac{\Omega_{ql}^2}{1 + \dfrac{k_s^2}{q^2}} = \frac{mZ}{3M} v_F^2 q^2 \tag{6-25a}$$

or

$$\omega_{ql} = \left(\frac{mZ}{3M}\right)^{1/2} v_F q \tag{6-25b}$$

From this expression we see that the dressed longitudinal phonons have a sound-wave type of dispersion law with sound velocity $(mZ/3M)^{1/2} v_F$, where v_F is the Fermi velocity. This result for jellium was first derived by Bohm and Staver.[119] The situation in real metals is considerably more complicated since the bare phonon frequencies $\Omega_{q\lambda}$ and the bare electron–phonon matrix elements depend on crystallographic orientation. In addition, umklapp processes are important. Despite these complications, screening reduces the bare ion frequencies to sound-wave modes (i.e., $\omega_q \propto q$) so that jellium is not an unrealistic model on which one can test methods.

Suppose we are interested in coupling the dressed phonons to the electrons. While the ion–ion interactions have been screened, the electron–ion interactions and therefore the electron–phonon interactions remain unscreened. Intuitively, one would expect the screened matrix element $\bar{g}$ to be given by

$$\bar{g}_{ql} = \frac{g_{ql}}{\kappa(q)} \tag{6-26}$$

that is, the bare matrix element divided by the wave number and frequency-dependent dielectric function of the conduction electrons. This result is easily obtained by noting that the screened interaction is given by the series shown in Figure 6-9, which is just the expansion of (6-26) in powers of the irreducible polarizability.

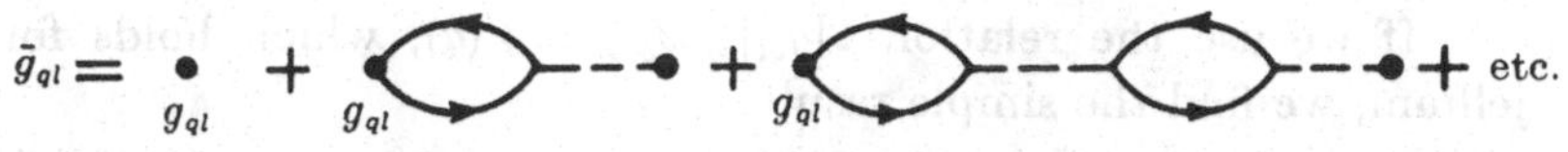

FIGURE 6-9 The random phase approximation for the screened interaction between electrons and longitudinal phonons.

If one uses the dressed D-function [given by (6-22)] and the screened electron–phonon matrix element in calculating other dynamical quantities, one must be careful not to double count by including vacuum polarization processes already accounted for in these functions. The prescription for using D and $\bar{g}$ is clear from their definition.

We now determine $G(p)$ within the screened exchange approximation of (6-18) but include longitudinal phonons as well as the Coulomb interaction in the calculation. The screened Coulomb piece of $\Sigma(p)$ is still given by (6-18), while the one-phonon process shown in Figure 6-10 gives

$$\Sigma^{\mathrm{ph}}(p) = i \int G_0(p + q)\{\bar{g}_{ql}\}^2 D_l(-q)\,\frac{d^4q}{(2\pi)^4} \qquad (6\text{-}27)$$

where $\{\bar{g}_{q,l}\}^2 \equiv \bar{g}_{ql}\bar{g}_{-ql}$. The total self-energy within this approximation is

$$\Sigma(p) = i \int G_0(p + q)[\mathscr{V}_c(q) + \{\bar{g}_{ql}\}^2 D_l(-q)]\,\frac{d^4q}{(2\pi)^4} \qquad (6\text{-}28)$$

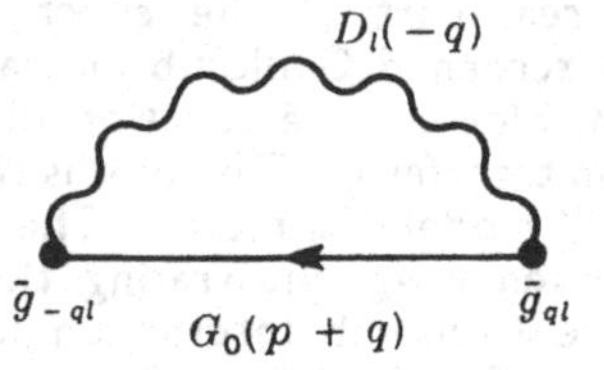

FIGURE 6-10 The lowest-order dressed phonon contribution to the electron self-energy.

If we use the relation $2|g_{ql}|^2/\Omega_{q,l} = V(q)$, which holds for jellium, we find the simple result

$$\Sigma(p) = i \int G_0(p+q)\mathscr{V}_c(q)\left[\frac{q_0^2}{q_0^2 - \dfrac{\Omega_{ql}^2}{\kappa(q)} + i\delta}\right]\frac{d^4q}{(2\pi)^4} \quad (6\text{-}29a)$$

or

$$\Sigma(p) = i \int G_0(p+q)\frac{V(q)}{1 + V(q)P(q) - \dfrac{\Omega_{ql}^2}{q_0^2} + i\delta}\frac{d^4q}{(2\pi)^4} \quad (6\text{-}29b)$$

The denominator in (6-29b) is just the total dynamical dielectric constant of the system including electronic and ionic polarizabilities, since the ionic polarizability is given by the high-frequency

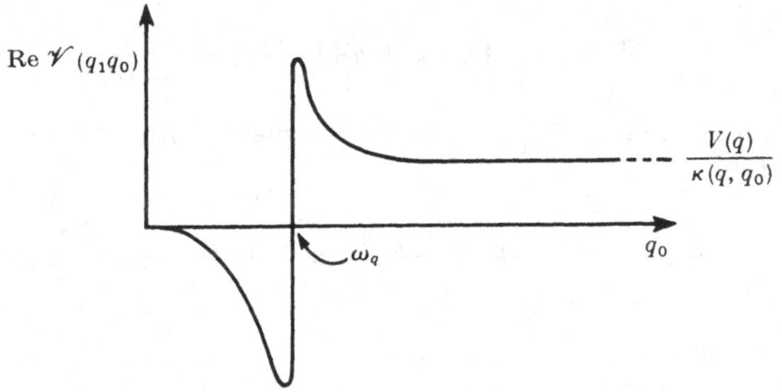

FIGURE 6-11 The real part of the effective interaction between electrons due to the screened Coulomb interaction and the exchange of a dressed phonon, plotted as a function of the energy transfer q_0 for a fixed momentum transfer q. The plot is shown for the RPA treatment of the "jellium" model of a metal. The resonance occurs at the dressed phonon frequency ω_q, illustrating the effect of ionic overscreening of the bare Coulomb interaction for $q_0 < \omega_q$ and underscreening for $q_0 > \omega_q$. For high-frequency $q_0 \gg \omega_q$, the ions do not respond and $\mathscr{V}(\mathbf{q}, q_0)$ approaches the bare Coulomb interaction reduced by the electronic dielectric function $\kappa(\mathbf{q}, q_0)$.

form $-\Omega_{q,i}^2/q_0^2$ in analogy with the limiting form (6-11) of the electronic polarizability. Thus, the effective screened potential between conduction electrons is the bare Coulomb potential $V(q)$ divided by the total dielectric function of the medium. From (6-29a) we see that if $q_0^2 < \omega_q^2$ the effective potential is attractive, corresponding to an *overscreening* of the Coulomb repulsion between electrons by the vibrating ions. For $q_0^2 > \omega_q^2$ the effective potential is *underscreened* by the ions vibrating out of phase with the electrons so that the effective repulsion is *stronger* than the bare interaction. This phenomenon is just a dielectric anomaly, familiar in the theory of dielectrics. For $q_0^2 \gg \omega_q^2$ the ionic polarization is negligible and the bare potential is screened only by the conduction electrons. A plot of this effective potential as a function of q_0 is shown in Figure 6-11 for fixed $\mathbf{q}$. It is the attractive region to the left of the dielectric anomaly which is primarily responsible for superconductivity.

It is perhaps best to insert a word of caution at this point, in that we have used the word "potential" in a broader sense than the formalism of Hamiltonian dynamics allows. In particular, one *cannot* use this "potential" in a straightforward Hamiltonian formalism since the strong frequency dependence of the effective interaction implies strong retardation effects. Since a two-body potential local must be in time in the Hamiltonian scheme, the retardation effects must be simulated by a velocity-dependent potential, a somewhat misleading procedure. In addition the effective interaction is not a real function so that H will not be Hermitian in general, unless damping effects are neglected. For these reasons it is best to treat the superconductor by Green's function methods, which allow retardation and damping effects to be included.

To obtain an explicit expression for $\Sigma(p)$ within the screened exchange approximation (6-28), we need only consider the phonon contribution (6-27) since the Coulomb piece (6-18) has been discussed above. The integrations in (6-27) are best performed by first carrying out the three-momentum integral. If the variables $|\mathbf{p}'| \equiv |\mathbf{p} + \mathbf{q}|$, $|\mathbf{q}|$, and φ (the azimuthal angle of $\mathbf{p}'$ around the polar axis **p**) are introduced, as shown in Figure 6-12, the phonon contribution becomes

$$\Sigma^{\mathrm{ph}}(p) = \frac{i}{(2\pi)^3|\mathbf{p}|}\int_{-\infty}^{\infty} dq_0 \int p'\,dp' \frac{1}{(p_0 + q_0)(1 + i\delta) - \epsilon_{p'}} \times \int q\,dq\{\bar{g}_{ql}\}^2 D_l(-q) \quad (6\text{-}30)$$

For convenience we measure all energies with respect to the Fermi energy E_F so that $\epsilon_{p_F} = 0$. Since D decreases as $1/q_0{}^2$ for large q_0, the dominant part of the integral comes from $|q_0| \gtrsim \omega_{\mathrm{av}}$ [a typical phonon energy, i.e., $\simeq (m/M)^{1/2}E_F \simeq 10^{-2}E_F$]. We shall be interested in electron energies $|p_0| \gtrsim \omega_{\mathrm{av}}$ so that the most important values of $|\epsilon_{p'}|$ are also of order ω_{av} or less. For this reason the p'-integral can be replaced by an integral over $\epsilon_{p'}$ with the limits extending from $-\infty$ to ∞. Thus,

$$\Sigma^{\mathrm{ph}}(p) \cong \frac{im}{(2\pi)^3 p}\int_{-\infty}^{\infty} dq_0 \int_{-\infty}^{\infty} d\epsilon_{p'} \frac{1}{(p_0 + q_0)(1 + i\delta) - \epsilon_{p'}} \times \int_0^{2k_F} q\,dq\{\bar{g}_{ql}\}^2 D_l(q) \quad (6\text{-}31)$$

The limits on the q integral have been simplified by using the fact that only states with $|p'| \simeq k_F$ contribute strongly to the

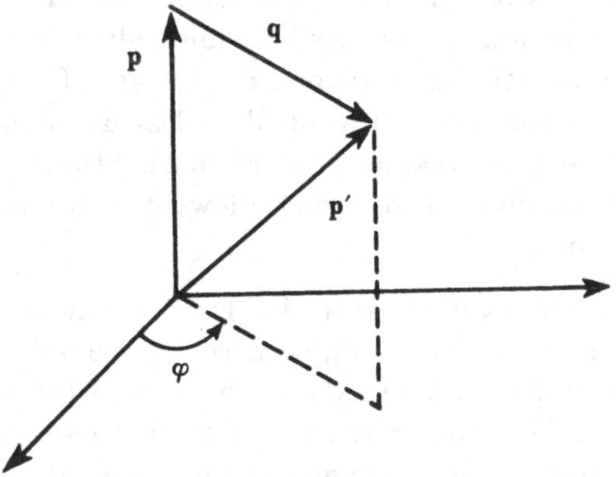

FIGURE 6-12 Coordinate system for carrying out the momentum integral in the expression for Σ^{ph}.

integral, as mentioned above. If the maximum phonon wave number q_m is less than $2k_F$ (as it usually is), transitions involving $q > q_m$ are to be interpreted as umklapp processes (which we include here) and the appropriate reduced-phonon wave vector is to be used in calculating D. After carrying out the $\epsilon_{p'}$-integral by the method of residues one finds

$$\Sigma^{\text{ph}}(p) = \frac{m}{8\pi^2 p}\int_{-\infty}^{\infty} \operatorname{sgn}(p_0 + q_0)\, dq_0 \int_0^{2k_F} q\, dq\{\bar{g}_{ql}\}^2 D_l(q)$$
$$= \frac{m}{4\pi^2 p}\int_0^{p_0} dq_0 \int_0^{2k_F} q\, dq\{\bar{g}_{ql}\}^2 D_l(q) \qquad (6\text{-}32)$$

where we have used the fact that $\{\bar{g}\}^2 D$ is an even function of q_0. The most important terms in $\operatorname{Im}\Sigma^{\text{ph}}(p)$ come from taking $\{\bar{g}_{ql}\}^2$ to be real and making a pole approximation (with a real frequency ω_{ql}) for $D_l(q)$. Within these approximations we have

$$\operatorname{Im}\Sigma^{\text{ph}}(p) = \frac{-m}{4\pi p}\operatorname{sgn} p_0 \int_0^{q(p_0)} q\, dq\{\bar{g}_{ql}\}^2 \frac{\Omega_{ql}}{\omega_{ql}} \qquad (6\text{-}33)$$

where $q(p_0)$ is the wave number such that $\omega_{ql} = |p_0|$. For the jellium model with $\kappa(q)$ given by the static long wavelength limit k_s^2/q^2 [see (6-8)], we find the damping rate of electrons, because of real phonon emission, is given by

$$2\operatorname{Im}\Sigma^{\text{ph}}(p) = -\frac{mM}{4\pi k_F n}\operatorname{sgn} p_0 \int_0^{|p_0|} \omega^2\, d\omega = -\frac{mM}{12\pi k_F n} p_0^{\,3} \qquad (6\text{-}34)$$

or

$$|\Gamma^{\text{ph}}(p)| = \frac{1}{3}\left(\frac{4}{9\pi}\right)^{1/3} r_s \left|\frac{p_0}{\Omega_p}\right|^3 \Omega_p$$

where $\Omega_p^{\,2} = 4\pi N Z e^2/M$ is the ionic plasma frequency and we have set $p = k_F$ since Σ varies slowly with p. While (6-34) holds for $p_0 \ll \Omega_p$, one obtains a reasonably good approximation (within the jellium model) for all p_0 by using (6-34) for $|p_0|$ less than the maximum phonon energy $\omega_{\max} \simeq \Omega_p$ and a constant given by setting $|p_0| = \omega_{\max}$ in (6-34) for $|p_0| > \omega_{\max}$.

By combining (6-19b) and (6-34) one obtains the total damping rate due to electron–hole pair production and real phonon emission:

$$\begin{aligned}|2\Gamma(k_F, p_0)| &= |2\Gamma^{\text{pair}}(k_F, p_0)| + |2\Gamma^{\text{ph}}(k_F, p_0)| \\ &\cong 2\Omega_p\left[0.252r_s^{1/2}\frac{\Omega_p}{E_F}\left(\frac{p_0}{\Omega_p}\right)^2 + \frac{1}{3}\left(\frac{4}{9\pi}\right)^{1/3} r_s\left\{\left|\frac{p_0}{\Omega_p}\right|^3, 1\right\}\right]\end{aligned} \tag{6-35}$$

where $\{a, b\}$ means the smaller quantity, a or b. Since $\Omega_p \sim 10^{-3}E_F$ phonon processes dominate the damping rate up to $|p_0| \sim 0.1E_F$, except for extremely small energies $|p_0| \sim 10^{-3}\Omega_p$, which are of no real importance in superconductivity.

From (6-32) the real part of $\Sigma_1{}^{\text{ph}}(p)$ is given by

$$\text{Re}\,\Sigma^{\text{ph}}(p) \cong -\frac{m}{4\pi^2|\mathbf{p}|}\int_0^{2k_F} q\,dq\{\bar{g}_{ql}\}^2\,\frac{\Omega_{ql}}{\omega_{ql}}\log\left|\frac{p_0 + \omega_{ql}}{p_0 - \omega_{ql}}\right| \tag{6-36}$$

For jellium this leads to the effective mass correction near the Fermi surface,

$$\delta m_{\text{ph}} = m\,\frac{4}{\pi}\left(\frac{\pi^2}{18}\right)^{1/3} r_s \ln\left[\frac{(2k_F)^2 + k_s{}^2}{k_s{}^2}\right] \tag{6-37}$$

From (6-37) we see that the phonon cloud surrounding the electron increases its effective mass. The numerical value of δm_{ph} given by this expression is not to be trusted in real metals [nor is the damping rate (6-34) to be trusted] since umklapp processes have been treated as if they were normal processes in our jellium model and transverse phonons have been neglected altogether. In addition, the bare electron–phonon matrix elements in real metals no doubt differ widely from the jellium results for large momentum transfer, as discussed in Chapter 4. Nevertheless, these expressions suggest that damping effects and mass corrections due to phonon processes are quite important, a result supported by experiment.

We turn now to an important observation made by Migdal.[15] Consider for the moment the class of graphs for Σ, in which no Coulomb interactions appear, other than those implicitly included by working with the screened electron–phonon vertex and the dressed-phonon line. Transverse phonons are neglected for convenience. This is essentially the model studied by Migdal. Within the framework of this model one has the exact relation

$$\Sigma^{\text{ph}}(p) = i \int \bar{g}_q G(p+q)\Gamma(p,q)D(q)\frac{d^4q}{(2\pi)^4} \tag{6-38}$$

where G is expressed in terms of the unknown function Σ by Dyson's equation. The vertex function $\Gamma(p, q)$ is given by the sum of all graphs in which an electron of four-momentum $p + q$ leaves and an electron and phonon of four-momenta p and q, respectively, enter, subject to the condition that the graph cannot be separated into two unconnected pieces by cutting one electron or one phonon line. A few low-order terms of Γ are shown in Figure 6-13a. External lines are connected to the graphs for clarity but are *not* to be included in the definition of Γ. By estimating the magnitude of the second and higher terms in this series Migdal argues that the corrections to the leading term $\bar{g}_q$ are of order $(m/M)^{1/2}\bar{g}_q \sim 10^{-2}\bar{g}_q$ and are therefore negligible.

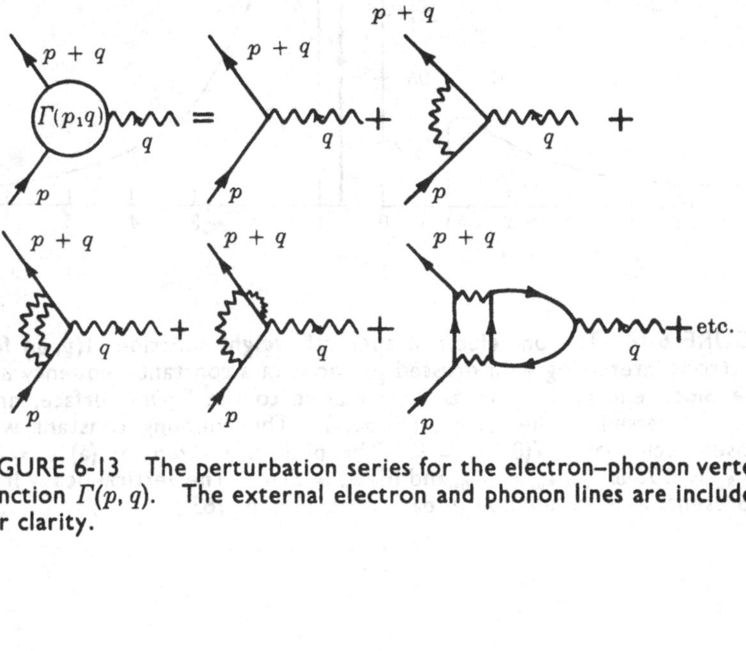

FIGURE 6-13 The perturbation series for the electron–phonon vertex function $\Gamma(p, q)$. The external electron and phonon lines are included for clarity.

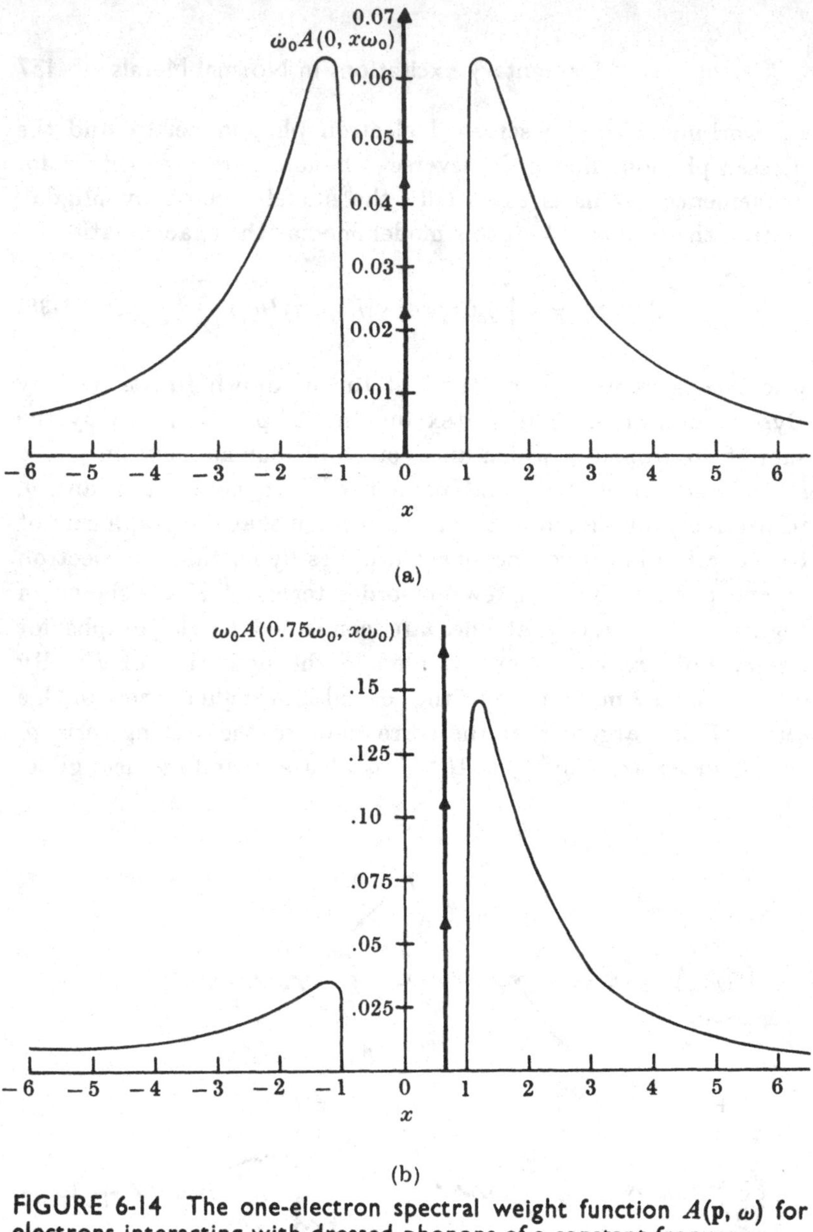

FIGURE 6-14 The one-electron spectral weight function $A(\mathbf{p}, \omega)$ for electrons interacting with dressed phonons of a constant frequency ω_0. The Bloch energy ϵ_p is measured relative to the Fermi surface, and A is expressed in the form $A(\epsilon_p, x\omega_0)$. The coupling constant was chosen such that $g^2N(0)/\omega_0 = \frac{1}{2}$. The plots are given for (a) $\epsilon_p = 0$, (b) $\epsilon_p = 0.75\omega_0$, (c) $\epsilon_p = 2\omega_0$, and (d) $\epsilon_p = 5\omega_0$. The vertical heavy line represents a delta function in each plot. (See p. 163.)

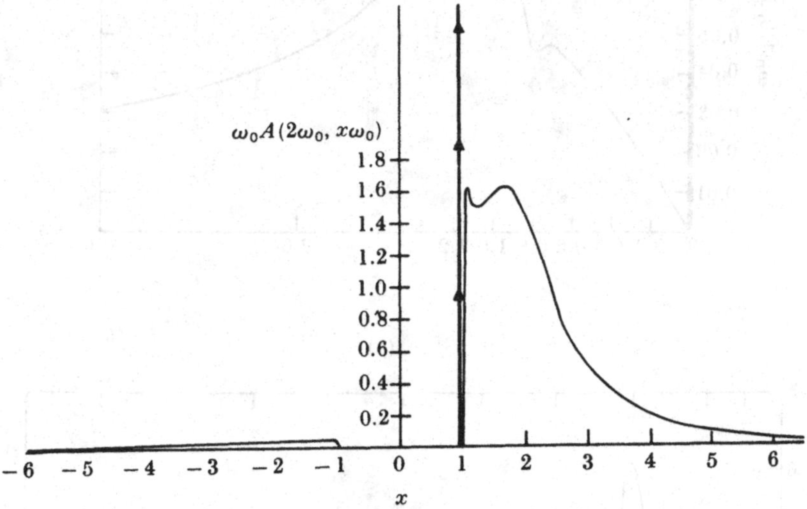

(c)

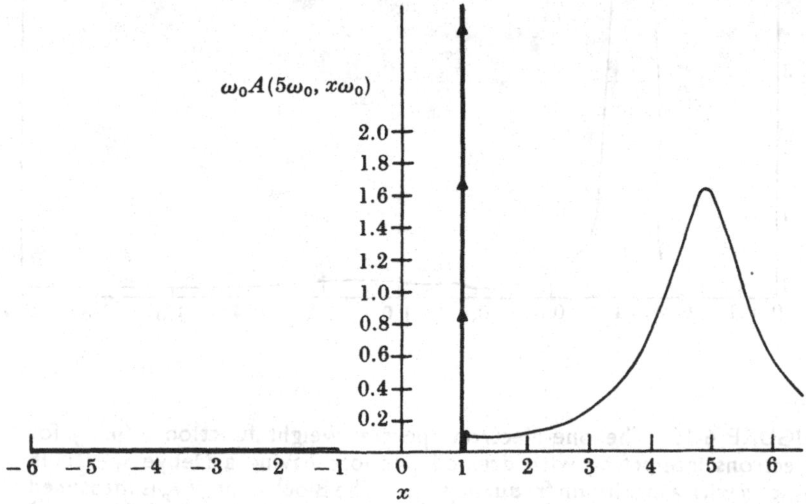

(d)

FIGURE 6-14 (*continued*)

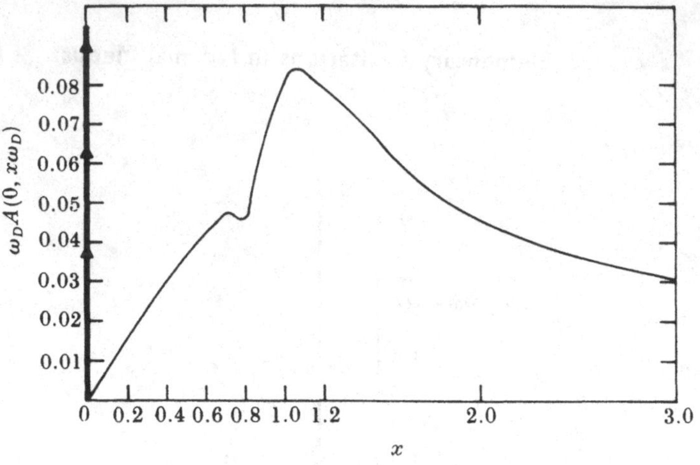

(a)

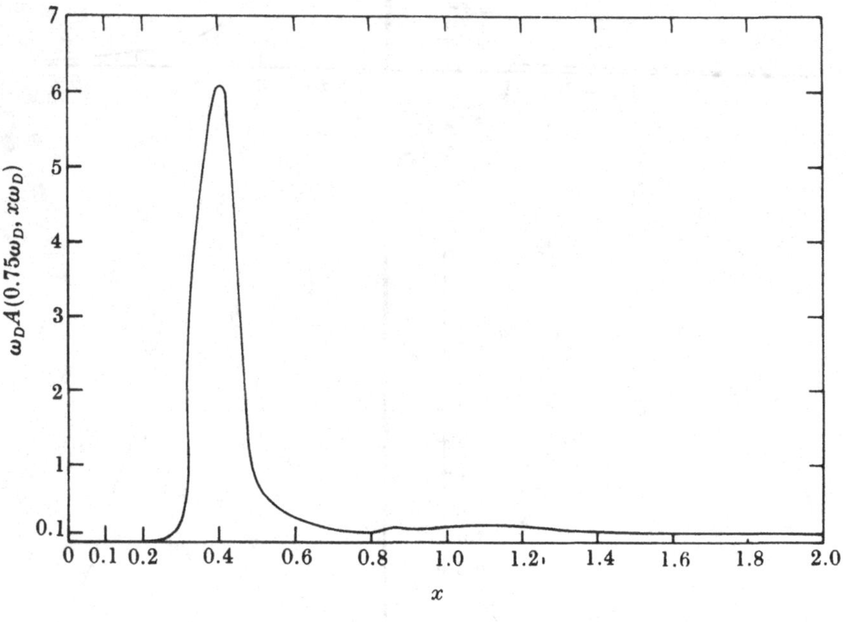

(b)

FIGURE 6-15 The one-electron spectral weight function $A(\mathbf{p}, \omega)$ for electrons interacting with dressed phonons having a Debye spectrum $\omega_q \propto q$ with a maximum frequency ω_D. The Bloch energy ϵ_p is measured relative to the Fermi surface. (a) Spectral weight function $A(\epsilon_p, x\omega_D)$ for $\epsilon_p = 0$. There is a delta-function contribution at $x = 0$. (b) Spectral weight function for $\epsilon_p = 0.75\omega_D$. (c) Spectral weight function for $\epsilon_p = 2\omega_D$. (d) Spectral weight function for $\epsilon_p = 5\omega_D$. (see p. 163.)

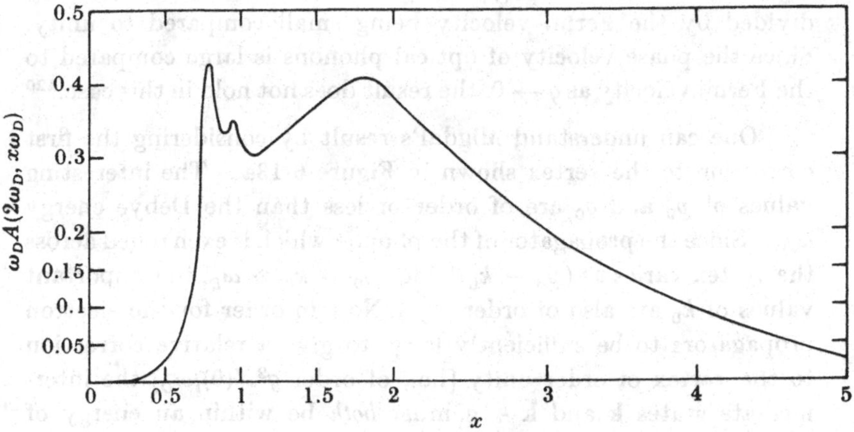

(c)

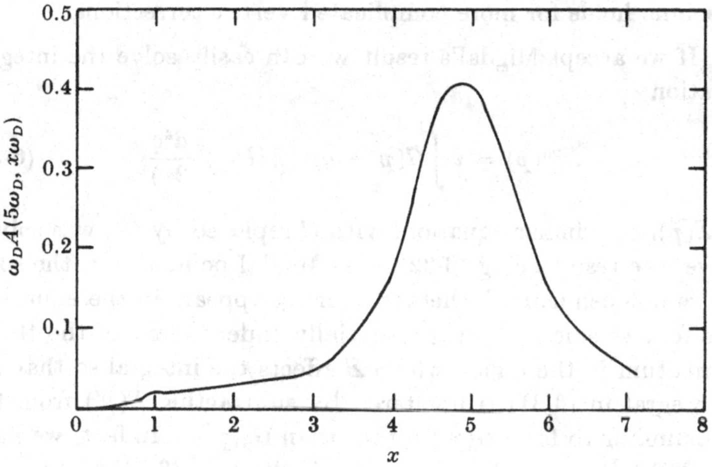

(d)

FIGURE 6-15 (*continued*)

His result depends strongly on the ratio of the sound velocity divided by the Fermi velocity being small compared to unity. Since the phase velocity of optical phonons is large compared to the Fermi velocity as $q \to 0$, the result does not hold in this case.[120]

One can understand Migdal's result by considering the first correction to the vertex shown in Figure 6-13a. The interesting values of p_0 and q_0 are of order or less than the Debye energy ω_D. Since the propagator of the phonon which is exchanged across the vertex varies as $(p_0 - k_0)^{-2}$ for $p_0 - k_0 \gg \omega_D$, the important values of k_0 are also of order ω_D. Now in order for the electron propagators to be sufficiently large to give a relative correction to the vertex of order unity [i.e., of order $g^2N(0)/\omega_D$] the intermediate states $\mathbf{k}$ and $\mathbf{k} + \mathbf{q}$ must *both* be within an energy of order ω_D of the Fermi surface (unless $q \ll k_F$; this case requires special care). Therefore, instead of a relative correction $g^2N(0)/\omega_D$, one has the result $[g^2N(0)/\omega_D](\omega_D/E_F) \sim (m/M)^{1/2}$. The same reasoning holds for more complicated vertex corrections.

If we accept Migdal's result we can easily solve the integral equation

$$\Sigma^{\text{ph}}(p) = i \int G(p + q)\{\bar{g}_q\}^2 D(q) \frac{d^4q}{(2\pi)^4} \tag{6-39}$$

for $\Sigma(p)$. A similar equation, with G replaced by G_0, was solved above, the result being (6-32). As Migdal pointed out, the solution is independent of whether G or G_0 appears in the equation. This follows since $\Sigma^{\text{ph}}(p)$ is essentially independent of the three-momentum in the region where Σ affects the integral so that the $\epsilon_{p'}$-integral in (6-31) is unaltered by subtracting $\Sigma(p')$ from the denominator to form $G(p')$ rather than $G_0(p')$. In fact, we have used Migdal's integration procedure in deriving (6-31) to illustrate this point. Thus, the solution of (6-39) is given by (6-32) and the results following from (6-32) still hold.

While the above result is attractive because of its simplicity, one becomes suspicious of Migdal's argument since it states that the phonon-induced electron self-energy is given correctly to order $(m/M)^{1/2} \sim 10^{-2}$ by (6-32). As we have seen, (6-32) leads to a continuous excitation spectrum without an energy gap, while superconductors exhibit an energy gap which in some cases is $10^{-1}\Omega_p$, i.e., ten times the error limit set by Migdal's argument. The difficulty is simply that certain sums of corrections to Γ, while formally being of order $(m/M)^{1/2}$, are actually divergent. The resolution of these singularities leads to a qualitatively new state of the system which exhibits an energy gap in the elementary excitation spectrum of the electrons and corresponds to the superconducting state of the metal.

We close this chapter by noting that the solution (6-32) of Migdal's problem (i.e., the system described by what is sometimes known as "Fröhlich's" Hamiltonian) allows one to determine the spectral weight function $A(\mathbf{p}, \omega)$ for this problem if g_q and ω_q are given. Engelsberg and the author[120] have studied two models for g_q and ω_q. The first assumes an Einstein spectrum of phonons, that is, $\omega_q = \omega_0$, a constant, and assumes a general function for g_q. The second model takes g_q and ω_q proportional to q, as in the Fröhlich model. The weight functions for these two models are shown in Figures 6-14 and 6-15 for several values of kinetic energy above the Fermi surface. The coupling was chosen to be roughly appropriate to the strong-coupling superconductors, although the situation is similar for weaker coupling. The main lesson to be learned is that, while the quasi-particle approximation is good for states very near to or very far from the Fermi surface, the weight function is far from a Lorentzian or a simple sum of Lorentzians for energies $\sim \omega_{av}/2 \rightarrow 2\omega_{av}$, where ω_{av} is the average phonon frequency. However, it is just this region in which most of the interaction bringing about superconductivity occurs, as we mentioned above. Fortunately, one can solve for most properties of a superconductor without explicitly making a quasi-particle (or pole) approximation for G.

CHAPTER 7

FIELD-THEORETIC METHODS APPLIED TO SUPERCONDUCTIVITY

In Chapters 2 and 3, the superconducting state was treated as resulting from nonretarded interactions between undamped quasi-particles of the normal phase. In real metals, retardation and damping effects play a strong role in a fundamental understanding of the superconducting phase. In addition, collective modes are important in understanding the electromagnetic properties of a superconductor in a gauge-invariant manner. The methods of quantum field theory outlined in Chapters 4 to 6 are useful in treating these effects.

7-1 INSTABILITY OF THE NORMAL PHASE

In Chapter 2 we saw how the normal state becomes unstable with respect to bound-pair formation in the presence of an attractive two-body potential. It is instructive to go back to Migdal's argument and see which graphs lead to the instability. We would like to isolate a set of graphs in which two electrons multiply-

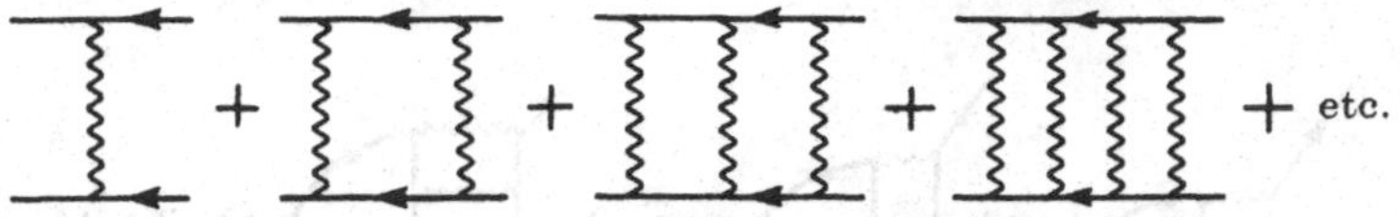

FIGURE 7-1 The ladder-graph series.

scatter each other corresponding to the two-particle problem discussed by Cooper.[41] It is clear that the series of ladder graphs shown in Figure 7-1 represents such an effect. If the effective potential is attractive, one would expect to find a bound state of the two-particle system, or an instability of the system as a whole. To understand how these graphs would enter $\Sigma(p)$, one need only close one of the electron lines onto itself as shown in Figure 7-2. If we reinterpret the graphs of this figure in terms of G, D, and Γ it follows that the instability is due to the singular behavior of Γ resulting from the graphs shown in Figure 7-3 [see (6-38)].

In order to understand the instability in more detail we consider the ladder graph series of Figure 7-1. Except for the external electron lines, the sum of the series is given by the "t"-matrix, defined by

$$\begin{aligned}\langle k' + q, -k'|t|k + q, -k\rangle &= \mathscr{V}(k' - k) \\ &+ i\int \mathscr{V}(k' - k'')G_0(k'' + q)G_0(-k'') \\ &\times \langle k'' + q, -k''|t|k + q, -k\rangle \frac{d^4k''}{(2\pi)^4} \qquad (7\text{-}1)\end{aligned}$$

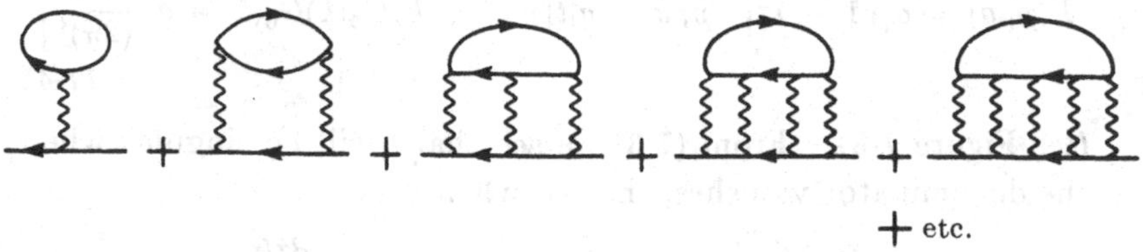

FIGURE 7-2 The ladder approximation for Σ.

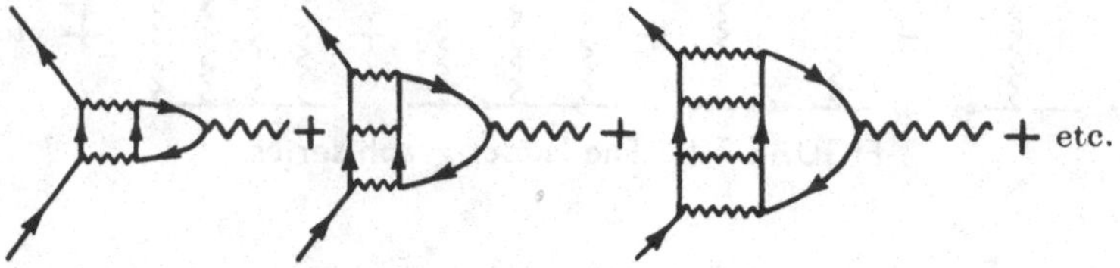

FIGURE 7-3 A class of ladder graphs which contributes to the electron–phonon vertex function Γ and leads to the superconductor instability.

One can readily establish this relation by performing an iteration solution of (7-1) in powers of $\mathscr{V}$ and noticing that one generates the desired series. While this integral equation cannot be solved in general, a solution is immediate if we replace $\mathscr{V}(q)$ by a non-retarded factorizable s-wave potential which is finite in a shell around the Fermi surface

$$\mathscr{V}(p-k) = \lambda_0 w_{\mathbf{p}}{}^* w_{\mathbf{k}} \tag{7-2}$$

where

$$w_{\mathbf{k}} = \begin{cases} 1 & |\epsilon_k| < \omega_c \\ 0 & \text{otherwise} \end{cases}$$

Then one finds the solution

$$\langle k' + q, -k'|t|k+q, -k\rangle$$
$$= \frac{\lambda_0 w_{\mathbf{k}'+\mathbf{q}}{}^* w_{\mathbf{k}+\mathbf{q}}}{1 - i\lambda_0 \int |w_{\mathbf{p}+\mathbf{q}}|^2 G_0(p+q)G_0(-p)\,\dfrac{d^4p}{(2\pi)^4}} \tag{7-3}$$

which can be checked by direct substitution.

If we are able to show that t is ill-behaved, then Γ will also be ill-behaved since in our approximation Γ is given by

$$\Gamma(p, q) = g_q\left[1 - i\int \langle p, k+q|t|p+q, k\rangle G_0(k)G_0(k+q)\,\frac{d^4k}{(2\pi)^4}\right] \tag{7-4}$$

(see Figure 7-3). From (7-3) we see that t will be singular when the denominator vanishes, that is, when

$$\frac{1}{\lambda_0} = i\int |w_{\mathbf{p}+\mathbf{q}}|^2 G_0(k+q)G_0(-k)\,\frac{d^4k}{(2\pi)^4} \tag{7-5}$$

For simplicity we consider the case of zero center-of-mass momentum for the interacting pair of electrons (i.e., a $\mathbf{q} = 0$ pair, as in the Cooper problem). We then seek the values of q_0 such that (7-5) is satisfied. By inserting the expression for G_0 and carrying out the k_0-integral by residues, one readily finds that (7-5) reduces to

$$\frac{1}{\lambda_0} = \sum_{|\mathbf{k}|>k_F} \frac{|w_k|^2}{q_0 - 2\epsilon_k} - \sum_{|\mathbf{k}|<k_F} \frac{|w_k|^2}{q_0 - 2\epsilon_k} \equiv \tilde{\Phi}(q_0) \tag{7-6}$$

where we have replaced $\int d^3k/(2\pi)^3$ by $\sum_k$ since we are interested in drawing an analogy with the solutions of the Cooper problem (2-11) for $\mathbf{q} = 0$. If $|w_k|^2$ were zero for states below the Fermi surface, (2-11) and (7-6) would be identical. By allowing the interaction to extend down into the Fermi sea, the states below the Fermi surface play a role in determining the singularities of t. A plot of the right-hand side of (7-6) is shown in Figure 7-4. For $\lambda_0 > 0$ (i.e., a repulsive potential) the perturbed states are again trapped between the unperturbed levels and no bound states

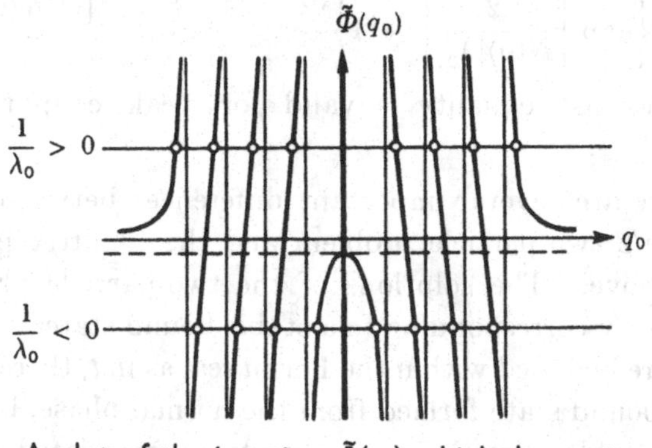

FIGURE 7-4 A plot of the function $\tilde{\Phi}(q_0)$ which determines the poles of the t-matrix in the many-body system. For a repulsive s-wave interaction ($\lambda_0 > 0$), all poles are real. For an attractive interaction ($\lambda_0 < 0$), two pure imaginary poles of t appear (regardless of the strength $|\lambda_0|$ if one considers the limit of a large system). These poles illustrate the instability of the normal phase at $T = 0$ for an attractive two-body interaction.

appear. For $\lambda_0 < 0$ and very small, again no bound states appear. However as λ_0 becomes a large negative number (i.e., stronger attractive interaction) the $1/\lambda_0$ line slides up toward the q_0-axis until it reaches the critical value such that it is tangent to the peak in $\tilde{\Phi}(q_0)$ near $q_0 = 0$. Thus far, no bound states have appeared. Beyond this point, however, two pure imaginary roots appear corresponding to the two roots which disappeared for λ_0 greater than the critical value. These pure imaginary roots[101] can be obtained by converting the sums in (7-6) to integrals again and assuming the density of Bloch states in energy near the Fermi surface is a constant $N(0)$. Then if we set $q_0 = i\alpha$, we have

$$\begin{aligned} \frac{1}{N(0)\lambda_0} &= \int_0^{\omega_c} \frac{d\epsilon}{i\alpha - 2\epsilon} - \int_{-\omega_c}^{0} \frac{d\epsilon}{i\alpha - 2\epsilon} \\ &= -\frac{1}{2}\log\left[\frac{(2\omega_c)^2 + \alpha^2}{\alpha^2}\right] \end{aligned} \tag{7-7}$$

For $N(0)\lambda_0 < 0$ (and larger than the critical value mentioned above, which is vanishingly small in the limit of a large system) one has the pair of pure imaginary roots

$$\alpha = \pm \frac{2\omega_c}{\left\{\exp\left[\dfrac{2}{N(0)|\lambda_0|}\right] - 1\right\}^{1/2}} \simeq \pm 2\omega_c \exp\left[-\frac{1}{N(0)|\lambda_0|}\right] \tag{7-8}$$

where the last equality is valid for weak coupling, that is, $|N(0)\lambda_0| < \frac{1}{4}$.

There are several important differences between the results of Cooper's two-particle problem and the t-matrix problem discussed above. The solutions of the two-particle problem gave a real energy corresponding to a stable bound state. When interactions are included within the Fermi sea, as in t, there is no longer a stable bound state formed from the normal phase, but there is a rapid unstable growth in amplitude of correlated pair states. Unfortunately, one cannot trace the time evolution of the unstable system long enough to discover directly what state it is tending toward, since the t-approximation is not accurate enough to treat the correlations in the superconducting phase.

Another difference is that the binding energy of the one-pair problem involves $e^{-2/N(0)|\lambda_0|}$, while the growth rate α involves the exponentially larger quantity $e^{-1/N(0)|\lambda_0|}$. As we saw in Chapter 2, the energy gap in the superconducting state involves the latter quantity, a result which depends on a consistent treatment of correlations both above and below the Fermi surface.

If one carries out a solution of (7-5) for finite center-of-mass momentum $\mathbf{q}$, one finds the growth rate $\alpha(q)$ decreases for increasing $\mathbf{q}$. Thus, the $\mathbf{q} = 0$ pairs are the most unstable if there is no net current in the normal state, and one would expect these pairs to play an important role in forming the ground state of the superconducting phase. (For an average drift velocity $\mathbf{v}_d$ of the electrons in the normal state the $\mathbf{q} = 2m\mathbf{v}_d$ pair state has the most rapid growth rate.)

7-2 NAMBU–GOR'KOV FORMALISM

While the perturbation series discussed in Chapter 5 for the Green's functions is not adequate for resolving the instability of the normal state, a modified scheme due to Nambu,[114] which preserves the simplicity of the Feynman–Dyson series, is applicable. A closely related scheme was developed by Gor'kov[121] prior to Nambu's work. Since the Nambu formalism is particularly convenient for carrying out calculations, we shall discuss this approach in some detail.

Perhaps the simplest way to understand the Nambu formalism is to consider a system of electrons interacting via a two-body nonretarded potential V. The Hamiltonian for the system is

$$\begin{aligned} H &= \sum_{k,s} \epsilon_k n_{ks} + \tfrac{1}{2} \sum_{\substack{k,k',q\\ s,s'}} \langle \mathbf{k} + \mathbf{q}, \mathbf{k}' - \mathbf{q}|V|\mathbf{k}, \mathbf{k}'\rangle \\ &\qquad\qquad \times c_{k+q,s}{}^{+} c_{k'-q,s'}{}^{+} c_{k's'} c_{ks} \\ &\equiv H_0 + H_{\text{int}} \end{aligned} \tag{7-9}$$

In the Hartree–Fock approximation one linearizes the interaction term with respect to a given state $|0\rangle$ so that the two-body operator H_{int} is replaced by a *one-body* operator. In the linearization one replaces a typical operator product $c_1{}^+ c_2{}^+ c_3 c_4$ by

$$\begin{aligned} c_1{}^+ c_2{}^+ c_3 c_4 \Rightarrow \langle 0|c_1{}^+ c_4|0\rangle c_2{}^+ c_3 &- \langle 0|c_1{}^+ c_3|0\rangle c_2{}^+ c_4 \\ &+ \langle 0|c_2{}^+ c_3|0\rangle c_1{}^+ c_4 - \langle 0|c_2{}^+ c_4|0\rangle c_1{}^+ c_3 \end{aligned} \tag{7-10}$$

The state $|0\rangle$ is then determined self-consistently in terms of the eigenstates of the linearized Hamiltonian. An equivalent, but for our purposes, more convenient way of formulating this approximation is to introduce a modified zero-order Hamiltonian

$$H_0' = H_0 + H_\chi - \mu N \tag{7-11}$$

where

$$H_\chi = \sum_{k,s} \chi_k n_{ks} \tag{7-12}$$

will turn out to be the Hartree–Fock potential which, for simplicity, we assume to be translationally invariant and spin independent. The factor

$$\mu N = \mu \sum_{k,s} n_{ks} \tag{7-13}$$

is included as a convenience to shift the origin of energy, where μ is the chemical potential. The modified interaction Hamiltonian is then

$$H'_{\text{int}} = H_{\text{int}} - H_\chi \tag{7-14}$$

so that

$$H' = H_0' + H'_{\text{int}} = H - \mu N \tag{7-15}$$

Therefore, the energy spectrums of H and H' are identical except for the energy shift implied by μN. In this language the Hartree–Fock (HF) approximation is equivalent to requiring that the elementary excitation spectrum of H_0' is to first order unaffected by the residual interactions H'_{int}.

To carry out this prescription, we note that the poles of the one-particle Green's function $G_0(\mathbf{p}, p_0)$ appropriate to H_0' give the elementary excitation energies $\bar{\epsilon}_p$ of H_0'. Restricting the discussion to zero temperature, we have

$$G_{0s}(\mathbf{p}, t) = -i\langle 0|T\{c_{ps}(t)c_{ps}^+(0)\}|0\rangle \tag{7-16}$$

where $|0\rangle$ is the N_0-particle ground state of H_0' and

$$c_{ps}(t) = e^{iH_0't}c_{ps}(0)e^{-iH_0't} = c_{ps}(0)e^{-i(\epsilon_p + \chi_p - \mu)t} \tag{7-17}$$

From (7-17) we find that

$$G(\mathbf{p}, p_0) = \frac{e^{i\delta p_0}}{p_0 - (\epsilon_p + \chi_p - \mu) + i\delta p_0} \tag{7-18}$$

where $\delta = 0^+$. The factor $e^{i\delta p_0}$ in (7-18) ensures that the p_0-integral of G_{0s} is the average occupation number:

$$-iG_{0s}(\mathbf{p}, t = 0) = -i \int G_{0s}(\mathbf{p}, p_0) \frac{dp_0}{2\pi} = \langle 0|n_{ps}|0\rangle \quad (7\text{-}19)$$

By inserting the expression (7-18) into (7-19) and closing the p_0-contour in the upper half-plane we find

$$\langle 0|n_{ps}|0\rangle = \begin{cases} 1 & \epsilon_p + \chi_p < \mu \\ 0 & \epsilon_p + \chi_p > \mu \end{cases} \quad (7\text{-}20)$$

Since the total number of electrons must be N_0, we find the restriction

$$\langle 0| \sum_{p,s} n_{ps}|0\rangle = 2 \int \frac{d^3p}{(2\pi)^3}\bigg|_{\epsilon_p + \chi_p < \mu} = N_0 \quad (7\text{-}21)$$

which determines the chemical potential μ for a given Hartree–Fock potential χ_p. The pole of $G_{0s}(\mathbf{p}, p_0)$ is at $p_0 = \epsilon_p + \chi_p - \mu \equiv \bar{\epsilon}_p$. As we saw in Chapter 5, $\bar{\epsilon}_p$ is the excitation energy of the $N_0 + 1$ particle system when an electron is added to the N_0-particle system, if $\bar{\epsilon}_p > 0$. If $\bar{\epsilon}_p < 0$, $-\bar{\epsilon}_p = |\bar{\epsilon}_p|$ is the excitation energy of the $N_0 - 1$ system when an electron is removed from the N_0-particle system.

To find the effect of H'_{int} on the excitation spectrum we require the self-energy $\Sigma(\mathbf{p}, p_0)$ due to these residual interactions. By using the standard Feynman–Dyson rules, one finds to first order in H'_{int}

$$\Sigma(\mathbf{p}, p_0) = -2i \int \langle \mathbf{p}, \mathbf{p}'|V|\mathbf{p}, \mathbf{p}'\rangle G_0(\mathbf{p}', p_0') \frac{d^4p'}{(2\pi)^4} + i \int \langle \mathbf{p}', \mathbf{p}|V|\mathbf{p}, \mathbf{p}'\rangle G_0(\mathbf{p}', p_0') \frac{d^4p'}{(2\pi)^4} - \chi_p \quad (7\text{-}22)$$

The terms correspond to the direct and exchange contributions and the subtracted Hartree–Fock potential, respectively, as shown in Figure 7-5. Since the matrix elements of the two-body potential are independent of p_0', we can use (7-19) to obtain the simple result

$$\Sigma(\mathbf{p}, p_0) = \int_{\bar{\epsilon}_{p'} < 0} \frac{d^3p'}{(2\pi)^3} \{2\langle \mathbf{p}, \mathbf{p}'|V|\mathbf{p}, \mathbf{p}'\rangle - \langle \mathbf{p}', \mathbf{p}|V|\mathbf{p}, \mathbf{p}'\rangle\} - \chi_p \quad (7\text{-}23)$$

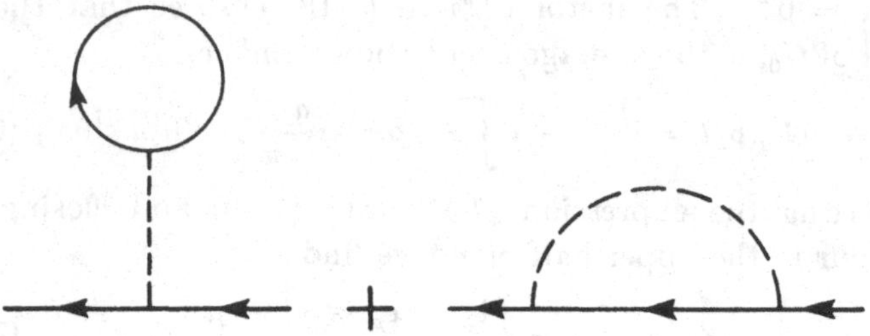

FIGURE 7-5 Contributions to Σ within the Hartree–Fock approximation. The electron lines include the self-consistently determined self-energy Σ.

If the poles of $G_0(\mathbf{p}, p_0)$ are to be unaffected by Σ, we must require the self-consistency condition

$$\Sigma(\mathbf{p}, \bar{\epsilon}_p) = 0 \tag{7-24}$$

to be satisfied. Since in our particular problem Σ is independent of p_0, we have the familiar Hartree–Fock relation

$$\chi_p = \int_{\bar{\epsilon}_{p'} < 0} \frac{d^3p'}{(2\pi)^3} \{2\langle \mathbf{p}, \mathbf{p}'|V|\mathbf{p}, \mathbf{p}'\rangle - \langle \mathbf{p}', \mathbf{p}|V|\mathbf{p}, \mathbf{p}'\rangle\} \tag{7-25}$$

determining χ_p.

While this appears to be a complicated way of phrasing the HF approximation, the scheme is easily extended to include the pairing correlations. The idea is simply to generalize the linearization (7-10) to include Hartree-like terms involving $\langle 0|c_1{}^+c_2{}^+|0\rangle$ and $\langle 0|c_3c_4|0\rangle$. The state $|0\rangle$ is then to be determined self-consistently as in the standard Hartree–Fock approach.

At this point one might argue that since the full Hamiltonian commutes with the total number of particles operator N_{op}, the exact eigenstates of H are eigenfunctions of N_{op}. One might then argue that our approximate state $|0\rangle$ should also be chosen to be an eigenfunction of N_{op} and therefore the Hartree-like terms $\langle 0|c_1{}^+c_2{}^+|0\rangle$ and $\langle 0|c_3c_4|0\rangle$ vanish identically. We can make two counter arguments. First, in the limit of a large system the ground states of the N_0 and the $N_0 + \nu$ particle system are degenerate for $|\nu| \ll N_0$, if the origin of energy is shifted by measuring the kinetic energies relative to the exact chemical potential μ. Therefore, we are free to choose the ground state of H as a linear combination of these degenerate states, and therefore the ground state (or more generally, the eigenstates) of H *need not* be chosen to be eigenstates of N_{op}. Furthermore, even if the exact ground states describing different numbers of particles were not degenerate, one might be able to obtain with a given amount of effort a more accurate ground-state energy and excitation spectrum if the approximate states were *not* required to be eigenfunctions of N_{op}. We saw an illustration of this in the BCS theory. In any event, one can view the generalized Hartree scheme as one which defines a zero-order Hamiltonian H_0' with the difference between the true Hamiltonian and H_0' remainng as a perturbation to be treated by other means. The essential point is that it is simpler to include the nonanalytic pairing correlations in zero-order (and violate the requirement that $|0\rangle$ be an eigenfunction of N_{op}) than to require the states to be strict eigenfunctions of N_{op} at each state of the calculation. A similar situation occurs in nuclear structure theory, where in lowest order one introduces a nonspherical Hartree potential which gives a ground state that is not an eigenfunction of angular momentum.

As in the HF scheme, an alternate way of viewing this generalized HF approximation is to consider the modified zero-order Hamiltonian

$$H_0' = H_0 + (H_\chi + H_\phi) - \mu N \tag{7-26}$$

where the added term H_ϕ is of the form

$$H_\phi = \sum_k \{\phi_k{}^* c_{k\uparrow}{}^+ c_{-k\downarrow}{}^+ + \text{H.c.}\} \tag{7-27}$$

Here one desires to describe pairing correlations between $k\uparrow$ and $-k\downarrow$. Other pairings can be handled in a similar manner. Owing to the presence of H_ϕ, H_0' is no longer a one-particle operator in the conventional sense, a fact which complicates a perturbation treatment of the modified interaction Hamiltonian,

$$H'_{\text{int}} = H_{\text{int}} - (H_\chi + H_\phi) \tag{7-28}$$

Nambu discovered a neat way of getting around this formal complication. He pointed out that if one introduces a two-component (spinor) field operator

$$\Psi_k = \begin{pmatrix} c_{k\uparrow} \\ c_{-k\downarrow}{}^+ \end{pmatrix} \qquad \Psi_{k_1} = c_{k\uparrow} \qquad \Psi_{k_2} = c_{-k\downarrow}{}^+ \tag{7-29}$$

and its Hermitian adjoint

$$\Psi_k{}^+ = (c_{k\uparrow}{}^+, c_{-k\downarrow}) \qquad \{\Psi_k, \Psi_{k'}{}^+\} = \delta_{kk'}\mathbf{1}, \{\Psi_k, \Psi_{k'}\} = 0 \tag{7-30}$$

H_0' can be formally written as a one-body operator in the Ψ_k field. To see this we introduce the four Pauli matrices

$$\tau_1 = \begin{pmatrix} 0 & 1 \\ 1 & 0 \end{pmatrix} \quad \tau_2 = \begin{pmatrix} 0 & -i \\ i & 0 \end{pmatrix} \quad \tau_3 = \begin{pmatrix} 1 & 0 \\ 0 & -1 \end{pmatrix} \quad \mathbf{1} = \begin{pmatrix} 1 & 0 \\ 0 & 1 \end{pmatrix} \tag{7-31}$$

from which it follows that the Hermitian operators $\Psi_k{}^+\tau_i\Psi_k$ are given by

$$\Psi_k{}^+\tau_1\Psi_k = c_{k\uparrow}{}^+c_{-k\downarrow}{}^+ + c_{-k\downarrow}c_{k\uparrow} \tag{7-32a}$$

$$\Psi_k{}^+\tau_2\Psi_k = -i[c_{k\uparrow}{}^+c_{-k\downarrow}{}^+ - c_{-k\downarrow}c_{k\uparrow}] \tag{7-32b}$$

$$\Psi_k{}^+\tau_3\Psi_k = n_{k\uparrow} + n_{-k\downarrow} - 1 \tag{7-32c}$$

$$\Psi_k{}^+\mathbf{1}\Psi_k = n_{k\uparrow} - n_{-k\downarrow} + 1 \tag{7-32d}$$

These relations allow one to write H_0' as

$$H_0' = \sum_k \Psi_k{}^+[\bar{\epsilon}_k\tau_3 + \phi_{k1}\tau_1 + \phi_{k2}\tau_2]\Psi_k + \sum_k \bar{\epsilon}_k \tag{7-33}$$

where ϕ_{k1} and ϕ_{k2} are the real and imaginary parts, respectively, of ϕ_k and $\bar{\epsilon}_k = \epsilon_k + \chi_k - \mu$ as before. The last term in (7-33) is included to compensate for the -1 occurring in (7-32c). We can easily remove this annoying (infinite) term once we go over to the Green's function scheme. The form (7-33) of H_0' is just that required to make the Feynman–Dyson perturbation series rules work if H_{int}' takes the form of a one- and/or two-body potential in the Ψ_k language. One readily finds that H'_{int} takes the form

$$H'_{\text{int}} = \tfrac{1}{2} \sum_{k,k',q} \langle \mathbf{k} + \mathbf{q}, \mathbf{k}' - \mathbf{q} | V | \mathbf{k}, \mathbf{k}' \rangle (\Psi_{k+q}{}^{+} \tau_3 \Psi_k)$$
$$\times (\Psi_{k'-q}{}^{+} \tau_3 \Psi_{k'}) - \sum_{k} \Psi_k{}^{+} (\chi_k \tau_3 + \phi_{k1} \tau_1 + \phi_{k2} \tau_2) \Psi_k \tag{7-34a}$$

if the matrix elements of V satisfy the symmetry requirements

$$\langle \mathbf{k}_1, \mathbf{k}_2 | V | \mathbf{k}_3, \mathbf{k}_4 \rangle = \langle -\mathbf{k}_3, -\mathbf{k}_4 | V | -\mathbf{k}_1, -\mathbf{k}_2 \rangle$$
$$= \langle -\mathbf{k}_3, \mathbf{k}_2 | V | -\mathbf{k}_1, \mathbf{k}_4 \rangle = \langle \mathbf{k}_1, -\mathbf{k}_4 | V | \mathbf{k}_3, -\mathbf{k}_2 \rangle \tag{7-34b}$$

These symmetry conditions are satisfied for potentials arising from the electron–phonon and Coulomb interactions. To make the first term of (7-34a) agree precisely with H_{int}, one must include extra terms, as in (7-33); however, these terms will be automatically included by our choice of G_0, and we shall disregard them for now.

To determine the chemical potential μ and the Hartree fields χ_k and ϕ_k we imitate the Hartree–Fock treatment given above. The one-particle Green's function (matrix) is defined as

$$\mathbf{G}_{0\alpha\beta}(\mathbf{p}, t) = -i\langle 0 | T\{\Psi_{p\alpha}(t) \Psi_{p\beta}{}^{+}(0)\} | 0 \rangle \tag{7-35}$$

where $|0\rangle$ is the ground state of $H_0{}'$ for an *average* number of electrons N_0 and

$$\Psi_p(t) = e^{iH_0{}'t} \Psi_p(0) e^{-iH_0{}'t} \tag{7-36}$$

The quantity

$$\mathbf{G}_{011}(\mathbf{p}, t) = -i\langle 0 | T\{c_{p\uparrow}(t) c_{p\uparrow}{}^{+}(0)\} | 0 \rangle \tag{7-37}$$

is the Green's function for spin-up electrons, while

$$\mathbf{G}_{022}(\mathbf{p}, t) = -i\langle 0 | T\{c_{-p\downarrow}{}^{+}(t) c_{-p\downarrow}(0)\} | 0 \rangle \tag{7-38}$$

can be thought of as the Green's function for spin-down holes. The off-diagonal matrix elements of G

$$\mathbf{G}_{012}(\mathbf{p}, t) = -i\langle 0 | T\{c_{p\uparrow}(t) c_{-p\downarrow}(0)\} | 0 \rangle \tag{7-39a}$$
$$\mathbf{G}_{021}(\mathbf{p}, t) = -i\langle 0 | T\{c_{-p\downarrow}{}^{+}(t) c_{p\uparrow}{}^{+}(0)\} | 0 \rangle \tag{7-39b}$$

are related to the amplitude for subtracting or adding a pair of particles to the system without creating excitations. In order to remove the infinite c-number term from H_0' it is convenient to define $G_0(p, t = 0)$ as

$$\mathbf{G}_{011}(\mathbf{p}, t = 0) = \lim_{t \to 0^-} \mathbf{G}_{011}(p, t) \qquad (7\text{-}40a)$$

$$\mathbf{G}_{022}(\mathbf{p}, t = 0) = \lim_{t \to 0^+} \mathbf{G}_{022}(\mathbf{p}, t) \qquad (7\text{-}40b)$$

With these conditions it follows that the time Fourier transform of $G_0(\mathbf{p}, t)$ is

$$\mathbf{G}_0(\mathbf{p}, p_0) = \frac{(p_0 \mathbf{1} + \bar{\epsilon}_p \tau_3 + \phi_{p1}\tau_1 + \phi_{p2}\tau_2)e^{i\delta p_0 \tau_3}}{p_0^2 - \bar{\epsilon}_p^2 - \phi_{p1}^2 - \phi_{p2}^2 + i\delta} \qquad (7\text{-}41)$$

Since the pole of $G(\mathbf{p}, p_0)$ for $p_0 > 0$ gives the quasi-particle energy E_p, we see that E_p is given by

$$E_p = (\bar{\epsilon}_p^2 + \phi_{p1}^2 + \phi_{p2}^2)^{1/2} \qquad (7\text{-}42)$$

More explicitly, one has the various matrix components of G_0 given by

$$\mathbf{G}_{011}(p) = \frac{(p_0 + \bar{\epsilon}_p)e^{i\delta p_0}}{p_0^2 - E_p^2 + i\delta} \qquad (7\text{-}43a)$$

$$\mathbf{G}_{022}(p) = \frac{(p_0 - \bar{\epsilon}_p)e^{-i\delta p_0}}{p_0^2 - E_p^2 + i\delta} \qquad (7\text{-}43b)$$

$$\mathbf{G}_{012}(p) = \frac{\phi_p^*}{p_0^2 - E_p^2 + i\delta} = \mathbf{G}_{021}^*(p) \qquad (7\text{-}43c)$$

As in the HF approximation we require the total (average) number of electrons to be N_0. Owing to the limiting forms (7-40), we have

$$\begin{aligned} \langle 0| \sum_{p,s} n_{ps} |0\rangle &= \sum_p (-i)[\mathbf{G}_{011}(p, t = 0) - \mathbf{G}_{022}(p, t = 0)] \\ &= \sum_p (-i) \operatorname{Tr}[\tau_3 \mathbf{G}_0(\mathbf{p}, t = 0)] = N_0 \end{aligned} \qquad (7\text{-}44)$$

From (7-41) we find that

$$\begin{aligned} -i\mathbf{G}_0(\mathbf{p}, t = 0) &= -i \int \mathbf{G}_0(\mathbf{p}, p_0) \frac{dp_0}{2\pi} \\ &= \frac{(E_p - \bar{\epsilon}_p)\tau_3 - \phi_{p_1}\tau_1 - \phi_{p_2}\tau_2}{2E_p} \end{aligned} \qquad (7\text{-}45)$$

Thus, the restriction determining the chemical potential μ is

$$\int\left(1-\frac{\epsilon_p}{E_p}\right)\frac{d^3p}{(2\pi)^3} \equiv 2\int v_p{}^2\,\frac{d^3p}{(2\pi)^3} = N_0 \tag{7-46}$$

where we have used $\tau_i{}^2 = \mathbf{1}$, $\operatorname{Tr}\tau_i = 0$, and $\operatorname{Tr}\mathbf{1} = 2$. The reader will recognize (7-46) as the result we found in Chapter 2.

We turn now to the problem of fixing χ_k and ϕ_k. Since we can use the Feynman–Dyson rules to calculate the effects of H'_{int}, we find the lowest-order self-energy contribution, in analogy with (7-22), is

$$\begin{aligned}\boldsymbol{\Sigma}(\mathbf{p}, p_0) = &-i\tau_3\int\langle\mathbf{p},\mathbf{p}'|V|\mathbf{p},\mathbf{p}'\rangle\operatorname{Tr}[\tau_3\mathbf{G}_0(p')]\,\frac{d^4p'}{(2\pi)^4}\\ &+ i\int\langle\mathbf{p}',\mathbf{p}|V|\mathbf{p},\mathbf{p}'\rangle\tau_3\mathbf{G}_0(p')\tau_3\,\frac{d^4p'}{(2\pi)^4}\\ &- (\chi_p\tau_3 + \phi_{p1}\tau_1 + \phi_{p2}\tau_2)\end{aligned} \tag{7-47}$$

The only change in the rules of Chapter 5 is that *each vertex* coupling an electron line to an interaction line of a two-body potential or to a phonon *includes* a factor of τ_3 because the coupling is of the form $\Psi^+\tau_3\Psi$ [see (7-34a)]. One must also remember to keep the matrices in their proper order as given by the graphs, and to interpret a closed loop as implying a trace over the free matrix scripts, as in the first term of (7-47). By using (7-45) we can reduce (7-47) to

$$\begin{aligned}\boldsymbol{\Sigma}(p, p_0) = &\,\tau_3\Bigg[\int\frac{d^3p'}{(2\pi)^3}\{2\langle\mathbf{p},\mathbf{p}'|V|\mathbf{p},\mathbf{p}'\rangle\\ &\qquad - \langle\mathbf{p}',\mathbf{p}|V|\mathbf{p},\mathbf{p}'\rangle\}v_{p'}{}^2 - \chi_p\Bigg]\\ &+ \tau_1\Bigg[\int\frac{d^3p'}{(2\pi)^3}\langle\mathbf{p}',\mathbf{p}|V|\mathbf{p},\mathbf{p}'\rangle\,\frac{\phi_{p'_1}}{2E_{p'}} + \phi_{p_1}\Bigg]\\ &+ \tau_2\Bigg[\int\frac{d^3p'}{(2\pi)^3}\langle\mathbf{p}',\mathbf{p}|V|\mathbf{p},\mathbf{p}'\rangle\,\frac{\phi_{p'_2}}{2E_{p'}} + \phi_{p_2}\Bigg]\end{aligned} \tag{7-48}$$

If we require that the quasi-particle energy E_p is unaffected to first order in H'_{int}, we obtain the self-consistency condition

$$\boldsymbol{\Sigma}(\mathbf{p}, E_p) = 0 \tag{7-49}$$

as a matrix equation. Since the Pauli matrices are linearly independent, (7-49) requires the coefficients of the τ's in (7-48) to be separately zero and we obtain three equations determining χ, ϕ_1, and ϕ_2:

$$\chi_p = \int \frac{d^3p'}{(2\pi)^3} \{2\langle \mathbf{p}, \mathbf{p}' | V | \mathbf{p}, \mathbf{p}' \rangle - \langle \mathbf{p}', \mathbf{p} | V | \mathbf{p}, \mathbf{p}' \rangle\} v_{p'}^2 \tag{7-50a}$$

$$\phi_{p_1} = - \int V_{pp'} \frac{\phi_{p'1}}{2E_{p'}} \frac{d^3p'}{(2\pi)^3} \tag{7-50b}$$

$$\phi_{p_2} = - \int \frac{d^3p'}{(2\pi)^3} V_{pp'} \frac{\phi_{p'2}}{2E_p'} \tag{7-50c}$$

where the pairing matrix $V_{pp'}$ is given by

$$V_{pp'} = \langle \mathbf{p}', -\mathbf{p}' | V | \mathbf{p}, -\mathbf{p} \rangle = \langle \mathbf{p}', \mathbf{p} | V | \mathbf{p}, \mathbf{p}' \rangle \tag{7-51}$$

[see (7-34b)]. Since the total Hamiltonian is invariant to rotations in τ-space about the τ_3-axis (τ_1 and τ_2 never enter H), we are free to choose phases so that $\phi_2 = 0$ and (7-50b) reduces to the BCS energy-gap equation, where ϕ_p is identified as the energy-gap parameter Δ_p. It is interesting to note that the Hartree–Fock potential χ_p is of the expected form since $v_{p'}^2$ gives the average occupancy of states $\mathbf{p}'$, s. Thus, this generalized HF scheme is equivalent to the BCS theory discussed in Chapter 2 if χ_p is lumped with the energy ϵ_p of that chapter.

To make connection with the spectral weight function $A(\mathbf{p}, \omega)$ of Chapter 5, we note that $G_{011}(p)$ can be written

$$\mathbf{G}_{011}(p) = \frac{p_0 + \bar{\epsilon}_p}{p_0^2 - E_p^2 + i\delta} = \frac{u_p^2}{p_0 - E_p + i\delta} + \frac{v_p^2}{p_0 + E_p - i\delta} \tag{7-52}$$

where

$$u_p^2 = \frac{1}{2}\left(1 + \frac{\bar{\epsilon}_p}{E_p}\right) \tag{7-53a}$$

$$v_p^2 = \frac{1}{2}\left(1 - \frac{\bar{\epsilon}_p}{E_p}\right) \tag{7-53b}$$

as usual. Since we are measuring all energies relative to μ, we have from (5-41)

$$A(\mathbf{p}, \omega) = -\frac{\operatorname{sgn} \omega}{\pi} \operatorname{Im} G_{011}(\mathbf{p}, \omega)$$
$$= u_p^2 \, \delta(\omega - E_p) + v_p^2 \, \delta(\omega + E_p) \qquad (7\text{-}54)$$

The physical interpretation of the two peaks in $A(\mathbf{p}, \omega)$ was given in connection with (5-53) and will not be repeated here.

Another way of viewing the generalized HF scheme, which is easily extended to include retardation and damping effects, etc., is that of self-consistent perturbation theory. In this approach one calculates $\mathbf{\Sigma}(p)$ as a perturbation series in which one uses one-particle propagators which themselves include the self-energy being calculated. Therefore, one obtains an approximate integral equation determining $\mathbf{\Sigma}$. In carrying out the procedure one must be careful not to double count graphs. If we continue to use the Nambu notation, the most general form for $\mathbf{\Sigma}(p)$ is

$$\mathbf{\Sigma}(p) = [1 - Z(p)]p_0\mathbf{1} + \chi(p)\tau_3 + \phi(p)\tau_1 + \bar{\phi}(p)\tau_2 \qquad (7\text{-}55)$$

We again choose phases so that the coefficient of τ_2 is equal to zero. In contrast to the above treatment, the coefficient of τ_1 need not be real; in fact, the imaginary part of ϕ contributes to the damping rate of quasi-particles, as we shall see. Furthermore, the quantities Z, χ, and ϕ in (7-55) are functions of the four-momentum $(\mathbf{p}, p_0)$ rather than the three-momentum $\mathbf{p}$ as in (7-50). This generality is required to treat retardation effects as we did in Chapter 6 for the normal state. The generalized HF scheme then corresponds to first-order self-consistent perturbation theory since within this approximation $\mathbf{\Sigma}$ (7-55) is given by

$$\mathbf{\Sigma}(p) = -i\tau_3 \int \frac{d^4p'}{(2\pi)^4} \langle \mathbf{p}, \mathbf{p}' | V | \mathbf{p}, \mathbf{p}' \rangle \operatorname{Tr} \tau_3 \mathbf{G}(p')$$
$$+ i \int \frac{d^4p'}{(2\pi)^4} \langle \mathbf{p}', \mathbf{p} | V | \mathbf{p}, \mathbf{p}' \rangle \tau_3 \mathbf{G}(p') \tau_3 \qquad (7\text{-}56)$$

From Dyson's equation (now a matrix equation)

$$\mathbf{G}^{-1}(p) = \mathbf{G}_0^{-1}(p) - \mathbf{\Sigma}(p) \qquad (7\text{-}57)$$

and (7-55) we have

$$\mathbf{G}(p) = \frac{Z(p)p_0 + \bar{\epsilon}(p)\tau_3 + \phi(p)\tau_1}{[Z(p)p_0]^2 - E(p)^2 + i\delta} \tag{7-58}$$

where by $\mathbf{G}_0(p)$ we now mean the true free-electron Green's function in the Nambu notation,

$$\mathbf{G}_0(p) = \frac{p_0\mathbf{1} + \epsilon_p\tau_3}{p_0{}^2 - \epsilon_p{}^2 + i\,\delta} = [p_0\mathbf{1} - \epsilon_p\tau_3 + i\,\delta p_0\mathbf{1}]^{-1} \tag{7-59}$$

and ϵ_p is measured relative to μ. As before we use the notation

$$\bar{\epsilon}(p) = \epsilon_p + \chi(p) \tag{7-60a}$$

$$E(p) = [\bar{\epsilon}(p)^2 + \phi(p)^2]^{1/2} \tag{7-60b}$$

Thus, (7-56) represents a set of coupled integral equations determining Z, χ, and ϕ. Since the coefficient of $\mathbf{1}$ vanishes on the right-hand side (7-56), one finds $Z(p) = 1$ within this approximation, and the remaining terms in (7-56) reduce to (7-50), as required.

In the general case, the Nambu formalism is equivalent to a self-consistent perturbation approach to determine the coefficients of the Pauli matrices in the expression for $\mathbf{\Sigma}$ (7-55).

7-3 ZERO-TEMPERATURE EXCITATION SPECTRUM

We are now in the position of being able to handle damping and retardation effects in determining the quasi-particle spectrum of a superconductor at zero temperature. Nambu[114] and Eliashberg[122] have treated the problem by self-consistent perturbation theory and have retained the lowest-order dressed-phonon and dressed-Coulomb contributions to $\mathbf{\Sigma}$, as shown in Figure 7-5. Within this approximation one finds the (matrix) equation

$$\mathbf{\Sigma}(p) = i\int \tau_3\mathbf{G}(p')\tau_3\left[\sum_\lambda \{g_{pp'\lambda}\}^2 D_\lambda(p - p') + \mathscr{V}_c(p - p')\right]\frac{d^4p'}{(2\pi)^4} \tag{7-61}$$

determining Σ, where G is given by (7-58). From the above equation one can determine the unknown complex functions

$Z(p)$, $\chi(p)$, and $\phi(p)$. If there are no further singularities in the electron–phonon vertex function (other than those accounted for in the resolution of the pair instability), this integral equation treats the electron–phonon interaction exactly to order $(m/M)^{1/2} \simeq 10^{-2}$. The quantity $\mathscr{V}_c(p - p')$ is the Coulomb potential screened by the electronic-dielectric function κ [see (6-12)]. While in principle we should construct an equation to determine the dressed-phonon propagator D, which enters (7-61), numerical solutions of this equation show that the general features of the solutions are insensitive to details of the phonon spectrum. Thus, one can expect to obtain reasonable results with a fairly simple form for D, which is best obtained from experimentally determined phonon spectra. Since the phonon frequencies are essentially the same in the normal and superconducting states,[66] one need not include this small shift (typically $\delta\omega_q/\omega_q \gtrsim 10^{-4}$) in the calculation of $\mathbf{\Sigma}$. The effective interaction in (7-61)

$$\sum_{\lambda} \{g_{pp'\lambda}\}^2 D_\lambda(p - p') + \mathscr{V}_c(p - p') \tag{7-62}$$

is the generalization of the interaction occurring in (6-28), the only difference being that (7-62) includes transverse phonons as well as umklapp processes.

In Chapter 6 we saw that in the normal state the phonon contribution to Σ could be obtained analytically since $\Sigma^{\text{ph}}(\mathbf{p}, p_0)$ was essentially independent of $\mathbf{p}$ in the region where Σ played a role in the form of G [see (6-39)]. As we saw, if the three-momentum integral is carried out first, the residue at the pole of G is independent of Σ so that the integral equation for Σ is reduced to a quadrature. The situation is distinctly different for the superconducting state, since Migdal's trick of integrating first with respect to the three-momentum does not lead to an expression independent of Σ as in the N-state. Nevertheless, the trick is highly useful and reduces (7-61) to a one-dimensional rather than a four-dimensional integral equation, once the screened Coulomb interaction $\mathscr{V}_c$ is handled properly. The one-dimensional integral equations can then be solved numerically, as we shall see below.

To reduce (7-61) to one-dimensional form we begin with the term coming from the phonon part of the interaction,

$$\boldsymbol{\Sigma}^{\text{ph}}(p) = i\int \tau_3 \mathbf{G}(p')\tau_3 \sum_\lambda \{\bar{g}_{pp'\lambda}\}^2 D_\lambda(p - p')\,\frac{d^4p'}{(2\pi)^4} \tag{7-63}$$

The right-hand side of this equation is a function of $\mathbf{p}$ only through the momentum transfer $|\mathbf{p}' - \mathbf{p}|$ (in the absence of crystalline anisotropy effects). Since the momentum transfer is averaged over in performing the angular part of the $\mathbf{p}'$-integral, $\boldsymbol{\Sigma}^{\text{ph}}$ varies slowly with $|\mathbf{p}|$ about p_F, the variation being appreciable when $|\mathbf{p}|$ varies by $\sim \frac{1}{2}p_F$. Now we shall be interested in $\boldsymbol{\Sigma}^{\text{ph}}(\mathbf{p}, p_0)$ for energies $|p_0| \gtrsim \omega_D \ll E_F$. Since the phonon propagator decreases as $1/(p_0' - p_0)^2$ for $|p_0' - p_0| \gg \omega_{\mathbf{p}'-\mathbf{p}}$, it is clear that the dominant contribution from the p_0'-integral comes from energies $p_0' \gtrsim \omega_c$, where the cutoff energy ω_c is several times the Debye energy $\omega_D \ll E_F$. However, this means that the major contribution to the integral over $|\mathbf{p}|$ comes from states with kinetic energy $\gtrsim \omega_c$, owing to the form of G. Therefore, in evaluating $\boldsymbol{\Sigma}^{\text{ph}}$ we can approximate $\boldsymbol{\Sigma}(\mathbf{p}, p_0)$ occurring under the integral by $\boldsymbol{\Sigma}(p_F, p_0)$, i.e., a function of p_0 alone. By using the variables p', q, and ϕ, as in (6-30), we obtain the expression

$$\begin{aligned}\boldsymbol{\Sigma}^{\text{ph}}(p) \cong \frac{im}{(2\pi)^3|\mathbf{p}|}\int_{-\infty}^{\infty} dp_0' \int_{-\infty}^{\infty} d\bar{\epsilon}_{p'} \\ \times \frac{[Z(p_0')p_0'\mathbf{1} - \phi(p_0')\tau_1]}{[Z(p_0')p_0']^2 - \bar{\epsilon}_{p'}{}^2 - \phi^2(p_0') + i\,\delta} \\ \times \sum_\lambda \int_0^{2k_F} q\,dq\{\bar{g}_{q\lambda}\}^2 D_\lambda(q, p_0 - p_0')\end{aligned} \tag{7-64}$$

where we have used the same approximations as in (6-31) and for simplicity have assumed particle-hole symmetry to be valid near the Fermi surface in order to make the τ_3-term in $\boldsymbol{\Sigma}^{\text{ph}}$ vanish. We use the integration procedure of Eliashberg[122] and break up $D(p - p')$ into two parts, D^u which is analytic in the upper half of the p_0'-plane and D^l analytic in the lower half-plane [see the spectral representation of D (5-59)]. For the term D^u, the portion of the p_0'-contour originally along the positive real axis is folded through the upper half-plane back along the negative real axis as shown in Figure 7-6, the cuts representing the singularities of $\mathbf{G}(p')$. Since D^u is analytic in the upper half-plane, it has no discontinuity across the left G-cut. By using the relation

$$\mathbf{G}(\mathbf{p}, p_0 + i\,\delta) = \mathbf{G}^*(\mathbf{p}, p_0 - i\,\delta) \tag{7-65}$$

that is, the values of $\mathbf{G}$ on opposite sides of the cut are related by complex conjugation (see Section 5-4), the deformed contour can be replaced by an integral along the lower side of the cut if $\mathbf{G}(p)$ is replaced by $\mathbf{G}(p) - \mathbf{G}^*(p) = 2i\,\mathrm{Im}\,\mathbf{G}(p)$. Therefore the D^u-piece of Σ^{ph} is given by

$$\begin{aligned}\mathbf{\Sigma}_u{}^{\mathrm{ph}}(p) &= \frac{-2m}{(2\pi)^3|\mathbf{p}|}\int_{-\infty}^{0} dp_0{}' \\ &\quad\times \mathrm{Im}\left\{\int_{-\infty}^{\infty} d\epsilon' \frac{[Z'p_0{}'\mathbf{1} - \phi'\tau_1]}{(Z'p_0{}')^2 - \phi^{12} - \epsilon'^2 + i\,\delta}\right\} \\ &\quad\times \sum_\lambda \int_0^{2k_F} q\,dq\{\bar{g}_{q\lambda}\}^2 D_\lambda{}^u(q, p_0 - p_0{}')\end{aligned} \tag{7-66}$$

On performing the ϵ'-integral one obtains the expression

$$\begin{aligned}\mathbf{\Sigma}_u{}^{\mathrm{ph}}(p) &= \frac{m}{(2\pi)^2|\mathbf{p}|}\int_{-\infty}^{0} dp_0{}'\,\mathrm{Re}\left\{\frac{Z'p_0{}'\mathbf{1} - \phi'\tau_1}{[(Z'p_0{}')^2 - \phi'^2]^{1/2}}\right\} \\ &\quad\times \sum_\lambda \int_0^{2k_F} q\,dq\{\bar{g}_{q\lambda}\}^2 D_\lambda{}^u(q, p_0 - p_0{}')\end{aligned} \tag{7-67}$$

In a similar manner the term coming from D^l can be handled by folding into the lower half-plane the portion of the $p_0{}'$-contour originally along the negative real axis, as shown in Figure 7-7.

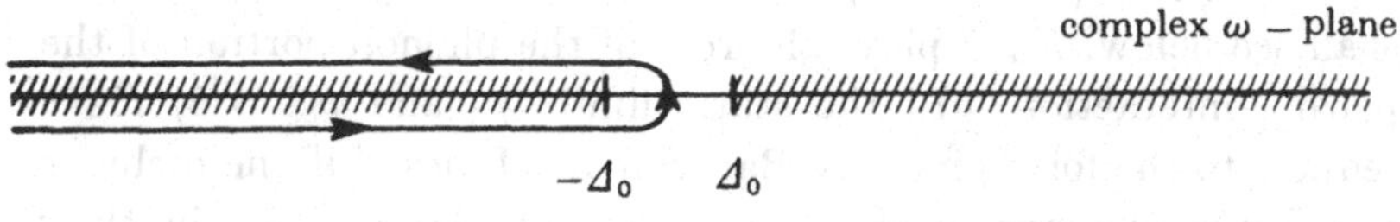

FIGURE 7-6 Folded contour for evaluating $\Sigma_u{}^{\mathrm{ph}}$.

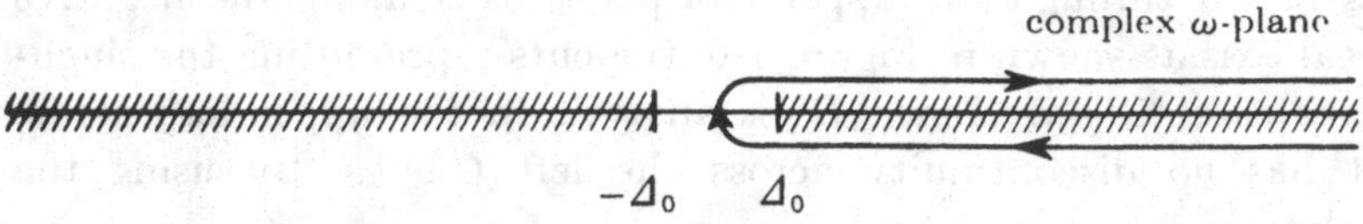

FIGURE 7-7 Folded contour for evaluating Σ_l^{ph}.

Again, D^l has no discontinuity across the right G-cut and one finds

$$\mathbf{\Sigma}_l^{\text{ph}}(p) = \frac{m}{(2\pi)^2|\mathbf{p}|}\int_0^\infty dp_0' \,\text{Re}\left\{\frac{Z'p_0'\mathbf{1} - \phi'\tau_1}{[(Z'p_0')^2 - \phi'^2]^{1/2}}\right\} \times \sum_\lambda \int_0^{2k_F} q\,dq\,\{\bar{g}_{q\lambda}\}^2 D_\lambda{}^l(q, p_0 - p_0') \quad (7\text{-}68)$$

By sending $p_0' \to -p_0'$ in (7-67) and using the fact that $Z(p)$ and $\phi(p)$ are even functions of p_0, a fact which follows from (7-61), one finds for $|\mathbf{p}| \sim p_F$,

$$\mathbf{\Sigma}^{\text{ph}}(p) = \mathbf{\Sigma}_l^{\text{ph}}(p) + \mathbf{\Sigma}_u^{\text{ph}}(p) = N(0)\int_0^\infty dp_0' \,\text{Re}\left\{\frac{Z'p_0'\mathbf{1} - \phi'\tau_1}{[(Z'p_0')^2 - \phi'^2]^{1/2}}\right\} K_\pm{}^{\text{ph}}(p_0, p_0') \quad (7\text{-}69)$$

The interaction kernels $K_+{}^{\text{ph}}$ and $K_-{}^{\text{ph}}$ are defined by

$$K_\pm{}^{\text{ph}}(p_0, p_0') = \sum_\lambda \int_0^{2k_F} \frac{q\,dq}{2k_F{}^2}\int_0^\infty d\omega B_\lambda(q,\omega)\{\bar{g}_{q\lambda}\}^2 \times \left[\frac{1}{p_0' + p_0 + \omega - i\,\delta} \pm \frac{1}{p_0' - p_0 + \omega - i\,\delta}\right] \quad (7\text{-}70)$$

where $K_-{}^{\text{ph}}$ is to be used with the $\mathbf{1}$-component of (7-69) and $K_+{}^{\text{ph}}$ is associated with the τ_1-component. In (7-70), $B_\lambda(q, \omega)$ is the phonon spectral weight function defined by (5-58). As we shall see below, $K_+{}^{\text{ph}}$ plays the role of the phonon portion of the pairing interaction. In the static limit (p_0 and $p_0' \to 0$) $K_+{}^{\text{ph}}$ reduces to the form given by Bardeen and Pines,[93] if one includes transverse phonons in their analysis and sets $\epsilon_p = \epsilon_{p'}$ in their expressions. For general p_0 and p_0', $K_+{}^{\text{ph}}$ differs from their velocity-dependent, nonretarded interaction. $K_+{}^{ph}$ gives the correct form in the region where the expressions differ, since as we mentioned above it arises from an equation which is exact to order $(m/M)^{1/2} \simeq 10^{-2}$.

We now turn to the problem of reducing the Coulomb term in the equation for Σ, (7-61), to one-dimensional form.[123] Unfortunately, the potential $\mathscr{V}_c(p - p')$ does not decrease rapidly for $|p_0 - p_0'| > \omega_D$, as is the case for the phonon interaction. For this reason the p_0'-integral is not limited to the region $|p_0'| < \omega_c$, as it was above, and the trick of integrating first with respect to the three-momentum does not work here in a straightforward way. To get around this complication we would like to introduce a pseudo-potential which accounts for the Coulomb interaction in (7-61) outside of the energy interval $-\omega_c < p_0' < \omega_c$. This general approach was first discussed by Bogoliubov, Tolmachev, and Shirkov[52] and reformulated by Morel and Anderson[124] to treat pairing correlations in He^3. To determine the pseudo-potential we consider the Coulomb part of the electron self-energy

$$\mathbf{\Sigma}^c(p) = i \int \tau_3 \mathbf{G}(p')\tau_3 \mathscr{V}_c(p - p') \frac{d^4p'}{(2\pi)^4} \tag{7-71}$$

If we define ϕ^c, χ^c, and $(1 - Z)^c p_0$ to be the coefficients of τ_1, τ_3, and $\mathbf{1}$, respectively, in $\mathbf{\Sigma}^c$ we have

$$\phi^c(p) = -i \int \frac{d^4p'}{(2\pi)^4} \frac{\phi'}{(Z'p_0')^2 - \bar{\epsilon}'^2 - \phi'^2} \mathscr{V}_c(p - p') \tag{7-72a}$$

$$\chi^c(p) = i \int \frac{d^4p'}{(2\pi)^4} \frac{\bar{\epsilon}'}{(Z'p_0')^2 - \bar{\epsilon}'^2 - \phi'^2} \mathscr{V}_c(p - p') \tag{7-72b}$$

$$[1 - Z(p)]^c p_0 = i \int \frac{d^4p'}{(2\pi)^4} \frac{Z'p_0'}{(Z'p_0')^2 - \bar{\epsilon}'^2 - \phi'^2} \mathscr{V}_c(p - p') \tag{7-72c}$$

For simplicity we assume that $\mathscr{V}_c(p - p')$ is well represented by a statically screened potential so that it is independent of p_0 and p_0'. Since the left-hand side of (7-72c) is antisymmetric in p_0 and the right-hand side is independent of p_0, we find $[1 - Z(p)]^c = 0$. We also neglect χ^c since its main effect is to give an unimportant shift to the chemical potential and change the effective mass associated with the Bloch states. The latter effect can be included in the density of states $N(\epsilon)$. Therefore we are interested in the Eq. (7-72a) for ϕ^c, which can be written as

$$\phi^c(p) = 2\int_{\Delta_0}^{\infty} \frac{dp_0'}{2\pi} \int \frac{d^3p'}{(2\pi)^3} \operatorname{Im}\left\{\frac{\phi'}{(Z'p_0')^2 - \bar{\epsilon}^2 - \phi'^2}\right\} \mathscr{V}_c(\mathbf{p} - \mathbf{p}') \tag{7-73}$$

In the reduction we have folded the p_0'-contour as in Figure 7-7 and have used the fact that $2 \operatorname{Im} G$ is the discontinuity across the G-cut. Since we want the pseudo-potential U_c to reduce the p_0'-interval in (7-73) to $\Delta_0 \rightarrow \omega_c$, we define

$$\phi^c(p) = 2\int_{\Delta_0}^{\omega_c} \frac{dp_0'}{2\pi} \int \frac{d^3p'}{(2\pi)^3} \operatorname{Im}\left\{\frac{\phi'}{(Z'p_0')^2 - \bar{\epsilon}'^2 - \phi'^2}\right\} U_c(p, p') \tag{7-74}$$

For (7-73) and (7-74) to agree, $U_c(p, p')$ must satisfy

$$U_c(p, p') = \mathscr{V}_c(\mathbf{p} - \mathbf{p}') + 2\int_{\omega_c}^{\infty} \frac{dp_0''}{2\pi} \int \frac{d^3p''}{(2\pi)^3} \mathscr{V}_c(p - p'') \times \operatorname{Im}\left\{\frac{1}{p_0''^2 - E''^2 + i\,\delta}\right\} U_c(p'', p') \tag{7-75}$$

where we have simplified this equation by using

$$\left.\begin{aligned} Z(p) &\rightarrow 1 \\ \phi(p) &\rightarrow \phi^c(p) \end{aligned}\right\} \quad \text{for } p_0 > \omega_c \tag{7-76}$$

which follows from the form of K_+^{ph} (7-70) and the integral equation (7-69). That U_c defined by (7-75) actually leads to the correct ϕ^c can be seen by substituting the formal solution for U_c into (7-74) and rearranging the resulting expression so that it agrees with (7-73).

To estimate the magnitude of U_c we assume s-wave pairing in the superconductor. In this case only the spherical average of $\mathscr{V}_c(\mathbf{p} - \mathbf{p}')$ enters:

$$\tfrac{1}{2}\int \mathscr{V}_c(p - p')\, d\mu \equiv V_c(p, p') \tag{7-77}$$

where μ is the cosine of the angle between $\mathbf{p}$ and $\mathbf{p}'$. The pseudo-potential equation (7-75) now becomes

$$U_c(p,p') = V_c(p,p') - \int \frac{d^3p''}{(2\pi)^3}\,\theta(E_{p''} - \omega_c)V_c(p,p'')\frac{1}{2E_{p''}}U_c(p'',p') \tag{7-78a}$$

where the θ-function is defined by

$$\theta(x) = \begin{cases} 1 & x > 0 \\ 0 & x < 0 \end{cases} \tag{7-78b}$$

The physical interpretation of the pseudo-potential is clear from the form of this equation since it is just the equation satisfied by the s-wave part of the t-matrix for two-particles scattering in the region outside of $-\omega_c \rightarrow \omega_c$. It is reasonable that the effective potential to use between particles near the Fermi surface is given by the sum of all multiple scatterings of the particles in the region away from the Fermi surface, plus the Born term for scattering near the Fermi surface, in agreement with (7-78a). The pseudo-potential can be obtained explicitly if $V_c(p, p')$ is approximated by a factorizable potential:

$$V_c(p,p') = \begin{cases} V_c & |\epsilon_p| \text{ and } |\epsilon_{p'}| < \omega_m \\ 0 & \text{otherwise} \end{cases} \tag{7-79}$$

where the maximum energy ω_m is of order the Fermi energy E_F. While this is a rough approximation, it should give the correct order of magnitude for U_c, and one finds

$$U_c(p,p') = \frac{V_c}{1 + N(0)V_c \ln\left(\dfrac{\omega_m}{\omega_c}\right)} \tag{7-80}$$

if the density of Bloch states is taken constant for $|\epsilon_p| < \omega_m$. This result, first given by Bogoliubov, Tolmachev, and Shirkov,[52] shows that the effective Coulomb repulsion to be used near the Fermi surface is weaker than the true screened Coulomb interaction due to the factor $[1 + N(0)V_c \ln(\omega_m/\omega_c)]$, which is typically of order 2 or 3. Physically, this reduction is associated with the fact that scatterings far from the Fermi surface lead to a smaller probability for two electrons being within the range of the screened Coulomb potential. Therefore, in the region $|p_0''| < \omega_c$, the matrix elements of the screened Coulomb interaction taken between these correlated states are smaller than the corresponding plane-wave matrix elements.

We are now in a position to perform the three-momentum integral since the dominant part of this integral comes from Bloch states for which $U_c(p, p')$ is a constant U_c. The reduction goes through just as for the phonon contribution and one finally obtains, for the total $\mathbf{\Sigma}$,

$$\mathbf{\Sigma}(p) = N(0) \int_0^{\omega_c} dp_0' \operatorname{Re}\left\{\frac{Z'p_0'\mathbf{1} + \phi'\tau_1}{[(Z'p_0')^2 - \phi'^2]^{1/2}}\right\} K_\pm(p_0, p_0') \quad (7\text{-}81)$$

As before K_+ is to be used with the τ_1-component and K_- with the $\mathbf{1}$-component, where

$$K_+(p_0, p_0') = K_+{}^{\mathrm{ph}}(p_0, p_0') - U_c \quad (7\text{-}82a)$$

$$K_-(p_0, p_0') = K_-{}^{\mathrm{ph}}(p_0, p_0') \quad (7\text{-}82b)$$

and $K_\pm{}^{\mathrm{ph}}$ are given by (7-70). Had we included the dynamic dielectric function $\kappa(\mathbf{q}, q_0)$ and not assumed particle-hole symmetry, the Coulomb interaction would enter K_- as well. These effects have been discussed by Schrieffer, Scalapino, and Wilkins.[78] The integral equations given by (7-81) can be simplified by introducing the energy-gap parameter Δ, defined by

$$\Delta(p_0) \equiv \frac{\phi(p_0)}{Z(p_0)} \quad (7\text{-}83)$$

It is this quantity which most closely corresponds to the BCS parameter Δ_p. We then obtain the integral equations

$$\Delta(p_0) = \frac{N(0)}{Z(p_0)} \int_{\Delta_0}^{\omega_c} dp_0' \times \operatorname{Re}\left\{\frac{\Delta(p_0')}{[p_0'^2 - \Delta^2(p_0')]^{1/2}}\right\} K_+(p_0, p_0') \quad (7\text{-}84a)$$

$$[1 - Z(p_0)]p_0 = N(0) \int_{\Delta_0}^{\omega_c} dp_0' \times \operatorname{Re}\left\{\frac{p_0'}{[p_0'^2 - \Delta^2(p_0')]^{1/2}}\right\} K_-(p_0, p_0') \quad (7\text{-}84b)$$

which determine Δ and Z where Δ_0, the value of the gap parameter at the edge of the gap, is defined by

$$\Delta_0 = \Delta(\Delta_0) \tag{7-84c}$$

Numerous approximate calculations have been made in an attempt to solve these equations. The first numerical calculations were performed by Swihart,[125] who set $Z = 1$ and in effect took K_+ to be a square-well potential in the variable $p_0 - p_0'$, attempting to approximate the Bardeen–Pines potential. From his results he concluded that $\Delta(p_0)$ decreases monotonically and changes sign as p_0 increases from zero. Morel and Anderson[124] and Culler et al.[126] used the correct Eliashberg potential K appearing in (7-84) but took $Z = 1$, as before. The former group treated analytically the case of constant-frequency phonons (an Einstein spectrum), while the latter group treated the Debye spectrum ($\omega_q \propto q$). In both cases it was found that Δ increased initially with increasing p_0 before changing sign near the average phonon frequency, the peak being due to the resonant nature of the Eliashberg interaction.

The most complete calculations at present are those of Scalapino, Wilkins, and the author.[78] In attempting to explain the tunnel-current anomalies observed in superconducting lead by Rowell, Anderson, and Thomas,[88] and earlier by Giaever, Hart, and Megerle,[87] they assumed that the weight function determining $K_\pm{}^{\text{ph}}$ (7-61) is of the form

$$N(0) \sum_\lambda \int_0^{2k_F} \frac{q\,dq}{2k_F{}^2} \{\bar{g}_{q\lambda}\}^2 B_\lambda(q, \omega) = \sum_\lambda \frac{w_\lambda \omega_2{}^\lambda/\pi}{(\omega - \omega_1{}^\lambda)^2 + (\omega_2{}^\lambda)^2} \tag{7-85}$$

The Lorentzian functions were adjusted to represent in a gross manner the phonon density of states for each polarization. The values $\omega_1{}^t = 4.4 \times 10^{-3}$ ev and $\omega_1{}^l = 8.5 \times 10^{-3}$ were used for the average transverse and longitudinal frequencies with the half-widths $\omega_2{}^t = 0.75 \times 10^{-3}$ ev and $\omega_2{}^l = 0.5 \times 10^{-3}$ ev. These values were chosen to give a rough fit to the phonon spectrum observed by Brockhouse in lead and to approximately reproduce the positions of the tunnel-current anomalies in this material. The coupling strengths w_λ were chosen to be independent of λ, a reasonable approximation since the dominant part of the interaction involves umklapp processes in this case. The value of w was adjusted to make Δ_0 be 1.34×10^{-3} ev, the value appropriate for lead. The real and imaginary parts of $\Delta(E)$ are shown in Figure 7-8. Notice that Re $\Delta \equiv \Delta_1$ goes through a maximum as E increases toward $\omega_1{}^t + \Delta_0$ or $\omega_1{}^l + \Delta_0$, reflecting the resonance in K_{ph} near these frequencies. The imaginary part of $\Delta \equiv \Delta_2$ is small until p_0 approaches the "emission threshold" for transverse phonons $\simeq \omega_1{}^t + \Delta_0$, a second rise appearing near $\omega_1{}^l + \Delta_0$ because of longitudinal phonon emission. In these calculations the Coulomb pseudo-potential was set equal to zero, while plots of $\Delta(E)$ and $Z(E)$ for $N(0)U_c = 0.11$, a value which is reasonable for lead, are shown in Figures 7-9a and 7-9b. To compare these results with experiment, we note that according to (3-43a) and (3-44a) the theoretical tunneling density of states is given by

$$\frac{N_T(E)}{N(0)} = -\frac{1}{\pi}\int_{-\infty}^{\infty} d\epsilon_k \operatorname{Im} G(\mathbf{k}, E) = \operatorname{Re}\left\{\frac{E}{[E^2 - \Delta^2(E)]^{1/2}}\right\} \tag{7-86}$$

where we have used the relation

$$G(\mathbf{k}, E) = \mathbf{G}_{11}(k, E) = \frac{1}{Z(E)}\left[\frac{E + \hat{\epsilon}}{E^2 - \hat{\epsilon}^2 - \Delta^2(E) + i\,\delta}\right] \tag{7-87}$$

for G where $\hat{\epsilon} = \epsilon_k/Z(E)$. The tunneling density of states is plotted using the above solutions for $\Delta(E)$, along with the experimental results of Rowell, Anderson, and Thomas. The agreement is remarkably good in spite of the simple model of the phonon spectrum and the electron–phonon coupling used. The gross structure of $N_T(E)$ can be understood by expanding (7-86) to first order in Δ^2:

$$\frac{N_T(E)}{N(0)} = 1 + \frac{[(\operatorname{Re}\Delta)^2 - (\operatorname{Im}\Delta)^2]}{2E^2} \tag{7-88}$$

As pointed out above, just below the phonon-emission thresholds Re Δ increases with increasing E, causing a tendency for N_T to rise, as seen by the knee just below $(E - \Delta_0)/\omega_1{}^t = 1$. After the threshold, Re Δ decreases while Im Δ becomes large, thereby

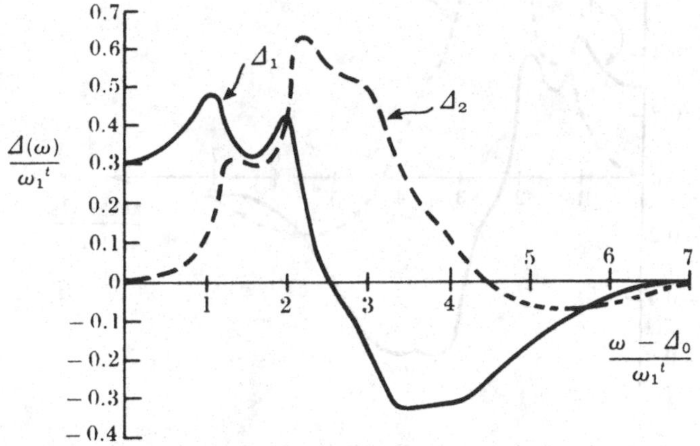

FIGURE 7-8 The real and imaginary parts of the energy-gap parameter Δ plotted as a function of energy for lead, setting the Coulomb pseudo-potential equal to zero.

producing a sharp drop in N_T. A similar situation obtains near the longitudinal phonon emission threshold. Thus a peak in the phonon density of states is reflected by a knee or peak in N_T followed by a rapid drop in this function. A good deal of detailed information about the phonon density of states can be gained in this manner from the I–V tunneling curves. In particular, van Hove singularities in the phonon spectrum along with more general singularities are reflected in d^2I/dV^2. These have been discussed by Scalapino and Anderson.[127]

It is interesting to note that the electron–phonon coupling is so strong for lead that the quasi-particle picture is meaningless over much of the energy spectrum. Nevertheless, the Green's function approach is sufficiently powerful and simple to allow this problem to be treated in detail. Furthermore, one finds structure in measurable quantities, e.g., $N_T(E)$, over energy intervals which are small compared to the level width one would calculate in perturbation theory. This is due to the fact that the spectral weight function $A(\mathbf{p}, \omega)$ is distinctly non-Lorentzian in form in this case.

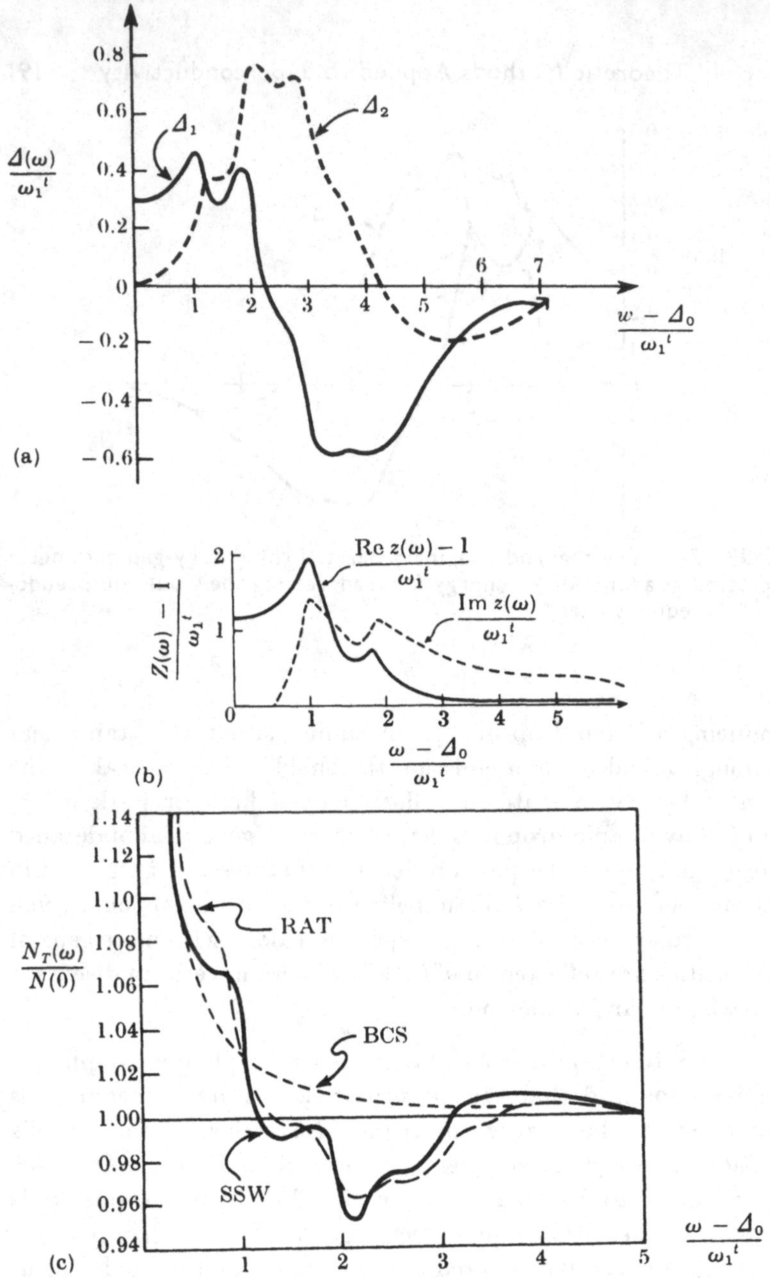

FIGURE 7-9 (a) The same as Figure 7-8 except the Coulomb pseudopotential is set equal to a value roughly appropriate for lead, $N(0)U_c = 0.11$. (b) The real and imaginary parts of the renormalization function $Z(\omega)$ for this case. (c) Tunneling density of states.

Thus far we have said nothing about the isotope effect emerging from these calculations. Estimates of the isotope-effect coefficient

$$\alpha \equiv -\frac{\partial \ln \Delta_0}{\partial \ln M} \tag{7-89}$$

by Morel and Anderson[124] are in reasonably good agreement with experiment for a large group of superconductors, with a few exceptions. Within their approximations they found the small value ~ 0.35 for zinc (a very weak coupling superconductor), while more complete calculations by Garland[24] removed the discrepancy. In addition, Garland argues that in "dirty" transition metals (i.e., those which are sufficiently impure that the single-particle states are best represented as admixtures of *s*- and *d*-states) the large *d*-band density of states strongly affects the effective Coulomb potential to be used in solving the gap equation and that very large deviations from $\alpha = \frac{1}{2}$ can be obtained even though the attraction is due to phonon exchange. This is a possible explanation of the vanishing isotope effect in ruthenium and the small value $\alpha \sim 0.2$ for osmium. For highly purified transition metals, an *s*–*s* Colomb interaction made attractive by *d*- or *f*-band polarization effects cannot be ruled out as a possible mechanism leading to superconductivity without an isotope effect; however, this mechanism appears to be highly unlikely from both a theoretical and experimental point of view.

7-4 EXTENSION TO FINITE TEMPERATURE

For finite-temperature problems one would like to define the one-particle Green's function as the statistical average of Green's functions defined for the exact excited states of the system.[91c, 99b] More generally, it is convenient to include an ensemble average over systems with different total number of particles and use the grand canonical ensemble as the appropriate weighting factor. Thus we define

$$G_s(\mathbf{r}_1, \mathbf{r}_2, \tau, \beta, \mu) = -i \operatorname{Tr} [u(\beta, \mu) T\{\psi_s(\mathbf{r}, \tau)\psi_s^+(\mathbf{r}_2, 0)\}] \tag{7-90}$$

where the grand canonical density matrix is given by

$$u(\beta, \mu) = \frac{e^{-\beta(H-\mu N)}}{\mathrm{Tr}\, e^{-\beta(H-\mu N)}} \qquad \beta = \frac{1}{k_B T} \tag{7-91}$$

and the T-product is defined as before. We suppress the β- and μ-dependence of G. For convenience we define $H - \mu N = K$ and let the time dependence develop by K rather than H,

$$\psi_s(\mathbf{r}, \tau) = e^{iK\tau}\psi_s(\mathbf{r}, 0)e^{-iK\tau} \tag{7-92}$$

For a translationally invariant system we are interested in the Green's function

$$G_s(p, \tau) = -i\,\mathrm{Tr}\,[u(\beta, \mu)T\{c_{ps}(\tau)c_{ps}^{+}(0)\}] \tag{7-93}$$

To gain insight into this function we note that a spectral representation for the time Fourier transform of $G(\mathbf{p}, \tau)$ can be derived just as in Chapter 5 and one finds

$$G(\mathbf{p}, p_0) = \int_{-\infty}^{\infty} d\omega \frac{\rho^{(+)}(\mathbf{p}, \omega)}{p_0 - \omega + i\,\delta} + \int_{-\infty}^{\infty} d\omega \frac{\rho^{(-)}(\mathbf{p}, \omega)}{p_0 + \omega - i\,\delta} \tag{7-94}$$

where the spectral functions are given by

$$\rho^{(+)}(\mathbf{p}, \omega) = \sum_{n,m} u_n |\langle m|c_p^{+}|n\rangle|^2\, \delta(E_m - E_n - \omega) \tag{7-95a}$$

$$\rho^{(-)}(\mathbf{p}, \omega) = \sum_{n,m} u_n |\langle m|c_p|n\rangle|^2\, \delta(E_m - E_n - \omega) \tag{7-95b}$$

The states $|n\rangle$ are the eigenstates

$$K|n\rangle = (H - \mu N)|n\rangle = E_n|n\rangle \tag{7-96a}$$

and u_n is the diagonal element of the density matrix,

$$u_n = \frac{e^{-\beta E_n}}{\sum_m e^{-\beta E_m}} \tag{7-96b}$$

It follows from (7-95) and (7-96) that $\rho^{(+)}$ and $\rho^{(-)}$ are related by

$$\rho^{(-)}(\mathbf{p}, \omega) = e^{\beta\omega}\rho^{(+)}(\mathbf{p}, -\omega) \tag{7-97}$$

as can be seen by interchanging dummy indices in (7-95b). Thus, the spectral representation (7-94) can be rewritten as

$$\operatorname{Re} G(\mathbf{p}, p_0) = P \int_{-\infty}^{\infty} \frac{A(\mathbf{p}, \omega)}{p_0 - \omega} \, d\omega \qquad \text{(7-98a)}$$

and

$$\operatorname{Im} G(\mathbf{p}, p_0) = -\pi A(\mathbf{p}, p_0) \tanh \frac{\beta p_0}{2} \qquad \text{(7-98b)}$$

The spectral weight function $A(\mathbf{p}, p_0)$ is defined by

$$A(\mathbf{p}, \omega) = \rho^{(+)}(\mathbf{p}, \omega) + \rho^{(-)}(\mathbf{p}, -\omega) = \rho^{(+)}(p, \omega)(1 + e^{-\beta\omega})$$

and reduces to the corresponding function introduced in Chapter 5 at zero temperature. As opposed to the zero-temperature case, however, $\rho^{(+)}$ and $\rho^{(-)}$ are in general finite over the entire ω-axis, since the system may lower its energy relative to μ when a particle is added to or subtracted from the system if $T \neq 0$. Therefore, the simple picture that the positive and negative frequency parts of $A(\mathbf{p}, \omega)$ correspond to particle- and hole-injection processes, respectively, no longer holds if $T \neq 0$.

As for $T = 0$, $A(\mathbf{p}, \omega)$ satisfies the sum rule

$$\int_{-\infty}^{\infty} A(\mathbf{p}, \omega) \, d\omega = 1 \qquad \text{(7-99a)}$$

This is seen by calculating

$$\begin{aligned} \int_{-\infty}^{\infty} A(\mathbf{p}, \omega) \, d\omega &= \int_{-\infty}^{\infty} [\rho^{(+)}(\mathbf{p}, \omega) + \rho^{(-)}(\mathbf{p}, \omega)] \, d\omega \\ &= \sum_{n, m} u_n \{\langle n | c_p c_p^+ | n \rangle + \langle n | c_p^+ c_p | n \rangle\} \\ &= \operatorname{Tr} u = 1 \qquad \text{(7-99b)} \end{aligned}$$

where we have used the anticommutation relations for the c's.

As pointed out in Chapter 5, the functions A, $\rho^{(+)}$, and $\rho^{(-)}$ are related to the functions A_{BK}, $G^{>}$, and $G^{<}$, discussed by Kadanoff and Baym,[91c] by

$$A_{BK}(\mathbf{p}, \omega) = 2\pi A(\mathbf{p}, \omega) \qquad \text{(7-100a)}$$

$$G^{>}(\mathbf{p}, \omega) = 2\pi \rho^{(+)}(\mathbf{p}, \omega) \qquad \text{(7-100b)}$$

$$G^{<}(\mathbf{p}, \omega) = 2\pi \rho^{(-)}(\mathbf{p}, -\omega) \qquad \text{(7-100c)}$$

The highly automatic nature of the zero-temperature perturbation series for G can be carried over to finite temperature by an elegantly simple procedure introduced by Abrikosov, Gor'kov, and Dzyaloshinskii,[128] who extended the pioneering work of Matsubara in this area. Similar techniques were advanced independently by Martin and Schwinger[99b] in their fundamental work on Green's function techniques in the many-body problem. The basic result of their development is that $G(\mathbf{p}, p_0)$ can be determined by the analytic continuation of a Green's function $\mathscr{G}(\mathbf{p}, i\omega_n)$ defined over a discrete set of pure imaginary frequencies $i\omega_n$. The function $\mathscr{G}(\mathbf{p}, i\omega_n)$ can be constructed by the usual Feynman–Dyson rules if all energy variables p_0 associated with fermion lines occurring in the zero-temperature expansion are formally replaced by

$$p_0 \rightarrow i\omega_n = \frac{i(2n+1)\pi}{\beta} \qquad (n = \text{integer}) \tag{7-101}$$

and the corresponding energy integrals are replaced by

$$\int_{-\infty}^{\infty} \frac{dp_0}{2\pi} \rightarrow \frac{i}{\beta} \sum_{n=-\infty}^{\infty} \tag{7-102}$$

The phonon energy variables are replaced by

$$q_0 \rightarrow i\nu_m = \frac{i2n\pi}{\beta} \tag{7-103}$$

that is, an *even* multiple of $\pi i/\beta$ (while the electronic energies are replaced by *odd* multiples of $\pi i/\beta$). The difference between the boson and fermion rules comes from the definition of the T-product. In analogy with (7-102), q_0-integrals are replaced by

$$\int_{-\infty}^{\infty} \frac{dq_0}{2\pi} \rightarrow \frac{i}{\beta} \sum_{m=-\infty}^{\infty} \tag{7-104}$$

The zero-order propagators to be used in the perturbation series are

$$\mathscr{G}_0(\mathbf{p}, i\omega_n) = \frac{1}{i\omega_n - \epsilon_p} \tag{7-105a}$$

$$\mathscr{D}_{0\lambda}(\mathbf{p}, i\nu_m) = \frac{2\Omega_{q\lambda}}{(i\nu_m)^2 - \Omega_{q\lambda}{}^2} \tag{7-105b}$$

Once the function $\mathcal{G}(\mathbf{p}, i\omega_n)$ defined on the integers $n = 0, \pm 1, \pm 2, \ldots$ is determined, this function is then to be analytically continued to the real-energy axis $i\omega_n \to p_0$ in such a way that $\mathcal{G}(\mathbf{p}, z)$ is a bounded function as $z \to \infty$ in the complex plane. In practice the continuation usually consists of setting functions such as $e^{\beta i\omega_n}$ and $e^{\beta i\nu_m}$ equal to -1 and $+1$, respectively, and replacing the remaining imaginary energies by real energies, i.e., $i\omega_n \to p_0$ (or $i\nu_m \to q_0$). Once the continued function $\mathcal{G}(\mathbf{p}, p_0)$ has been determined, the actual Green's function $G(\mathbf{p}, p_0)$ is given by

$$\text{Re}\, G(\mathbf{p}, p_0) = \text{Re}\, \mathcal{G}(\mathbf{p}, p_0) \tag{7-106a}$$

and

$$\text{Im}\, G(\mathbf{p}, p_0) = \tanh\left(\frac{\beta p_0}{2}\right) \text{Im}\, \mathcal{G}(\mathbf{p}, p_0 + i\,\delta) \tag{7-106b}$$

where $\delta = 0^+$ as usual.

The procedure is made particularly convenient by using the Poisson summation formulas

$$\begin{aligned}\sum_{n=-\infty}^{\infty} F(i\omega_n) &= -\frac{\beta}{2\pi i}\int_c \frac{F(\omega)}{e^{\beta\omega} + 1}\, d\omega \\ &= \frac{\beta}{2\pi i}\int_c \frac{F(\omega)}{e^{-\beta\omega} + 1}\, d\omega \qquad \left[\omega_n = \frac{(2n+1)\pi}{\beta}\right]\end{aligned} \tag{7-107}$$

where the contour c encircles the imaginary ω-axis as shown in Figure 7-10 and does not enclose any singularities of $F(\omega)$. This relation follows from Cauchy's theorem since the denominators $(e^{\beta\omega} + 1)$ and $(e^{-\beta\omega} + 1)$ have first-order poles at $\omega = i\omega_n$. For sums involving phonon energies one has

$$\begin{aligned}\sum_{m=-\infty}^{\infty} F(i\nu_m) &= \frac{\beta}{2\pi i}\int_c \frac{F(\nu)}{e^{\beta\nu} - 1}\, d\nu \\ &= -\frac{\beta}{2\pi i}\int_c \frac{F(\nu)\, d\nu}{e^{-\beta\nu} - 1} \qquad \left(\nu_m = \frac{2m\pi}{\beta}\right)\end{aligned} \tag{7-108}$$

The trick is then to expand the ω- or ν-contour to infinity, picking up contributions from singularities of $F(\omega)$ as the contour moves outward. In most cases the remaining integral around the

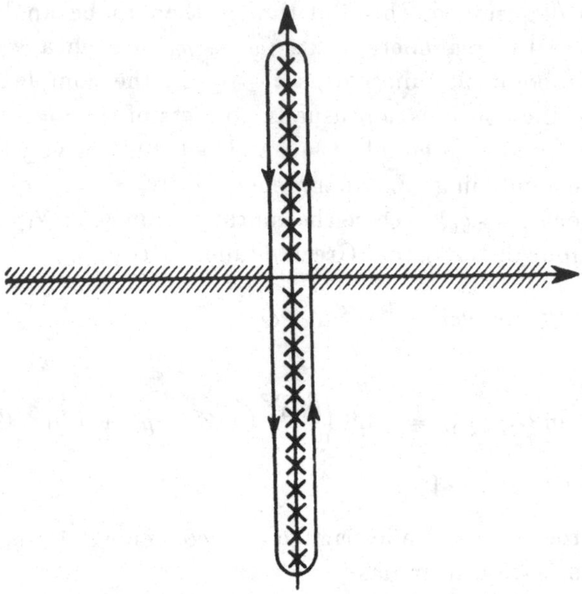

FIGURE 7-10 The contour of integration to be used with finite-temperature calculations.

infinite circle vanishes because of the vanishing of F on this contour.

To illustrate the procedure we calculate G for the Fröhlich model of the electron–phonon system in the normal state, keeping only the lowest-order contribution to Σ. For zero temperature we had

$$\Sigma(\mathbf{p}, p_0) = i \int \frac{d^4p'}{(2\pi)^4} |g_q|^2 G_0(\mathbf{p}', p_0') D_0(\mathbf{q}, p_0 - p_0') \quad (7\text{-}109)$$

where $\mathbf{q} = \mathbf{p}' - \mathbf{p}$ [see (6-27)]. To calculate $\mathcal{G}$ we use the prescription given above to transcribe this expression for Σ into one for the self-energy $\bar{\Sigma}$ appropriate to $\mathcal{G}$; thus

$$\bar{\Sigma}(\mathbf{p}, i\omega_n) = -\frac{1}{\beta} \sum_{n'=-\infty}^{\infty} \int \frac{d^3p'}{(2\pi)^3} |g_q|^2 \mathcal{G}_0(\mathbf{p}', i\omega_{n'}) \mathcal{D}_0(\mathbf{q}, i\omega_n - i\omega_{n'}) \quad (7\text{-}110)$$

By using (7-107) the n'-sum can be written as

$$-\frac{1}{\beta}\sum_{n'=-\infty}^{\infty} \mathscr{G}_0(\mathbf{p}', i\omega_{n'})\mathscr{D}_0(\mathbf{q}, i\omega_n - i\omega_{n'})$$
$$= \frac{1}{2\pi i}\int_C \mathscr{G}_0(p', \omega)\mathscr{D}_0(\mathbf{q}, i\omega_n - \omega)f(\omega)\, d\omega \quad (7\text{-}111)$$

where $f(\omega) = 1/(e^{\beta\omega} + 1)$ is the Fermi function. As the ω-contour is expanded to infinity one picks up contributions to the integral from the poles of $\mathscr{G}_0$ and $\mathscr{D}_0$ at $\omega = \epsilon_{p'}$ and $\omega = i\omega_n \pm \Omega_q$, respectively. (In the Fröhlich model Ω_q is assumed to include electron-screening effects so that $\Omega_q \propto q$.) The deformed contour is shown in Figure 7-11 for $\epsilon_{p'}$ and $\omega_n > 0$. Notice that the poles are encircled in the clockwise direction over the entire plane. By Cauchy's theorem one obtains

$$\frac{-2\Omega_q f(\epsilon_{p'})}{(\epsilon_{p'} - i\omega_n)^2 - \Omega_q{}^2} - \frac{f(i\omega_n + \Omega_q)}{i\omega_n + \Omega_q - \epsilon_{p'}} + \frac{f(i\omega_n - \Omega_q)}{i\omega_n - \Omega_q - \epsilon_{p'}} \quad (7\text{-}112)$$

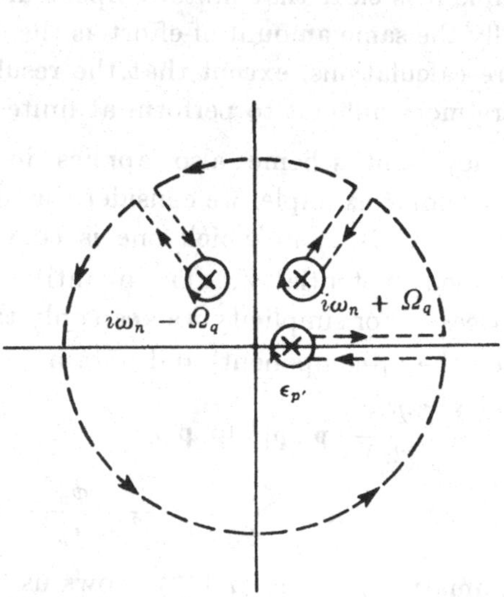

FIGURE 7-11 The deformed version of the contour shown in Figure 7-10. Notice that *all* poles are encircled in the clockwise direction. The poles shown are for the lowest-order phonon contribution to the electron self-energy in the normal state.

for the sum (7-111). In order that $\mathscr{G}(\mathbf{p}, z)$ be bounded for large z we must set

$$f(i\omega_n \pm \Omega_q) = \frac{1}{e^{i\beta\omega_n}e^{\pm\beta\Omega_q} + 1} = \frac{1}{1 - e^{\pm\beta\Omega_q}} \qquad (7\text{-}113)$$

before replacing $i\omega_n$ by z. Therefore, the properly continued self-energy is given by

$$\bar{\Sigma}(\mathbf{p}, z) = \int \frac{d^3p'}{(2\pi)^3} |g_q|^2 \left\{ \frac{1 - f_{p'} + N_q}{z - \epsilon_{p'} - \Omega_q} + \frac{f_{p'} + N_q}{z - \epsilon_{p'} + \Omega_q} \right\} \qquad (7\text{-}114)$$

As $T \to 0$, the first term in the brackets contributes only for $\epsilon_{p'} > 0$, that is, states above the Fermi surface. The second term enters for $\epsilon_{p'} < 0$, in agreement with our earlier result (5-78). The physically interesting Green's function G can now be obtained from the relations (7-112) and Dyson's equation for $\mathscr{G}$,

$$\mathscr{G}(\mathbf{p}, z) = \frac{1}{z - \epsilon_p - \bar{\Sigma}(\mathbf{p}, z)} \qquad (7\text{-}115)$$

From this example it is clear that finite-temperature calculations require essentially the same amount of effort as the corresponding zero-temperature calculations, except that the resulting momentum integrals are more difficult to perform at finite temperature.

The frequency sum scheme also applies in the Nambu formalism. As a simple example, we consider the finite temperature generalization of (7-56) in which one is concerned with a nonretarded two-body potential V. The quantities $\bar{\Sigma}$ and $\mathscr{G}$ are now 2×2 matrices. For simplicity we keep only the energy-gap portion of $\bar{\Sigma}$ (i.e., the τ_1-component) and obtain

$$\phi(\mathbf{p}, i\omega_n) \equiv \phi_p = \frac{1}{\beta} \int \frac{d^3p'}{(2\pi)^3} \langle \mathbf{p}', \mathbf{p} | V | \mathbf{p}, \mathbf{p}' \rangle \times \sum_{n'=-\infty}^{\infty} \frac{\phi_{p'}}{(i\omega_{n'})^2 - \epsilon_{p'}^2 - \phi_{p'}^2} \qquad (7\text{-}116)$$

The Poisson summation formula (7-113) allows us to convert the n'-sum to the form

$$\frac{1}{\beta} \sum_{n'=-\infty}^{\infty} \frac{\phi_{p'}}{(i\omega_{n'})^2 - \epsilon_{p'}^2 - \phi_{p'}^2} = -\frac{1}{2\pi i} \int_c \frac{\phi_{p'} f(\omega)\, d\omega}{\omega^2 - \epsilon_{p'}^2 - \phi_{p'}^2} \qquad (7\text{-}117)$$

where the contour encircles the imaginary ω-axis. By expanding the contour to infinity we pick up residues from poles at $\omega = \pm(\epsilon_{p'}{}^2 + \phi_{p'}{}^2)^{1/2} \equiv \pm E_{p'}$ as shown in Figure 7-12 and obtain

$$\frac{\phi_{p'}}{2E_{p'}}[f(E_{p'}) - f(-E_{p'})] = -\frac{\phi_{p'}}{2E_{p'}}[1 - 2f(E_{p'})]$$
$$= -\frac{\phi_{p'}}{2E_{p'}}\tanh\left(\frac{\beta E_{p'}}{2}\right) \qquad (7\text{-}118)$$

for the sum (7-117). Therefore the energy-gap equation becomes

$$\phi_{p'} = -\int \frac{d^3p'}{(2\pi)^3} V_{pp'} \frac{\phi_{p'}}{2E_{p'}} \tanh\left(\frac{\beta E_{p'}}{2}\right) \qquad (7\text{-}119)$$

where $V_{pp'} = \langle \mathbf{p}', \mathbf{p} | V | \mathbf{p}, \mathbf{p}' \rangle$, in agreement with the result of BCS and of the linearized equation of motion treatment (2-76). The finite-temperature generalization of the retarded interaction problem discussed in Section 7-3 is also straightforward to carry out by these techniques.[78]

Although we have emphasized the calculation of G at finite temperature, the phonon Green's function follows in a similar manner.

As a general conclusion of this chapter, one can say that, on refining the BCS theory to include noninstantaneous interactions and damping effects, only details of the quasi-particle spectrum are altered. In particular, the energy gap in the elementary

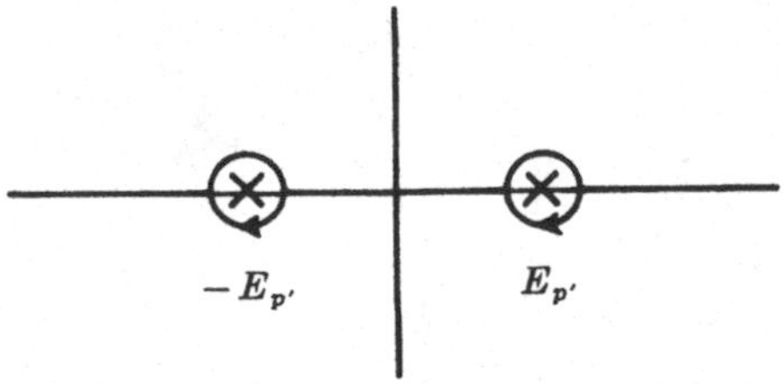

FIGURE 7-12 The poles contributing to the electron self-energy in the superconducting state if one assumes an instantaneous two-body potential.

excitation spectrum continues to exist within the above approximations. The effective density of quasi-particle states in energy is well approximated by the simple BCS model near the edge of the gap with small deviations of order or less than 1 to 5 per cent occurring in the vicinity of the Debye frequency. In the next chapter we shall see that collective effects do not alter these conclusions; however, collective modes can exist in the energy gap under suitable conditions.

CHAPTER 8

ELECTROMAGNETIC PROPERTIES OF SUPERCONDUCTORS

Among the many surprising properties of superconductors, their response to externally applied electric and magnetic fields is the most striking. In 1911, while measuring the electrical resistivity of metals at liquid helium temperature, Kamerlingh Onnes[7] discovered that certain metals passed into a radically new phase in which the voltage drop across the metallic specimen vanished even though a finite current was flowing through it. He characterized the new state as one of infinite electrical conductivity or "superconductivity." Equally striking was Meissner and Ochenfeld's[6] discovery that under ideal conditions a superconductor is a perfect diamagnet, that is, the magnetic field strength B vanishes within the bulk of the superconductor.

8-1 LONDON RIGIDITY

In Chapter 1 we gave a qualitative discussion of the origin of these unique properties. Here we take a more formal approach and show how these effects follow from microscopic considerations. While the mathematical aspects of the discussion become involved at points, the underlying physics which accounts for the phenomena was clearly stated by London[5] in 1935. He suggested that the wave function Ψ_s of the "superfluid" electrons is "rigid" or "stiff" with respect to perturbations due to the presence of a weak magnetic field. Then, as in the problem of diamagnetism of atoms, the vector potential leads to a finite-current density

$$\langle \mathbf{j} \rangle = \langle -ne\mathbf{v} \rangle = -\frac{ne}{m}\left\langle \mathbf{p} + \frac{e\mathbf{A}}{c} \right\rangle = -\frac{ne^2}{mc}\mathbf{A} \tag{8-1}$$

since $\langle \mathbf{p} \rangle = 0$, owing to the rigidity of Ψ_s. This induced current then gives rise to a magnetic field which screens out the external field and leads to perfect diamagnetism in a large system.

The microscopic implications of London's interpretation of the Meissner effect can be seen in the following manner. Since Maxwell's equations ensure that the magnetic field is necessarily a transverse field ($\nabla \cdot \mathbf{B} = 0$), the magnetic perturbation H' only affects the transverse excitations of the system. If Ψ_s is to be essentially unaffected by this (weak) perturbation, the sum of the squares of the first-order perturbation series amplitudes

$$\sum_\alpha |a_\alpha|^2 = \sum_\alpha \left| \frac{\langle \Psi_\alpha | H' | \Psi_s \rangle}{E_\alpha - E_s} \right|^2 \tag{8-2}$$

must be vanishingly small. Here Ψ_α is the state with transverse excitation α present. Presumably, this anomalously small value of the sum will occur if the matrix elements $\langle \Psi_\alpha | H' | \Psi_s \rangle$ tend to zero while the excitation energies $E_\alpha - E_s$ remain finite. Clearly, this will not be the situation for magnetic fields which vary rapidly in space. Such fields create excitations involving electronic states far from the Fermi surface. These states are presumably unaffected by superconducting correlations and therefore lead to finite matrix elements as in the normal state. Fortunately, in establishing the Meissner effect, one only requires the response of the system to magnetic fields which vary slowly in space. In this limit only electronic states near the Fermi surface enter, and there is no reason to expect the matrix elements not to vanish in this case. Therefore, London's interpretation of the Meissner effect leads one to suspect that (1) the matrix elements for creating transverse excitations from the superfluid by a magnetic field tend to zero for fields which vary slowly in space; (2) there is an energy gap in the transverse excitation spectrum of the superfluid. As we shall see below, these conditions are satisfied by the BCS theory. It is possible, however, that the matrix elements and the energy denominators *both* vanish in the long wavelength limit, in such a way as to give a finite sum. If this sum does not exactly cancel the diamagnetic current in this limit, a Meissner effect is still obtained, as in superconductors with $l \neq 0$ pairing[170] and "gapless" superconductors.[172] The essential difference between the normal and superconducting states in metals is that the paramagnetic and diamagnetic currents do not exactly cancel in the long wavelength limit in the latter.

On the basis of these arguments one might wonder why an insulator, which has as an energy gap for creating transverse electronic excitations, is not also a perfect diamagnet. The point here is that the energy gap arises from the one-body crystal potential in this case, rather than the effective electron-electron interaction. One can derive a sum rule which shows that for insulators the wave function shifts just enough to make $\langle \mathbf{p} \rangle$ cancel the diamagnetic term $e\mathbf{A}/c$ in (8-1) except for a weak diamagnetism, as in normal metals.

If one can explain the Meissner effect, the "infinite conductivity" observed by Onnes can also be explained. This follows because one can show the currents flowing in his configuration (i.e., a superconducting section in an otherwise normal circuit) were diamagnetic in the superconducting section.[1] That is, the currents in the superconductor were due to electrons described by a wave function which was essentially the same as in the absence of the current. The finite current then arose self-consistently from the magnetic field which was generated by the current itself.

In addition to the Meissner effect, one must understand the stability of persistent currents in a multiply connected body, for example, a superconducting ring. While the Meissner effect plays a role in the details of the phenomenon, the currents are not primarily diamagnetic in origin in this case. On the contrary, the wave function for the current-carrying state differs greatly from that in the absence of currents, in contrast with the situation described above. However, the effect is again due to a "rigidity" of the wave function with respect to all fluctuations which occur with a finite thermodynamic probability. That is, essentially all fluctuations lead to states of higher free energy and therefore regress without leading to decay of the current.

8-2 WEAK-FIELD RESPONSE

In the beginning of Chapter 2 we argued that transverse electromagnetic fields need not be included directly in calculating the detailed pairing interactions which bring about superconductivity. Their effect can be taken into account in terms of a space- and time-dependent average field which is calculated self-consistently from the external field and the currents flowing in the material. While the externally applied magnetic field generally represents a large perturbation on the system, the induced field arising from the supercurrents cancels the external field over most of the material, as we know from the Meissner effect. Therefore, the net field acts only very near the surface and can often be treated as a weak perturbation on the system as a whole. Thus, we shall formally treat the total transverse electromagnetic field as an externally applied field and solve for self-consistency as a separate problem.

As we have seen above, the Coulomb potential plays an essential role in the pairing theory. It cannot be treated by the self-consistent field scheme we use for the transverse field and therefore we include the total Coulomb interaction in the zero-order Hamiltonian.

We begin by considering a simply connected bulk superconductor of unit volume in the presence of a weak externally applied electromagnetic field described by the vector and scalar potentials $\mathbf{A}(\mathbf{r}, t)$ and $\varphi(\mathbf{r}, t)$, respectively. As usual, we use periodic boundary conditions. For convenience, we write

$$A_\mu(x) = \begin{cases} A_i(x) & (\mu = i = 1, 2, 3) \\ c\varphi(x) & (\mu = 0) \end{cases} \tag{8-3}$$

where $x \equiv (\mathbf{r}, t)$. To first order in A_μ, the coupling of the electrons to the electromagnetic field is

$$\begin{aligned} H^p &= -\frac{1}{c}\int \sum_\mu j_\mu{}^p(x)A_\mu(x)\, d^3r \\ &= -\frac{1}{c}\int [\mathbf{j}^p(x)\cdot \mathbf{A}(x) - \rho_e(x)c\varphi(x)]\, d^3r \end{aligned} \tag{8-4}$$

where we use the metric (1, 1, 1, − 1) in μ-sums for μ = 1, 2, 3, and 0, respectively. We call H^p the paramagnetic coupling. The paramagnetic four-current is defined by

$$j_\mu{}^p(x) = \begin{cases} j_i{}^p(x) \equiv -\dfrac{e}{2mi}\displaystyle\sum_s \{\psi_s{}^+(x)\nabla_i\psi_s(x) - [\nabla_i\psi_s{}^+(x)]\psi_s(x)\} \\ \qquad\qquad\qquad\qquad\qquad\qquad (\mu = i = 1, 2, 3) \\ \rho_e(x) = -e\displaystyle\sum_s \psi_s{}^+(x)\psi_s(x) = -e\rho(x) \qquad (\mu = 0) \end{cases} \tag{8-5}$$

the (1, 2, 3) components giving the electronic current density operator in the absence of A and the last component being the electronic charge density operator. The physical current density $\mathcal{J}_\mu(x)$ in the presence of A is the sum

$$\mathcal{J}_\mu(x) = j_\mu{}^p(x) + j_\mu{}^d(x) \tag{8-6}$$

where the diamagnetic current density j^d is given by

$$j_\mu{}^d(x) = \begin{cases} \dfrac{e}{mc}\,\rho_e(x)A_i(x) & (\mu = i = 1, 2, 3) \\ 0 & (\mu = 0) \end{cases} \tag{8-7}$$

The full coupling of the electrons to the perturbing electromagnetic field is then

$$H' = H^p + H^d$$

where the diamagnetic coupling is defined by

$$H^d = -\frac{e}{2mc^2}\int \rho_e(x)\sum_{i=1}^{3} A_i{}^2(x)\, d^3r \tag{8-8}$$

Therefore, the total system Hamiltonian is

$$\mathscr{H} = H + H'$$

If we work in an interaction representation where H' is taken to be the perturbation, and assume that $A_\mu \to 0$ as $t \to -\infty$, the ground state of the system in the presence of A evolves in time according to

$$|\Phi(t)\rangle = T \exp\left[-i\int_{-\infty}^{t} H'(t')\,dt'\right]|0\rangle \equiv U(t, -\infty)|0\rangle \quad (8\text{-}9)$$

Here $|0\rangle$ is the ground state of H and all quantities are expressed in the interaction representation. Therefore, the expectation value of the current density in the state $|\Phi(t)\rangle$ is given by

$$J_\mu(x) = \langle\Phi(t)|j_\mu(\mathbf{r}, t)|\Phi(t)\rangle = \langle 0|U^+(t, -\infty)j_\mu(\mathbf{r}, t)U(t, -\infty)|0\rangle \quad (8\text{-}10)$$

Since we are interested in the terms of J_μ which are first order in A_μ we have

$$J_\mu(x) = \frac{e}{mc}\langle 0|\rho_e(x)|0\rangle A_\mu(x)[1 - \delta_{\mu,0}] - i\langle 0|\left[j_\mu^p(\mathbf{r}, t), \int_{-\infty}^{t} H'(t')\,dt'\right]|0\rangle \quad (8\text{-}11)$$

The zeroth-order terms in J_μ vanish except for the average electronic-charge density $\langle j_0(x)\rangle$, which does not interest us here. By using the expressions (8-4) and (8-7) we find that the linear response J_μ and the externally applied potential A_μ are related by a nonlocal kernel $K_{\mu\nu}$:

$$J_\mu(x) = -\frac{c}{4\pi}\sum_\nu \int K_{\mu\nu}(\mathbf{r}, t; \mathbf{r}', t')A_\nu(\mathbf{r}', t')\,d^3r'\,dt' \quad (8\text{-}12)$$

where the spatial integral runs over the unit volume and the time integral extends from $-\infty$ to ∞. The electromagnetic response kernel $K_{\mu\nu}$ is given by

$$K_{\mu\nu}(x; x') = -\frac{4\pi i}{c^2}\langle 0|[j_\mu^p(x), j_\nu^p(x')]|0\rangle\theta(t - t') - \frac{4\pi e}{mc^2}\langle 0|\rho_e(x)|0\rangle\,\delta^4(x - x')\,\delta_{\mu\nu}[1 - \delta_{\nu,0}] \quad (8\text{-}13a)$$

where the theta function is defined by

$$\theta(t - t') = \begin{cases} 1 & (t > t') \\ 0 & (t < t') \end{cases} \tag{8-13b}$$

If the system is translationally invariant, $K_{\mu\nu}$ depends only on the difference $x - x' = (\mathbf{r} - \mathbf{r}', t - t')$. In this case it is convenient to work with the spatial Fourier transform of $K_{\mu\nu}$ defined by

$$\begin{aligned} K_{\mu\nu}(\mathbf{q}, t - t') &= \int K_{\mu\nu}(x; x') e^{-i\mathbf{q}\cdot(\mathbf{r}-\mathbf{r}')}\, d^3r\, d^3r' \\ &= -\frac{4\pi i}{c^2} \langle 0|[j_\mu{}^p(\mathbf{q}, t), j_\nu{}^p(-\mathbf{q}, t')]|0\rangle \theta(t - t') \\ &\quad + \frac{4\pi n e^2}{mc^2} \delta(t - t')\, \delta_{\mu\nu}(1 - \delta_{\nu,0}) \end{aligned} \tag{8-14}$$

where n is the number of electrons per unit volume. Since the diamagnetic (second) term in (8-14) is known explicitly, we concentrate on the paramagnetic (first) term in this expression, and define

$$R_{\mu\nu}(\mathbf{q}, \tau) = -i\langle 0|j_\mu{}^p(\mathbf{q}, \tau), j_\nu{}^p(-\mathbf{q}, 0)]|0\rangle \theta(\tau) \tag{8-15}$$

If the ground-state wave function were "rigid" with respect to *all* perturbations (rather than only those which lead to transverse excitations) $R_{\mu\nu}$ would be identically zero and (8-12) would reduce to London's equation

$$J_i(x) = -\frac{ne^2}{mc} A_i(x) \qquad (\mu = i = 1, 2, 3) \tag{8-16}$$

This relation is clearly not gauge-invariant since the predicted current depends upon the choice of gauge. In London's equation, only the transverse part of **A** is to be used[1] and therefore **J** is properly gauge-invariant. Since the longitudinal part of **A** couples to longitudinal excitations, the wave function is not "rigid" with respect to this type of perturbation and the paramagnetic term does not vanish in this case. In fact, if **A** is purely a gauge potential, the paramagnetic and diamagnetic terms exactly cancel as required by gauge invariance. In carrying out an approximate evaluation of $K_{\mu\nu}$, one may be able to accurately treat only excitations which enter the transverse response of the system. In this case the longitudinal part of the paramagnetic and diamagnetic currents will not cancel in general and the resultant current will not be manifestly gauge invariant. Nevertheless, if one recognizes the difficulty and uses only that part of $K_{\mu\nu}$ which is accurately calculated, correct physical predictions would be obtained for transverse fields. This is exactly the situation we shall meet when $K_{\mu\nu}$ is evaluated within the pairing (BCS) approximation. The inclusion of longitudinal collective modes or superfluid flow then restores gauge invariance by correcting the longitudinal part of the paramagnetic term.

In our discussion thus far, we have always dealt with time-ordered products of operators rather than retarded commutators of operators as appear in (8-15). It is the former that we can more readily handle by the Green's function scheme. Fortunately, $R_{\mu\nu}$ can be expressed in terms of a time-ordered product of current densities if one works with the time Fourier transforms of these quantities. To see this we note that $R_{\mu\nu}(q, q_0)$ defined by

$$R_{\mu\nu}(\mathbf{q}, \tau) = \int_{-\infty}^{\infty} R_{\mu\nu}(\mathbf{q}, q_0) e^{-iq_0\tau} \frac{dq_0}{2\pi} \tag{8-17}$$

can be expressed in the spectral form

$$R_{\mu\nu}(\mathbf{q}, q_0) = \int_{-\infty}^{\infty} \frac{C_{\mu\nu}(\mathbf{q}, \omega)\, d\omega}{q_0 - \omega + i\,\delta} \tag{8-18}$$

The spectral weight function $C_{\mu\nu}$ $(\mathbf{q}, \omega)$ is given by

$$\begin{aligned} C_{\mu\nu}(\mathbf{q}, \omega) = {} & \sum_n \langle 0|j_\mu{}^p(\mathbf{q})|n\rangle\langle n|j_\nu{}^p(-\mathbf{q})|0\rangle\, \delta(E_n - E_0 - \omega) \\ & - \sum_n \langle 0|j_\nu{}^p(-\mathbf{q})|n\rangle\langle n|j_\mu{}^p(\mathbf{q})|0\rangle\, \delta(E_n - E_0 + \omega) \end{aligned} \tag{8-19}$$

where

$$H|n\rangle = E_n|n\rangle \tag{8-20}$$

This spectral representation can be checked by inserting the complete set of intermediate states $|n\rangle$ between the operators in (8-15) and comparing this result with the expression given by combining (8-17) to (8-19).

Consider the corresponding time-ordered product expression

$$P_{\mu\nu}(\mathbf{q}, \tau) = -i\langle 0|T\{j_\mu{}^p(\mathbf{q}, \tau)j_\nu{}^p(-\mathbf{q}, 0)\}|0\rangle \tag{8-21}$$

Its time Fourier transform, defined by

$$P_{\mu\nu}(\mathbf{q}, \tau) = \int_{-\infty}^{\infty} P_{\mu\nu}(\mathbf{q}, q_0)e^{-iq_0\tau}\frac{dq_0}{2\pi} \tag{8-22}$$

is given by the spectral representation

$$P_{\mu\nu}(\mathbf{q}, q_0) = \int_{-\infty}^{\infty} \frac{C_{\mu\nu}(\mathbf{q}, \omega)\, d\omega}{q_0 - \omega + i\,\delta\omega} \tag{8-23}$$

as one can check by direct calculation. By comparing the spectral forms (8-18) and (8-23) we see that in the case that $C_{\mu\nu}(\mathbf{q}, \omega)$ is real, the real parts of $P_{\mu\nu}$ and $R_{\mu\nu}$ are identical while the imaginary parts differ by a minus sign for $q_0 < 0$; thus

$$\operatorname{Re} P_{\mu\nu}(\mathbf{q}, q_0) = \operatorname{Re} R_{\mu\nu}(\mathbf{q}, q_0) \tag{8-24a}$$

$$\operatorname{Im} P_{\mu\nu}(\mathbf{q}, q_0) = \operatorname{sgn} q_0 \operatorname{Im} R_{\mu\nu}(\mathbf{q}, q_0) \tag{8-24b}$$

More generally, the discontinuity of $P_{\mu\nu}(\mathbf{q}, q_0)$ across the cut determines $C_{\mu\nu}$ from which $R_{\mu\nu}$ can be obtained with the use of (8.18). Therefore, $R_{\mu\nu}$ is known once $P_{\mu\nu}$ is determined. (Since the expression for $K_{\mu\nu}$ involves only the system in the absence of A, the operators j_μ and $j_\mu{}^p$ are identical in this case and we shall often suppress the script p in the operator $j_\mu{}^p$.)

Summarizing the results obtained thus far, we find the response of the system to a weak externally applied potential $A_\mu(q) = [A(q), c\varphi(q)]$ is given by

$$\begin{aligned} J_\mu(q) &= -\frac{c}{4\pi}\sum_\nu K_{\mu\nu}(q)A_\nu(q) \\ &= -\frac{c}{4\pi}\left[\sum_{i=1}^{3} K_{\mu i}(q)A_i(q) - K_{\mu 0}(q)A_0(q)\right] \end{aligned} \tag{8-25}$$

where $q \equiv (\mathbf{q}, q_0)$. The kernel $K_{\mu\nu}$ is given by combining the expressions (8-14) and (8-15),

$$K_{\mu\nu}(q) = \frac{4\pi}{c^2} R_{\mu\nu}(q) + \frac{1}{\lambda_L{}^2}\,\delta_{\mu,\nu}[1 - \delta_{\nu,0}] \tag{8-26}$$

the two terms giving rise to the paramagnetic and diamagnetic currents, respectively. The quantity $\lambda_L{}^2 = mc^2/4\pi ne^2$ is the square of the London penetration depth. The paramagnetic kernel $R_{\mu\nu}$ is given in terms of the time-ordered quantity

$$P_{\mu\nu}(q) = \int_{-\infty}^{\infty} (-i)\langle 0|T\{j_\mu(q,\tau)j_\nu(-q,0)\}|0\rangle e^{iq_0\tau}\, d\tau \qquad (8\text{-}27)$$

by the relations (8-24).

8-3 THE MEISSNER–OCHSENFELD EFFECT

As Schafroth has shown,[14] the Meissner effect requires that the transverse part of the kernel $K_{\mu\nu}$ remain finite in the long wavelength limit ($\mathbf{q} \to 0$) for zero frequency ($q_0 = 0$). Now gauge invariance and charge conservation require that

$$\sum_{j=1}^{3} K_{ij}q_j = 0 \qquad \text{(gauge invariance)} \qquad (8\text{-}28a)$$

and

$$\sum_{i} q_i K_{ij} = 0 \qquad \text{(charge conservation)} \qquad (8\text{-}28b)$$

for $q_0 = 0$. When these relations are combined with the rotational invariance of the ground state $|0\rangle$, it follows that K_{ij} is of the form

$$K_{ij}(\mathbf{q}, 0) = \left[\delta_{ij} - \frac{q_i q_j}{\mathbf{q}^2}\right] K(\mathbf{q}^2) \qquad (8\text{-}29)$$

The Meissner effect then requires that

$$K(\mathbf{q}^2) > 0 \qquad \text{as} \qquad \mathbf{q}^2 \to 0 \qquad (8\text{-}30)$$

since the factor $[\delta_{ij} - q_i q_j/\mathbf{q}^2]$ guarantees that K_{ij} is purely transverse in this case.

The original BCS calculation of J_i was carried out in the transverse gauge, that is, $\mathbf{q} \cdot \mathbf{A}(\mathbf{q}) = 0$. In this gauge only the transverse part of K_{ij} is calculated, and one does not try to ensure that the longitudinal part of K_{ij} vanishes, as required by (8-29). It is, however, instructive to calculate the entire kernel within the BCS approximation so that we can understand the role of collective modes or superfluid flow in giving the correct result for the longitudinal part of K. To calculate the paramagnetic portion of K we require

$$P_{ij}(\mathbf{q}, \tau) = -i\langle 0|T\{j_i(\mathbf{q}, \tau)j_j(-\mathbf{q}, 0)\}|0\rangle \tag{8-31}$$

The current-density operator $\mathbf{j}(\mathbf{q})$ is given by the Fourier transform of the expression (8-5) and one finds

$$\mathbf{j}(\mathbf{q}) = -\frac{e}{m}\sum_{\mathbf{k},s}\left(\mathbf{k} + \frac{\mathbf{q}}{2}\right)c_{\mathbf{k}s}{}^{+}c_{\mathbf{k}+\mathbf{q},s} \tag{8-32}$$

Therefore P_{ij} becomes

$$P_{ij}(\mathbf{q}, \tau) = -\frac{ie^2}{m^2}\sum_{\mathbf{k},\mathbf{k}',s,s'}\left(\mathbf{k} + \frac{\mathbf{q}}{2}\right)_i\left(\mathbf{k}' + \frac{\mathbf{q}}{2}\right)_j \times \langle 0|T\{c_{\mathbf{k}s}{}^{+}(\tau)c_{\mathbf{k}+\mathbf{q},s}(\tau)c_{\mathbf{k}'+\mathbf{q},s'}{}^{+}(0)c_{\mathbf{k}'s'}(0)\}|0\rangle \tag{8-33}$$

We could evaluate this expression within the pairing approximation by replacing the exact ground state by the BCS ground state (2-33)

$$|\psi_0\rangle = \prod_{\mathbf{k}}(u_{\mathbf{k}} + v_{\mathbf{k}}b_{\mathbf{k}}{}^{+})|0\rangle \tag{8-34}$$

and expressing the c's in terms of the quasi-particle operators by the Bogoliubov–Valatin transformation (2-56). The time dependence is then approximated by that of free quasi-particles and the vacuum expectation value is evaluated in the standard way.

An equivalent procedure, which is more easily generalized beyond the pairing scheme, is to express (8-31) in terms of the Nambu field Ψ_p.[114] One then makes a Hartree factorization of the expectation value. In the Nambu notation, $\mathbf{j}(\mathbf{q})$ takes the form

$$\mathbf{j}(\mathbf{q}) = -\frac{e}{m}\sum_{\mathbf{k}}\left(\mathbf{k} + \frac{\mathbf{q}}{2}\right)(\Psi_{\mathbf{k}}{}^{+}\mathbf{1}\Psi_{\mathbf{k}+\mathbf{q}}) \tag{8-35}$$

and P_{ij} becomes

$$P_{ij}(\mathbf{q}, \tau) = -\frac{ie^2}{m^2}\sum_{\mathbf{k},\mathbf{k}'}\left(\mathbf{k} + \frac{\mathbf{q}}{2}\right)_i\left(\mathbf{k}' + \frac{\mathbf{q}}{2}\right)_j \times \langle 0|T\{\Psi_{\mathbf{k}}{}^{+}(\tau)\mathbf{1}\Psi_{\mathbf{k}+\mathbf{q}}(\tau)\Psi_{\mathbf{k}'+\mathbf{q}}{}^{+}(0)\mathbf{1}\Psi_{\mathbf{k}'}(0)\}|0\rangle \tag{8-36}$$

Within the Hartree factorization, the expectation value becomes

$$-\mathrm{Tr}\,[\langle 0|T\{\Psi_{\mathbf{k}+\mathbf{q}}(\tau)\Psi_{\mathbf{k}+\mathbf{q}}^{+}(0)\}|0\rangle\langle 0|T\{\Psi_{\mathbf{k}}(-\tau)\Psi_{\mathbf{k}}^{+}(0)\}|0\rangle]\,\delta_{\mathbf{k},\mathbf{k}'} = \mathrm{Tr}\,[\mathbf{G}(\mathbf{k}+\mathbf{q},\tau)\mathbf{G}(\mathbf{k},-\tau)]\,\delta_{\mathbf{k},\mathbf{k}'} \quad (8\text{-}37)$$

To make connection with Gor'kov's formulation,[121] we note that in carrying out the trace in (8-37), terms of the form $G_{11}G_{11}'$ and $G_{22}G_{22}'$ correspond to GG' in Gor'kov's notation, while $G_{12}G_{21}'$ and $G_{21}G_{12}'$ correspond to products of his F-functions.

Within this Hartree-like approximation, the time Fourier transform of P_{ij} is given by

$$P_{ij}(q) = -\frac{ie^2}{m^2}\int\frac{d^4k}{(2\pi)^4}\left(\mathbf{k}+\frac{\mathbf{q}}{2}\right)_i\left(\mathbf{k}+\frac{\mathbf{q}}{2}\right)_j \mathrm{Tr}\,[\mathbf{G}(k+q)\mathbf{G}(k)] \quad (8\text{-}38)$$

where $q \equiv (\mathbf{q}, q_0)$. If the pairing potential is nonretarded, we saw in Chapter 7 that within the pairing approximation $\mathbf{G}(k)$ is given by

$$\mathbf{G}(k) = \frac{k_0\mathbf{1} + \epsilon_k\tau_3 + \Delta_k\tau_1}{k_0^2 - E_k^2 + i\,\delta} \qquad [E_k = (\epsilon_k^2 + \Delta_k^2)^{1/2}] \quad (8\text{-}39)$$

[see (7-41)]. Since we are interested in the static Meissner effect, we set $q_0 = 0$ and P_{ij} reduces to

$$P_{ij}(\mathbf{q}, 0) = -2\left(\frac{e}{m}\right)^2\int\frac{d^3k}{(2\pi)^3}\left(\mathbf{k}+\frac{\mathbf{q}}{2}\right)_i\left(\mathbf{k}+\frac{\mathbf{q}}{2}\right)_j L(\mathbf{k},\mathbf{q}) \quad (8\text{-}40a)$$

where the function $L(\mathbf{k}, \mathbf{q})$ is defined by

$$\begin{aligned} L(\mathbf{k},\mathbf{q}) &= \frac{i}{2}\int_{-\infty}^{\infty}\frac{dk_0}{2\pi}\,\mathrm{Tr}\,[\mathbf{G}(k+q)\mathbf{G}(k)] \\ &= i\int_{-\infty}^{\infty}\frac{dk_0}{2\pi}\,\frac{(k_0^2 + \epsilon_k\epsilon_{k+q} + \Delta_k\Delta_{k+q})}{(k_0^2 - E_k^2 + i\,\delta)(k_0^2 - E_{k+q}^2 + i\,\delta)} \end{aligned} \quad (8\text{-}40b)$$

In the reduction, we have used the relations

$$\begin{aligned} \tau_i^2 &= \mathbf{1} \\ \mathrm{Tr}\,\mathbf{1} &= 2 \\ \mathrm{Tr}\,\tau_i &= 0 = \mathrm{Tr}\,\tau_i\tau_j \qquad (i \neq j) \end{aligned} \quad (8\text{-}40c)$$

The integral is performed by closing the contour in the upper (or lower) half-plane and one finds for $L(\mathbf{k}, \mathbf{q})$ the real quantity

$$L(\mathbf{k}, \mathbf{q}) = \frac{1}{2}\left(1 - \frac{\epsilon_k \epsilon_{k+q} + \Delta_k \Delta_{k+q}}{E_k E_{k+q}}\right) \frac{1}{E_k + E_{k+q}}$$
$$= \frac{p^2(\mathbf{k}, \mathbf{k} + \mathbf{q})}{E_k + E_{k+q}} \tag{8-41}$$

where $p(\mathbf{k}, \mathbf{k} + \mathbf{q})$ is the coherence factor we met in Chapter 3. This is the result of BCS.[8, 9]

To establish the Meissner effect within the pairing approximation, we note that $L(\mathbf{k}, \mathbf{q}) \to 0$ as $\mathbf{q} \to 0$, owing to the coherence factor $p^2(\mathbf{k}, \mathbf{q})$ vanishing in this limit and the energy denominator remaining finite ($E_k + E_{k+q} \geqslant 2\Delta_0$). Therefore,

$$\lim_{\mathbf{q} \to 0} P_{ij}(\mathbf{q}, 0) = \lim_{\mathbf{q} \to 0} R_{ij}(\mathbf{q}, 0) = 0 \tag{8-42}$$

and the electromagnetic kernel reduces to the London kernel

$$\lim_{\mathbf{q} \to 0} K_{ij}(\mathbf{q}, 0) = \frac{1}{\lambda_L^2} \delta_{ij} \qquad (i, j = 1, 2, 3) \tag{8-43}$$

The transverse part of this expression is

$$\lim_{q \to 0} K_{ij}(\mathbf{q}, 0) = \frac{\delta_{ij} - q_i q_j}{\mathbf{q}^2} \frac{1}{\lambda_L^2} \tag{8-44}$$

By comparing this result with the general form (8-29) we see that

$$\lim_{\mathbf{q}^2 \to 0} K(\mathbf{q}^2) = \frac{1}{\lambda_L^2} > 0 \tag{8-45}$$

which establishes the Meissner effect at zero temperature. Unfortunately, the longitudinal part of K does not vanish in this approximation, but is given by $(q_i q_j/q^2)(1/\lambda_L^2)$; however, this unphysical longitudinal response will be eliminated below.

The above derivation emphasizes the role of the energy gap in bringing about the Meissner effect. Aside from scale factors, the quantity $L(\mathbf{k}, \mathbf{q})$, given by (8-41), is the square of magnetic perturbation matrix element taken between the ground state and a transverse excited state, divided by the excitation energy for the two quasi-particles which are excited. In the superconducting state, the matrix element vanishes as $\mathbf{q}^2 \to 0$ [i.e., $p^2(\mathbf{k}, \mathbf{q}) \to 0$] and the energy denominator remains finite, in accordance with the discussion in the beginning of this chapter. Therefore, only the diamagnetic term in $K_{\mu\nu}$ survives in this long wavelength limit. In the normal metal $L(\mathbf{k}, \mathbf{q})$ goes over to the conventional result of second-order perturbation theory:

$$L_N(\mathbf{k}, \mathbf{q}) = \frac{1}{|\epsilon_{k+q} - \epsilon_k|} \tag{8-46}$$

if the $\mathbf{k}$ and $\mathbf{k} + \mathbf{q}$ are on opposite sides of the Fermi surface, and zero otherwise, as required by the Pauli principle. As we argued above, a finite value of $P_{ij}(\mathbf{q}, 0)$ arises in the normal metal as $\mathbf{q}^2 \to 0$ despite the fact that most of the matrix elements vanish, because of the vanishingly small-energy denominator in this case. If one calculates the magnitude of P_{ij} one finds that it almost exactly cancels the diamagnetic term, leaving the weak Landau diamagnetism of the normal state.

In "gapless" superconductors[172] both the matrix elements and energy denominators vanish, but the density of states near the Fermi surface is small enough to ensure that P_{ij} does not cancel the diamagnetic term as $\mathbf{q} \to 0$.

To extend this calculation to finite temperature, we use the prescription discussed in Chapter 7 to convert the zero temperature form (8-38) to one involving the discrete frequency sums. The only change is that $L(\mathbf{k}, \mathbf{q})$ becomes

$$L(\mathbf{k}, \mathbf{q}) = -\frac{1}{2\beta} \sum_{n=-\infty}^{\infty} \operatorname{Tr}\,[\mathbf{G}(\mathbf{k} + \mathbf{q}, i\omega_n)\mathbf{G}(\mathbf{k}, i\omega_n)] \tag{8-47}$$

where $\omega_n = (2n + 1)\pi/\beta$. As before we convert the sum to an integral using (7-107) and obtain

$$L(\mathbf{k}, \mathbf{q}) = -\frac{i}{2} \int_c \frac{d\omega}{2\pi} \operatorname{Tr}\,[\mathbf{G}(\mathbf{k} + \mathbf{q}, \omega)\mathbf{G}(\mathbf{k}, \omega)] f(\omega) \tag{8-48}$$

where the integral encircles the entire imaginary axis in a counterclockwise sense and $f(\omega) \equiv [e^{\beta\omega} + 1]^{-1}$ is the Fermi function. By expanding the contour to infinity we pick up residues at the four poles $\pm E_k \equiv \pm E$ and $\pm E_{k+q} \equiv \pm E'$, as shown in Figure 8-1. These residues lead to the real expression

$$L(\mathbf{k}, \mathbf{q}) = \frac{E^2 + \epsilon\epsilon' + \Delta\Delta'}{2E(E^2 - E'^2)}[1 - 2f(E)] + \frac{E'^2 + \epsilon\epsilon' + \Delta\Delta'}{2E'(E'^2 - E^2)}[1 - 2f(E')] \quad (8\text{-}49)$$

which can be written as

$$L(\mathbf{k}, \mathbf{q}) = \frac{p^2(\mathbf{k}, \mathbf{q})}{E_\mathbf{k} + E_{\mathbf{k}+\mathbf{q}}}[1 - f(E_\mathbf{k}) - f(E_{\mathbf{k}+\mathbf{q}})] + \frac{l^2(\mathbf{k}, \mathbf{q})}{E_\mathbf{k} - E_{\mathbf{k}+\mathbf{q}}}[f(E_{\mathbf{k}+\mathbf{q}}) - f(E_\mathbf{k})] \quad (8\text{-}50)$$

The coherence factors p^2 and l^2 are given by

$$\frac{1}{2}\left(1 \mp \frac{\epsilon_\mathbf{k}\epsilon_{\mathbf{k}+\mathbf{q}} + \Delta_\mathbf{k}\Delta_{\mathbf{k}+\mathbf{q}}}{E_\mathbf{k}E_{\mathbf{k}+\mathbf{q}}}\right) \quad (8\text{-}51)$$

the upper and lower signs applying to p and l, respectively. Since we are interested in establishing the Meissner effect at finite temperature we consider the limit of (8-50) as $\mathbf{q} \to 0$. The first

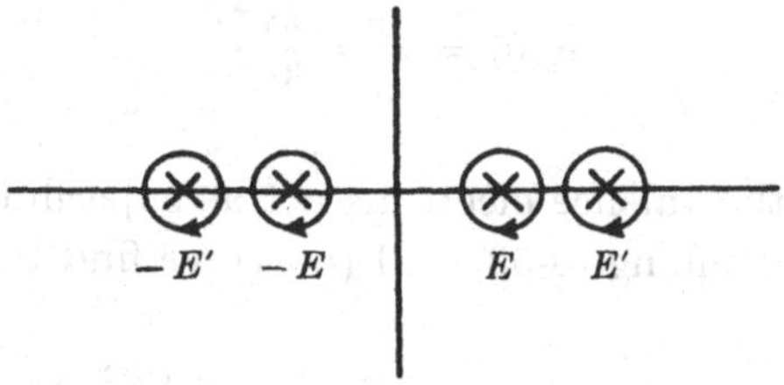

FIGURE 8-1 Poles contributing to the static electromagnetic kernel.

term gives the contribution of the "superfluid" electrons (i.e., the superfluid component in the two-fluid model), and vanishes in this limit as it did at zero temperature. The second term gives the contribution of the thermally excited quasi-particles (i.e., the "normal" fluid component) and does not vanish as $q \to 0$ since the denominator vanishes in this limit. The essential physical difference between the two terms is that the superfluid term involves creation of two quasi-particles, with the minimum excitation energy being $2\Delta_0$. On the other hand, the normal fluid term involves scattering of quasi-particles *already present* and the excitation energy in this case can be arbitrarily small, as in the normal metal. Therefore $L(\mathbf{k}, 0)$ becomes

$$\lim_{\mathbf{q}\to 0} L(\mathbf{k}, \mathbf{q}) = -\frac{\partial f(E_k)}{\partial E_k} = \frac{\beta e^{\beta E_k}}{(e^{\beta E_k} + 1)^2} = \beta f_k(1 - f_k) \tag{8-52}$$

and from (8-40a) we find

$$\lim_{\mathbf{q}\to 0} P_{ij}(\mathbf{q}, 0) = -2\left(\frac{e}{m}\right)^2 \int \frac{d^3k}{(2\pi)^3} k_i k_j \beta f_k(1 - f_k) \tag{8-53}$$

It is convenient to define the effective density $\rho_s(T)$ of superfluid electrons at temperature T by

$$\frac{\rho_s(T)}{\rho_s(0)} = 1 - \frac{2\beta E_F}{k_F{}^5} \int_0^\infty k^4 \frac{e^{\beta E_k}\, dk}{(e^{\beta E_k} + 1)^2} \tag{8-54a}$$

where the Fermi energy is given by $E_F = k_F{}^2/2m$. The relation

$$\rho_s(0) = n = \frac{k_F{}^3}{3\pi^2} \tag{8-54b}$$

states that all the valence electrons act as superfluid electrons at $T = 0$. On combining (8-53) and (8-54) we find the simple form

$$\lim_{\mathbf{q}\to 0} P_{ij}(\mathbf{q}, 0) = -\frac{ne^2}{m}\left[1 - \frac{\rho_s(T)}{\rho_s(0)}\right]\delta_{ij} \tag{8-55}$$

By using this result in the expression (8-24) for K_{ij}, one finds within the pairing approximation that

$$\lim_{\mathbf{q}\to 0} K_{ij}(\mathbf{q}, 0) = \frac{1}{\lambda_L^2(0)} \left[\frac{\rho_s(T)}{\rho_s(0)}\right] \delta_{ij} \qquad (8\text{-}56)$$

Thus, as long as $\rho_s(T)$ is nonzero, $K(q^2)$ [defined by (8-29)], is nonzero as $\mathbf{q}^2 \to 0$ and the Meissner effect is obtained. A plot of $\rho_s(T)/\rho_s(0)$ is shown in Figure 8-2. As $T \to T_c$, the density of superfluid electrons vanishes and one goes over to the normal state with its weak Landau diamagnetism.[92]

In summary, we find within the pairing approximation the following phenomena:

1. The Meissner effect is obtained for all $T \leqslant T_c$.

2. The transverse part of the electromagnetic response kernel K goes to the London form in the long wavelength limit (8-56).

3. By recognizing that only the superfluid electrons give a finite contribution to the transverse part of K as $\mathbf{q} \to 0$, we obtain an expression for the density of superfluid electrons $\rho_s(T)$ as a function of T (8-54).

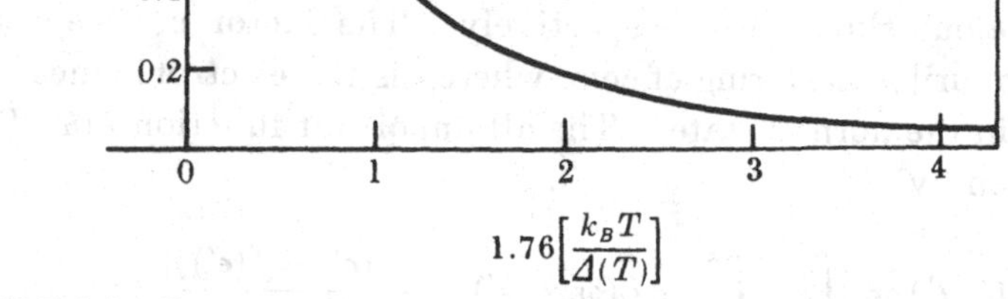

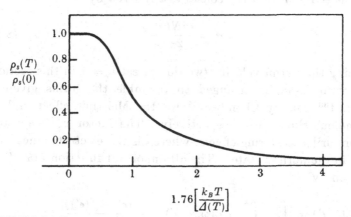

FIGURE 8-2 The superfluid density as a function of temperature. At $T = 0$ all the electrons are in the superfluid, while at $T \geqslant T_c$ all electrons are in the normal fluid.

4. The kernel K is not manifestly gauge invariant within this approximation.

8-4 ELECTROMAGNETIC PROPERTIES FOR FINITE q AND ω

Although it is gratifying to see the Meissner effect emerge within the pairing approximation, one would like to know the kernel $K_{ij}(\mathbf{q}, \omega, T)$ for general values of $\mathbf{q}$, ω, and T, as well as the effect of impurities on this function. This problem has been worked out by Mattis and Bardeen.[69a] Rather than re-deriving their results by the Green's function formalism,[69b, c] we simply state their conclusions. For many purposes it is more convenient to express the kernel in coordinate space rather than in q-space. If one works in the transverse gauge

$$\nabla \cdot \mathbf{A}(\mathbf{r}, \omega) = 0 \tag{8-57}$$

they find

$$\mathbf{J}(\mathbf{r}, \omega) = -\alpha \int d^3r' \frac{\mathbf{R}[\mathbf{R} \cdot \mathbf{A}(\mathbf{r}')]}{R^4} I(\omega, R, T) e^{-R/l} \tag{8-58}$$

Here $\mathbf{R} \equiv \mathbf{r} - \mathbf{r}'$ and the constant α is given by

$$\alpha = \frac{e^2 N(0) v_F}{2\pi^2 \hbar c} \tag{8-59}$$

v_F being the Fermi velocity (we do not set $\hbar = 1$ in this section). The form (8-58) is arranged to resemble the forms given by Pippard[33] and by Chambers[34] for the Meissner effect and the anomalous skin effect, respectively. The factor $e^{-R/l}$ accounts for impurity scattering effects, where l is the electronic mean free path in the normal state. The all important function $I(\omega, R, T)$ is given by

$$I(\omega, R, T) = \int_{-\infty}^{\infty} \int_{-\infty}^{\infty} \left\{ L(\omega, \epsilon, \epsilon') - \frac{[f(\epsilon) - f(\epsilon')]}{\epsilon' - \epsilon} \right\} \times \cos\left[\frac{R(\epsilon - \epsilon')}{\hbar v_F}\right] d\epsilon \, d\epsilon' \tag{8-60}$$

where as usual f is the Fermi function. The function $L(\omega, \epsilon, \epsilon')$ is the generalization of the function $L(\mathbf{k}, \mathbf{q})$ of the previous section, and is given by

$$\begin{aligned} L(\omega, \epsilon, \epsilon') = {} & \frac{1}{2} p^2(\epsilon, \epsilon')\left[\frac{1}{E + E' + \hbar\omega - i\,\delta} + \frac{1}{E + E' - \hbar\omega + i\,\delta}\right] \\ & \times [1 - f(E) - f(E')] \\ & + \frac{1}{2} l^2(\epsilon, \epsilon')\left[\frac{1}{E - E' + \hbar\omega - i\,\delta} \right. \\ & \left. + \frac{1}{E - E' - \hbar\omega + i\,\delta}\right] \\ & \times [f(E') - f(E)] \end{aligned} \tag{8-61}$$

The coherence factors are defined by

$$\begin{aligned} p^2(\epsilon, \epsilon') &= \frac{1}{2}\left(1 - \frac{\epsilon\epsilon' + \Delta\Delta'}{EE'}\right) \\ l^2(\epsilon, \epsilon') &= \frac{1}{2}\left(1 + \frac{\epsilon\epsilon' + \Delta\Delta'}{EE'}\right) \end{aligned} \tag{8-62}$$

and $E = (\epsilon^2 + \Delta^2)^{1/2}$. In several limiting cases $I(\omega, R, T)$ takes a simple form:

1. $\hbar\omega \gg \Delta_0$: In this limit, which includes the normal metal as a special case, I becomes

$$I(\omega, R, T) = i\pi\hbar\omega e^{iR\omega/v_F} \tag{8-63}$$

and (8-58) reduces to Chambers' expression for the anomalous skin effect. This allows the coefficient α to be evaluated in terms of the surface impedance of the normal metal in the extreme anomalous limit.

2. $\omega = 0$: In this low-frequency limit (8-58) reduces to a form closely related to Pippard's equation. It is conventional to introduce the function $J(R, T)$ (not to be confused with the current density $\mathbf{J}$) by the relation

$$I(0, R, T) = \left[\frac{\rho_s(T)}{\rho_s(0)}\right] \frac{\pi\hbar v_F}{\xi_0} J(R, T) \tag{8-64}$$

where Pippard's coherence length ξ_0 is defined in terms of microscopic parameters by

$$\xi_0 = \frac{\hbar v_F}{\pi \Delta_0} \tag{8-65}$$

In this limit (8-58) becomes

$$\mathbf{J}(\mathbf{r}) = \frac{-3}{4\pi c \Lambda(T)\xi_0} \int \frac{\mathbf{R}[\mathbf{R}\cdot\mathbf{A}(\mathbf{r}')]}{R^4} J(R, T) e^{-R/l}\, d^3r' \tag{8-66}$$

where $\Lambda(T) \equiv m/\rho_s(T)e^2$ is London's parameter.[1] This expression agrees with Pippard's equation except for the factor e^{-R/ξ_0} being replaced by $J(R, T)$ in (8-66). The definition (8-64) ensures that the $J(R, T)$ and e^{-R/ξ_0} have the same integral

$$\int_0^\infty J(R, T)\, dR = \xi_0 = \int_0^\infty e^{-R/\xi_0}\, dR \tag{8-67}$$

for all $T \leqslant T_c$. One finds that not only are the integrals of the two functions the same, but also the functions themselves resemble each other over the entire range of R and T. For example $J(R, 0)$ is within 5 per cent of e^{-R/ξ_0} for all R and $J(0, 0) = 1$, $J(0, T_c) = 1.33$.

3. $q\xi_0 \ll 1$, $\omega = 0$: In this long wavelength, zero-frequency limit we have already seen that one has the London expression

$$\mathbf{J}_s(\mathbf{r}) = -\frac{\rho_s(T)e^2}{mc}\mathbf{A}(\mathbf{r}) = -\frac{1}{c\Lambda(T)}\mathbf{A}(\mathbf{r}) \tag{8-68}$$

for the pure superconductor, in agreement with (8-66) as can be seen by taking $\mathbf{A}$ outside of the integral in this limit. For a short mean free path $l \ll \xi_0$, one obtains an extra factor $J(0, T)l/\xi_0 \simeq l/\xi_0$, which shows that the London penetration depth increases with impurity concentration.

4. $R/\xi_0 \ll 1$ ($q\xi_0 \gg 1$): If the field is well localized on space, for example, by a skin depth $\lambda \ll \xi_0$ or by the geometry of a thin film where $d \ll \xi_0$, one can evaluate $I(\omega, R, T)$ at $R = 0$ and take it outside of the integral in (8-58). Since the remaining integral is the same as in the normal state of the metal in this limit (i.e., the factor $e^{i\omega R/v_F} \simeq 1$), we can normalize the current

to that in the normal state and express the ratio in terms of the complex surface conductivities σ in the two states. Thus,

$$\frac{\sigma_s}{\sigma_n} = \frac{\sigma_1 + i\sigma_2}{\sigma_n} = \frac{I(\omega, 0, T)}{i\pi\hbar\omega} \tag{8-69}$$

The expression for σ_1/σ_n was given in Chapter 3 while the expression for σ_2 is

$$\frac{\sigma_2}{\sigma_n} = \frac{1}{\hbar\omega}\int_{\Delta-\hbar\omega,\,-\Delta}^{\Delta} \frac{[1 - 2f(E + \hbar\omega)][E^2 + \hbar\omega E + \Delta^2]\,dE}{\{[\Delta^2 - E^2][(E + \hbar\omega)^2 - \Delta^2]\}^{1/2}} \tag{8-70}$$

the lower limit being the larger of the two quantities $\Delta - \hbar\omega$ and $-\Delta$. At zero temperature, the ratio σ_2/σ_n is

$$\frac{\sigma_2}{\sigma_n} = \frac{1}{2}\left(1 + \frac{2\Delta}{\hbar\omega}\right)E(k') - \frac{1}{2}\left(1 - \frac{2\Delta}{\hbar\omega}\right)K(k') \tag{8-71}$$

while the absorptive part is given by

$$\frac{\sigma_1}{\sigma_n} = \left(1 + \frac{2\Delta}{\hbar\omega}\right)E(k) - \frac{4\Delta}{\hbar\omega}K(k) \tag{8-72}$$

In these expressions E and K are the complete elliptic integrals and

$$k' = (1 - k^2)^{1/2} \qquad \text{where} \qquad k = \left|\frac{2\Delta - \hbar\omega}{2\Delta + \hbar\omega}\right| \tag{8-73}$$

The functions σ_1 and σ_2 have been calculated by Tinkham[129] for $T = 0$. For $T \neq 0$ numerical calculations are necessary to determine the surface conductivity; however, a simple low-frequency limit is

$$\frac{\sigma_2}{\sigma_n} = \frac{\pi\Delta}{\hbar\omega}\tanh\frac{\Delta}{2k_BT} \tag{8-74}$$

Calculations for a wide range of frequency and temperature have been carried out by Miller.[130a]

In general one finds remarkably good quantitative agreement between these predictions of the pairing theory and experiment.

As mentioned in Ch. 3, a precursor absorption had been observed below the gap edge $\omega/2\Delta \sim 0.85$. While $\ell \neq 0$ collective modes give absorption in this region, the absorption is too weak to account for these experimental results, which were later found to be spurious.

8-5 GAUGE INVARIANCE

While the simple pairing approximation gives an accurate account of the response of the system to transverse electromagnetic fields, it does not in general give the correct response to longitudinal fields. In particular, we saw in a previous section that it predicts unphysical longitudinal currents which depend on the choice of gauge of the electromagnetic potentials. The physical origin of this difficulty was first recognized by Bardeen,[131] who pointed out that a (longitudinal) gauge potential couples primarily to the collective density fluctuation mode of the electron system (i.e., the plasmons of the charged electron gas). He argued that if one generalizes the pairing scheme to include this mode in a consistent way, a gauge-invariant theory would be obtained. While a number of authors have contributed to the detailed resolution of this problem, the pioneering work of Anderson[47] followed by that of Rickayzen[132] gave the essentials of a generalized pairing scheme which includes these effects. In essence, their approach is to extend the random phase approximation to include pairing correlations.

It is well known that a gauge-invariant response is a consequence of local charge conservation in the system. By local charge conservation we mean that the electronic current and charge density operators satisfy the continuity equation at each point in space and time,

$$\nabla \cdot \mathbf{j}(\mathbf{r}, t) + \frac{\partial \rho_e(\mathbf{r}, t)}{\partial t} = 0 \tag{8-75}$$

In Fourier transform variables this becomes

$$\mathbf{q} \cdot \mathbf{j}(\mathbf{q}, q_0) - q_0 \rho_e(\mathbf{q}, q_0) = 0 \tag{8-76a}$$

With the definitions (8-5) and (8-6) plus the metric (1, 1, 1, −1) used previously, the continuity equation for the four-current becomes

$$\sum_{\mu=0}^{3} q_\mu j_\mu(q) = 0 \tag{8-76b}$$

where $q = (\mathbf{q}, q_0)$ as usual. From these relations, it follows that the expected current

$$J_\mu(\mathbf{r}, t) = \langle j_\mu(\mathbf{r}, t)\rangle \tag{8-77}$$

satisfies the continuity equation

$$\nabla \cdot \mathbf{J}(\mathbf{r}, t) + \frac{\partial J_0(\mathbf{r}, t)}{\partial t} = 0 \tag{8-78a}$$

or in Fourier transform space,

$$\sum_{\mu=0}^{3} q_\mu J_\mu(q) = 0 \tag{8-78b}$$

If we concentrate on the linear response of the system to the potential A_μ, we define [see (8-25)]

$$J_\mu(q) = -\frac{c}{4\pi} \sum_{\nu=0}^{3} K_{\mu\nu}(q) A_\nu(q) \tag{8-79}$$

It follows from the continuity equation (8-78b) that the response kernel $K_{\mu\nu}(q)$ must satisfy the equation

$$\sum_{\mu=0}^{3} q_\mu K_{\mu\nu}(q) = 0 \tag{8-80}$$

[For $q_0 = 0$, this condition reduces to the condition (8-28b), used in discussing the static Meissner effect.]

Turning now to the restrictions imposed on K by gauge invariance, we note that the most general gauge transformation is of the form

$$\begin{aligned} \mathbf{A}(\mathbf{r}, t) &\Rightarrow \mathbf{A}(\mathbf{r}, t) + \nabla \Lambda(\mathbf{r}, t) \\ c\varphi(\mathbf{r}, t) &\Rightarrow c\varphi(\mathbf{r}, t) - \frac{\partial \Lambda(\mathbf{r}, t)}{\partial t} \end{aligned} \tag{8-81a}$$

The observable fields

$$\mathbf{E}(\mathbf{r}, t) = -\frac{1}{c}\frac{\partial \mathbf{A}(\mathbf{r}, t)}{\partial t} - \nabla\varphi(\mathbf{r}, t)$$
$$\mathbf{B}(\mathbf{r}, t) = \nabla \times \mathbf{A}(\mathbf{r}, t) \tag{8-81b}$$

are invariant under this transformation. In Fourier-transform space, the gauge transformation becomes

$$A_\mu(q) \Rightarrow A_\mu(q) + iq_\mu \Lambda(q) \tag{8-81c}$$

If the observable current is to be unaffected by the gauge transformation, we must require that K satisfies

$$\sum_{\nu=0}^{3} K_{\mu\nu}(q)q_\nu = 0 \tag{8-82}$$

To show the equivalence between the restrictions of local charge conservation (8-80) and of gauge invariance (8-82), we note that $K_{\mu\nu}(q)$ satisfies the symmetry relations

$$\operatorname{Re} K_{\mu\nu}(q) = \operatorname{Re} K_{\nu\mu}(-q) \tag{8-83a}$$
$$\operatorname{Im} K_{\mu\nu}(q) = -\operatorname{Im} K_{\nu\mu}(-q) \tag{8-83b}$$

as a consequence of the definition (8-26) and the spectral representation (8-18) for the retarded commutator $R_{\mu\nu}$. Therefore by changing dummy indices, the real part of the charge-conservation restriction (8-80) can be written as

$$\sum_\nu q_\nu \operatorname{Re} K_{\nu\mu}(q) = \sum_\nu \operatorname{Re} K_{\mu\nu}(-q)q_\nu = 0 \tag{8-84a}$$

or sending $q_\mu \to -q_\mu$

$$\sum_\nu \operatorname{Re} K_{\mu\nu}(q)q_\nu = 0 \tag{8-84b}$$

which is the real part of the gauge-invariance restriction (8-82). In a similar way one finds for imaginary part

$$\sum_\nu q_\nu \operatorname{Im} K_{\nu\mu}(q) = -\sum_\nu \operatorname{Im} K_{\mu\nu}(-q)q_\nu = 0 \tag{8-84c}$$

or

$$\sum_\nu \operatorname{Im} K_{\mu\nu}(q)q_\nu = 0 \tag{8-84d}$$

which agrees with the imaginary part of the gauge-invariance restriction (8-82). Therefore, in an exact calculation, gauge invariance would follow as a consequence of local charge conservation.

Unfortunately, approximate calculations of $K_{\mu\nu}$ may not maintain local charge conservation and therefore not lead to gauge-invariant results. This difficulty, however, is *not* peculiar to the superconducting state. It is a commonly held view that the gauge-invariance problem of the simple pairing approximation is due to the use of wave functions which do not describe a system with a fixed number of particles. That this is not the source of difficulty is seen by realizing that the matrix elements entering the kernel K involve the operator j_μ, which only connects states $|\alpha, N\rangle$ with the *same* number of particles. Therefore, if the states $|\alpha\rangle$ used in calculating these matrix elements are an ensemble average of states $|\alpha, N\rangle$, each with a fixed number of particles, as in the BCS approach, one simply obtains an ensemble average of matrix elements, each of which is evaluated between two states with the same number of particles. Since these fixed N matrix elements are slowly varying functions of N, the ensemble average does not affect the over-all result.

The actual source of error is that the quasi-particle excitations are not treated accurately enough in the simple pairing scheme to ensure local charge conservation under all conditions. Exactly the same situation exists in the normal state if one does not work within a "conserving" approximation, as Baym and Kadanoff[133] put it, even if one uses states which explicitly describe N-particles. In treating the motion of an electron in the medium, one must include the "backflow" of other electrons around the electron in question,[132, 134] as Feynman and Cohen[135] stressed in their work on excitations in superfluid helium. This backflow has a dipolar form at large distance. As one can show, the backflow cancels itself out if the quasi-particles are excited by a transverse field.[9] Therefore, in calculating the response of the system to transverse fields, the backflow currents play no role and one can obtain correct results by an approximation which neglects these complicating effects, as we did in the previous sections.

On the other hand, for the longitudinal response of the system, the backflow around a quasi-particle has a coupling to the external potential which in the long wavelength limit is equal in magnitude but opposite in sign to that given by the bare quasi-particle. Thus the dressed quasi-particle (i.e., the bare quasi-particle plus its associated backflow cloud) is very weakly coupled to a slowly varying longitudinal potential, as Pines and the author have discussed.[134] There is however an extra longitudinal mode of the system which occurs once backflow is properly taken into account and this is the collective density fluctuation mode.[47, 52] Physically, one can think of the collective mode simply as a compressional wave in the superfluid. From this physical picture it is reasonable that the current and particle densities associated with this mode will satisfy the continuity equation. Since in the long wavelength limit only the density fluctuation mode is coupled to a longitudinal potential, it is reasonable that a gauge-invariant response will be obtained once these effects are included.

Kadanoff and Ambegaokar have shown that in the long wavelength limit the collective mode can be described as a state in which the phase of the energy-gap parameter varies periodically in space and time, while the magnitude of the gap parameter remains fixed. Since the phase of the gap parameter gives the mean local center-of-mass momentum of the superfluid pairs, a periodically varying phase is exactly what one would expect if the superfluid momentum density varies periodically.

There are now a number of formalisms for including the backflow and the collective mode. One of the simplest ways to handle the problem is to make use of a "generalized Ward's identity," which is the Green's function analog of the continuity equation. By making approximations which are consistent with this identity one can ensure local charge conservation and therefore gauge invariance. This approach was first discussed by Nambu,[114] and we follow his line of argument below.

We consider the time-ordered quantity $\Lambda_\mu(x, y, z)$ defined in terms of the Nambu field Ψ by

$$\Lambda_\mu(x, y, z) = \langle 0|T\{j_\mu(z)\Psi(x)\Psi^+(y)\}|0\rangle \qquad (8\text{-}85)$$

The four-current density j_μ is defined by (8-6). It is clear that the paramagnetic kernel $R_{\mu\nu}$ can be calculated from Λ_μ by taking the appropriate gradients and traces of (8-85), as we shall see below. We define the vertex function $\mathbf{\Gamma}_\mu(x', y', z)$ by the integral relation

$$\mathbf{\Lambda}(x, y, z) = e \int \mathbf{G}(x, x')\mathbf{\Gamma}_\mu(x', y', z)\mathbf{G}(y', y)\, d^4x'\, d^4y' \tag{8-86}$$

where

$$\mathbf{G}(x, x') = -i\langle 0|T\{\Psi(x)\Psi^+(x')\}|0\rangle \tag{8-87}$$

is Nambu's one-particle Green's function. We assume the system to be translationally invariant so that we can write

$$\mathbf{\Gamma}_\mu(x', y', z) = \int \mathbf{\Gamma}_\mu(p + q, p)e^{i[p(x'-y')+q(x'-z)]} \frac{d^4p\, d^4q}{(2\pi)^8} \tag{8-88}$$

The generalized Ward's identity for the superconductor is then

$$\begin{aligned}\sum_\mu q_\mu \mathbf{\Gamma}_\mu(p + q, p) &= \sum_{i=1}^{3} q_i\mathbf{\Gamma}_i(p + q, p) - q_0\mathbf{\Gamma}_0(p + q, p) \\ &= \tau_3\mathbf{G}^{-1}(p) - \mathbf{G}^{-1}(p + q)\tau_3\end{aligned} \tag{8-89}$$

To prove this identity, we take the four-divergence of Λ_μ with respect to $z \equiv (\mathbf{z}, z_0 = t_z)$:

$$\begin{aligned}\sum_{i=1}^{3} \frac{\partial \mathbf{\Lambda}_i}{\partial z_i} + \frac{\partial \mathbf{\Lambda}_0}{\partial z_0} &= \langle 0|T\left\{\left[\sum_{i=1}^{3} \frac{\partial j_i(z)}{\partial z_i} + \frac{\partial j_0(z)}{\partial z_0}\right]\Psi(x)\Psi^+(y)\right\}|0\rangle \\ &\quad + \langle 0|T\{[j_0(z), \Psi(x)]\Psi^+(y)\}|0\rangle\, \delta(z_0 - x_0) \\ &\quad + \langle 0|T\{\Psi(x)[j_0(z), \Psi^+(y)]\}|0\rangle\, \delta(z_0 - y_0)\end{aligned} \tag{8-90}$$

The last two terms on the right-hand side arise from differentiating the time dependence due to the time-ordering symbol T. Now the first term on the right-hand side of this expression vanishes by virtue of the continuity equation (8-75). If we use the equal time anticommutation relations of the Ψ's (7-21), the commutators in (8-90) can be reduced to

$$[j_0(z), \Psi(x)]\, \delta(z_0 - x_0) = e\tau_3\Psi(z)\, \delta^4(z - x) \tag{8-91a}$$

and

$$[j_0(z), \Psi^+(y)]\, \delta(z_0 - y_0) = -e\Psi^+(y)\tau_3\, \delta^4(z - y) \tag{8-91b}$$

By inserting these expressions into (8-90) and using the definition (8-86), and (8-87) we find

$$i\mathbf{G}(x - z)\tau_3\, \delta^4(z - y) - i\tau_3\mathbf{G}(z - y)\, \delta^4(z - x)$$
$$= - \int \mathbf{G}(x - x')\left[\sum_{i=1}^{3} \frac{\partial \mathbf{\Gamma}_i}{\partial z_i} + \frac{\partial \mathbf{\Gamma}_0}{\partial z_0}\right]\mathbf{G}(y' - y)\, d^4x'\, d^4y' \quad (8\text{-}92)$$

Going over to Fourier transform variables we find (8-92) reduces to

$$\mathbf{G}(p + q)\tau_3 - \tau_3\mathbf{G}(p) = \mathbf{G}(p + q) \sum_{\mu=0}^{3} q_\mu \mathbf{\Gamma}_\mu(p + q, p)\mathbf{G}(p) \quad (8\text{-}93a)$$

or finally by operating with $\mathbf{G}^{-1}(p + q)$ on the left and with $\mathbf{G}^{-1}(p)$ on the right we obtain the generalized Ward's identity

$$\sum_{\mu=0}^{3} q_\mu \mathbf{\Gamma}_\mu(p + q, p) = \tau_3\mathbf{G}^{-1}(p) - \mathbf{G}^{-1}(p + q)\tau_3 \quad (8\text{-}93b)$$

as stated.

What is the physical significance of $\mathbf{\Gamma}_\mu$ and why is this identity of interest? The significance of $\mathbf{\Gamma}_\mu$ can be best understood by noticing that the four-current density operator $j_\mu{}^p(\mathbf{q})$ can be written in the Nambu notation as

$$j_\mu{}^p(\mathbf{q}) = \begin{cases} \sum_p \Psi_{\mathbf{p}}{}^+\left[-e\left(\mathbf{p} + \dfrac{\mathbf{q}}{2}\right)_i \mathbf{1}\right]\Psi_{\mathbf{p}+\mathbf{q}} & (\mu = i = 1, 2, 3) \\ \sum_p \Psi_{\mathbf{p}}{}^+\left[-e\tau_3\right]\Psi_{\mathbf{p}+\mathbf{q}} & (\mu = 0) \end{cases} \quad (8\text{-}94)$$

If we define the "free" vertex function $\boldsymbol{\gamma}_\mu(\mathbf{p} + \mathbf{q}, \mathbf{p})$ as

$$\boldsymbol{\gamma}_\mu(\mathbf{p} + \mathbf{q}, \mathbf{p}) = \begin{cases} \dfrac{1}{m}\left(\mathbf{p} + \dfrac{\mathbf{q}}{2}\right)_i \mathbf{1} & (\mu = i = 1, 2, 3) \\ \tau_3 & (\mu = 0) \end{cases} \quad (8\text{-}95)$$

then $j_\mu{}^p(\mathbf{q})$ can be written as

$$j_\mu{}^p(\mathbf{q}) = -e \sum_p \Psi_{\mathbf{p}}{}^+\boldsymbol{\gamma}_\mu(\mathbf{p} + \mathbf{q}, \mathbf{p})\Psi_{\mathbf{p}+\mathbf{q}} \quad (8\text{-}96)$$

The vertex function $\mathbf{\Gamma}_\mu(p + q, p)$ is then a "*dressed*" version of the *free* vertex $\gamma_\mu(p + q, p)$. To make this plausible, we apply the generalized Ward's identity to a system of noninteracting electrons. Since $G^{-1}(p)$ becomes

$$\mathbf{G}_0{}^{-1}(p) = p_0\mathbf{1} - \epsilon_p\tau_3 \tag{8-97}$$

in this limit, where $\epsilon_p = (p^2/2m) - \mu$, (8-93b) reduces to

$$\sum_\mu q_\mu \mathbf{\Gamma}_\mu(p + q, p) = (\epsilon_{\mathbf{p}+\mathbf{q}} - \epsilon_{\mathbf{p}})\mathbf{1} - q_0\tau_3 \tag{8-98}$$

However, this relation is identically satisfied if $\mathbf{\Gamma}_\mu$ is the free vertex γ_μ. Therefore, we can think of the dressed electrons as interacting with the electromagnetic field through the dressed vertex $(-e\mathbf{\Gamma}_\mu)$.

As to why the generalized Ward's identity is of interest, we shall now see that the paramagnetic kernel can be simply expressed in terms of $\mathbf{G}$ and $\mathbf{\Gamma}_\mu$. Furthermore, if we approximate $\mathbf{G}$ and $\mathbf{\Gamma}$ in a way which maintains Ward's identity, the full electromagnetic kernel $K_{\mu\nu}$ will be manifestly gauge-invariant. From the definitions of the time-ordered kernel $P_{\mu\nu}$ (8-21) and the vertex function $\mathbf{\Gamma}_\mu$ (8-85) and (8-86), it is straightforward to show that $P_{\mu\nu}(q)$ is given by

$$P_{\mu\nu}(q) = -ie^2 \int \mathrm{Tr}\,[\gamma_\mu(p, p + q)\mathbf{G}(p + q)\mathbf{\Gamma}_\nu(p + q, p)\mathbf{G}(p)]\,\frac{d^4p}{(2\pi)^4} \tag{8-99}$$

If we consider only the components of $P_{\mu\nu}$ with μ and $\nu \neq 0$, and approximate the vertex function $\mathbf{\Gamma}_\nu$ by the bare vertex γ_ν, we retrieve the expression (8-38) given by the pairing approximation. The lack of gauge invariance within this approximation is a consequence of calculating $P_{\mu\nu}$ with dressed G's but bare vertex functions, thereby violating the generalized Ward's identity.

It is convenient to represent the relation (8-99) in graphical form. In Figure 8-3 we show $P_{\mu\nu}(q)$ represented in terms of the dressed electron lines and the vertex parts. While $P_{\mu\nu}$ appears

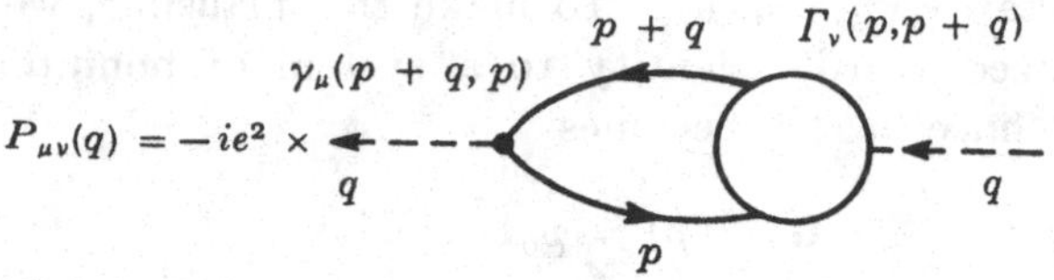

FIGURE 8-3 The polarizability kernel $P_{\mu\nu}$ represented in terms of the bare and dressed vertices γ_μ and Γ_ν, respectively.

to be unsymmetrical in μ and ν, the expression (8-99) is equally valid if $\boldsymbol{\gamma}_\mu \to \boldsymbol{\Gamma}_\mu$ and $\boldsymbol{\Gamma}_\nu \to \boldsymbol{\gamma}_\nu$.

We now prove that $K_{\mu\nu}$ is gauge-invariant if the generalized Ward's identity is satisfied. This follows by writing

$$\sum_\nu P_{\mu\nu}(q)q_\nu = -ie^2 \int \mathrm{Tr}\left[\boldsymbol{\gamma}_\mu(p, p+q)\mathbf{G}(p+q) \sum_\nu q_\nu \boldsymbol{\Gamma}_\nu(p+q, p) \times \mathbf{G}(p)\right] \frac{d^4p}{(2\pi)^4}$$

$$= ie^2 \int \mathrm{Tr}\,[\boldsymbol{\gamma}_\mu(p, p+q)\mathbf{G}(p+q)\{\mathbf{G}^{-1}(p+q)\tau_3 - \tau_3\mathbf{G}^{-1}(p)\}\mathbf{G}(p)]\frac{d^4p}{(2\pi)^4}$$

$$= ie^2 \int \mathrm{Tr}\,[\boldsymbol{\gamma}_\mu(p, p+q)\{\tau_3\mathbf{G}(p) - \mathbf{G}(p+q)\tau_3\}] \times \frac{d^4p}{(2\pi)^4} \quad (8\text{-}100)$$

In the second equality we have used Ward's identity (8-93b). Since τ_3 commutes with γ_μ, the last equality in (8-100) reduces to

$$\sum_\mu P_{\mu\nu}(q)q_\nu = ie^2 \int \mathrm{Tr}\,[\{\boldsymbol{\gamma}_\mu(p+q, p) - \boldsymbol{\gamma}_\mu(p, p-q)\} \times \tau_3\mathbf{G}(p)]\frac{d^4p}{(2\pi)^4} \quad (8\text{-}101)$$

when we use cyclic invariance of the trace. From the definition of γ_μ (8-95) we see that

$$\gamma_\mu(p + q, p) - \gamma_\mu(p, p - q) = \frac{q_\mu}{m}(1 - \delta_{\mu,0}) \qquad (8\text{-}102)$$

Therefore, we obtain the simple expression

$$\sum_\nu P_{\mu\nu}(q)q_\nu = -\frac{ne^2}{m} q_\mu(1 - \delta_{\mu,0}) \qquad (8\text{-}103)$$

where we have used the relation (7-35) giving the number of electrons per unit volume in terms of **G**. Since the right-hand side of this equation is real, (8-24) allows us to replace $P_{\mu\nu}$ in (8-103) by the physically relevant kernel $R_{\mu\nu}$ and we finally obtain

$$\sum_\nu R_{\mu\nu}(q)q_\nu = -\frac{ne^2}{m} q_\mu[1 - \delta_{\mu,0}] \qquad (8\text{-}104)$$

To check that $K_{\mu\nu}$ is manifestly gauge-invariant due to this result, we use (8-26) to write the gauge-invariance condition as

$$\sum_\nu K_{\mu\nu}(q)q_\nu = 0 = \frac{4\pi}{c^2}\sum_\nu R_{\mu\nu}(q)q_\nu + \frac{1}{\lambda_L^2} q_\mu[1 - \delta_{\mu,0}] \qquad (8\text{-}105)$$

Since $1/\lambda_L^2 = 4\pi ne^2/mc^2$, we see from (8-104) that the gauge-invariance condition is satisfied identically,

$$\sum_\nu K_{\mu\nu}(q)q_\nu = \left[-\frac{4\pi ne^2}{mc^2} + \frac{1}{\lambda_L^2}\right] q_\mu[1 - \delta_{\mu,0}] = 0 \qquad (8\text{-}106)$$

In the next section we shall discuss the generalization of the pairing scheme which is required in order that the generalized Ward's identity is to be satisfied.

8-6 THE VERTEX FUNCTION AND COLLECTIVE MODES

In seeking a gauge-invariant generalization of the pairing scheme, Nambu used a well-known prescription of quantum-field theory for constructing approximations which satisfy the generalized Ward's identity (GWI) (8-89).[137] If **G** is described

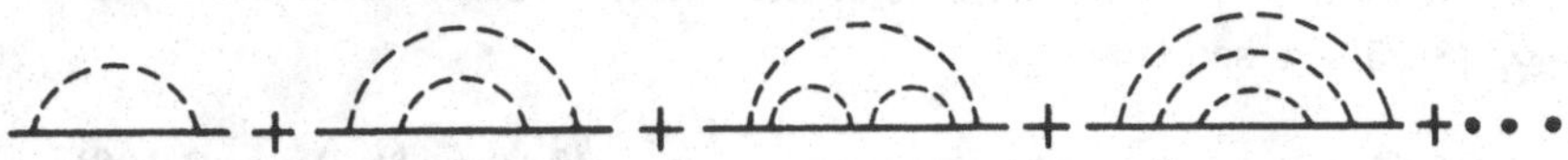

FIGURE 8-4 The pairing approximation sums all no-line-crossing graphs contributing to Σ.

by a certain set of perturbation series graphs, the corresponding vertex function $\boldsymbol{\Gamma}_\mu$ (which satisfies the GWI) is given by the sum of all graphs in which the free vertex $\boldsymbol{\gamma}_\mu$ is inserted in each bare electron line in this set. The pairing approximation for $\mathbf{G}$,

$$\mathbf{G}^{-1}(p) = p_0\mathbf{1} - \epsilon_p\tau_3 - \boldsymbol{\Sigma}(p) \tag{8-107}$$

$$\boldsymbol{\Sigma}(p) = i\int \tau_3\mathbf{G}(p')\tau_3\mathscr{V}(p - p')\,\frac{d^4p'}{(2\pi)^4} \tag{8-108}$$

can be formally thought of as the sum of all graphs in which no two interaction lines cross, as shown in Figure 8-4. If the vertex $\boldsymbol{\gamma}_\mu$ is inserted at all possible places in these graphs, the resultant series is summed by the ladder graph approximation for $\boldsymbol{\Gamma}_\mu$, as shown in Figure 8-5. Therefore $\boldsymbol{\Gamma}_\mu$ satisfies the linear integral equation

$$\begin{aligned}\boldsymbol{\Gamma}_\mu(p + q, p) &= \boldsymbol{\gamma}_\mu(p + q, p)\\ &\quad + i\int \tau_3\mathbf{G}(k + q)\boldsymbol{\Gamma}_\mu(k + q, k)\mathbf{G}(k)\tau_3\mathscr{V}(p - k)\,\frac{d^4k}{(2\pi)^4}\end{aligned} \tag{8-109}$$

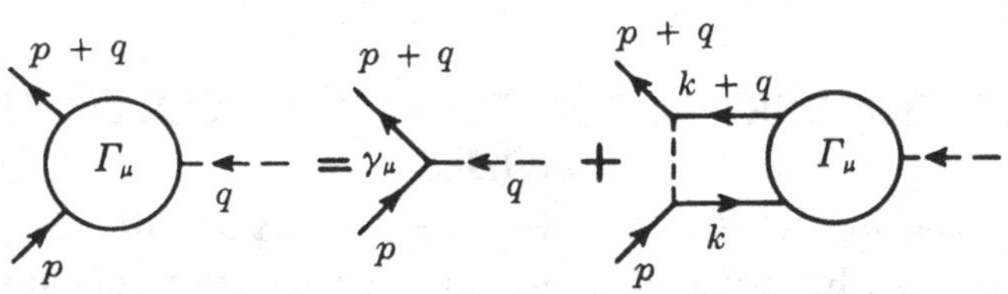

FIGURE 8-5 An equation for the vertex function $\boldsymbol{\Gamma}_\mu$ which leads to manifestly gauge-invariant results for the electromagnetic kernel within the pairing scheme.

where the $\mathbf{G}$'s are the Nambu functions evaluated within the pairing approximation (8-108). To check that the solution of this vertex equation is in fact consistent with the GWI, we form the quantity

$$\sum_\mu q_\mu \mathbf{\Gamma}_\mu(p+q,p) = \sum_\mu q_\mu \boldsymbol{\gamma}_\mu(p+q,p) + i\int \tau_3 \mathbf{G}(k+q) \sum_\mu q_\mu \mathbf{\Gamma}_\mu(k+q,k) \times \mathbf{G}(k)\tau_3 \mathscr{V}(p-k)\frac{d^4k}{(2\pi)^4} \quad (8\text{-}110)$$

The right-hand side of this equation can be reduced by use of the assumed GWI. By using the relation

$$\sum_\mu q_\mu \mathbf{\Gamma}_\mu(k+q,k) = \tau_3\mathbf{G}^{-1}(k) - \mathbf{G}^{-1}(k+q)\tau_3 \quad (8\text{-}111)$$

we see that the second term becomes

$$\left[i\int \tau_3\mathbf{G}(k+q)\tau_3\mathscr{V}(p-k)\frac{d^4k}{(2\pi)^4}\right]\tau_3 - \tau_3\left[i\int \tau_3\mathbf{G}(k)\tau_3\mathscr{V}(p-k)\frac{d^4k}{(2\pi)^4}\right] = \mathbf{\Sigma}(p+q)\tau_3 - \tau_3\mathbf{\Sigma}(p) \quad (8\text{-}112)$$

where we have used the equation determining $\mathbf{\Sigma}(p)$ (8-108). In addition, from (8-95) we see that the free vertex satisfies

$$\sum_\mu q_\mu \boldsymbol{\gamma}_\mu(p+q,p) = (\epsilon_{p+q} - \epsilon_p)\mathbf{1} - q_0\tau_3 \quad (8\text{-}113)$$

On combining these results, (8-93b) reduces to

$$\begin{aligned}\sum_\mu q_\mu \mathbf{\Gamma}_\mu(p+q,p) &= \tau_3[p_0\mathbf{1} - \epsilon_p\tau_3 - \mathbf{\Sigma}(p)] - [(p_0+q_0)\mathbf{1} - \epsilon_{p+q}\tau_3 - \mathbf{\Sigma}(p+q)]\tau_3 \\ &= \tau_3\mathbf{G}^{-1}(p) - \mathbf{G}^{-1}(p+q)\tau_3 \end{aligned} \quad (8\text{-}114)$$

which is the required GWI. Therefore, if $\mathbf{G}$ and $\mathbf{\Gamma}_\mu$ are given by solutions of Eqs. (8-107), (8-108), and (8-109), the electromagnetic kernel $K_{\mu\nu}$ determined through (8-99) will be *manifestly* gauge-invariant.

To understand the mechanism by which gauge invariance has been restored in this rather formal scheme, we again look at the

GWI. If we assume that the dressed vertex Γ_μ is a well-behaved function in the limit $\mathbf{q}$ and q_0 go to zero, the left-hand side of (8-114) vanishes while the right-hand side becomes the finite quantity

$$\tau_3\mathbf{\Sigma}(p) - \mathbf{\Sigma}(p)\tau_3 = 2i\tau_2\phi(p) = 2i\tau_2\,\Delta_p \tag{8-115}$$

(the second equality holding for a nonretarded pairing potential). Therefore, $\mathbf{\Gamma}_\mu(p + q, p)$ must be singular for $q = 0$. If one thinks of the coupling as going through a set of excited states of the system (i.e., one thinks of Γ_μ as written in a spectral form), one is tempted to argue that the $q = 0$ singularity in Γ reflects the existence of a low-lying collective mode whose frequency Ω_q goes to zero in the long-wavelength limit. To check this idea we would like to obtain an explicit solution of the vertex equation (8-109) and see if Γ_μ is actually singular for $q_0 = \Omega_q$ and $\mathbf{q} \neq 0$. As in Chapter 7, this t-matrix-like equation can be solved if $\mathscr{V}(k - p)$ is approximated by a factorizable potential

$$\mathscr{V}(k - p) = \lambda w^*(\mathbf{k})w(\mathbf{p}) \tag{8-116}$$

In solving (8-109) it is convenient to think of the 2×2 matrices $\mathbf{\Gamma}_\mu$ and $\boldsymbol{\gamma}_\mu$ as being represented by four-component column vectors. Thus we replace the matrix component $\langle l|\mathbf{\Gamma}_\mu|r\rangle$ by the column vector component $(\mathbf{\Gamma}_\mu)_{lr}$ and (8-109) becomes

$$\begin{aligned}\mathbf{\Gamma}_\mu(p + q, p) = \boldsymbol{\gamma}_\mu(p + q, p) &+ i\lambda w^*(\mathbf{p}) \int \tau_3{}^l\mathbf{G}^l(k + q)\tau_3{}^r\tilde{\mathbf{G}}^r(k) \\ &\times \mathbf{\Gamma}_\mu(k + q, k)w(\mathbf{k})\,\frac{d^4k}{(2\pi)^4}\end{aligned} \tag{8-117}$$

The scripts l and r indicate which part of the double script (lr) the matrices act and $\tilde{\mathbf{G}}$ means the transpose of $\mathbf{G}$. Since the last term in this expression is a constant (matrix) multiple of $w^*(\mathbf{p})$, we have

$$\mathbf{\Gamma}_\mu(p + q, p) = \boldsymbol{\gamma}_\mu(p + q, p) + w^*(\mathbf{p})C_q \tag{8-118}$$

where the constant C_q is defined by

$$C_q = i\lambda \int \tau_3{}^l\mathbf{G}^l(k + q)\tau_3{}^r\tilde{\mathbf{G}}^r(k)\mathbf{\Gamma}_\mu(k + q, k)w(\mathbf{k})\,\frac{d^4k}{(2\pi)^4} \tag{8-119}$$

By inserting the relation (8-118) into (8-119) and solving for C_q, one finds the explicit solution

$$\mathbf{\Gamma}_\mu(p + q, p) = \boldsymbol{\gamma}_\mu(p + q, p) + [1 - \lambda\boldsymbol{\phi}(q)]^{-1}\boldsymbol{\chi}_\mu(q)w^*(\mathbf{p}) \quad (8\text{-}120)$$

where $\boldsymbol{\phi}(q)$ is a generalization of the function $\tilde{\Phi}(q)$ [see (5-21)] used in discussing the instability of the normal state. The matrix function $\boldsymbol{\phi}(q)$ is given by

$$\boldsymbol{\phi}(q) = i \int \tau_3{}^l\mathbf{G}^l(k + q)\tau_3{}^r\mathbf{G}^r(k)|w(\mathbf{k})|^2 \frac{d^4k}{(2\pi)^4} \quad (8\text{-}121)$$

The matrix $\boldsymbol{\chi}(q)$ is defined by

$$\boldsymbol{\chi}_\mu(q) = i\lambda \int \tau_3{}^l\mathbf{G}^l(k + q)\tau_3{}^r\mathbf{G}^r(k)\boldsymbol{\gamma}_\mu(k + q, k)w(k) \frac{d^4k}{(2\pi)^4} \quad (8\text{-}122)$$

Since $\boldsymbol{\chi}_\mu$ is regular for $q_0 < 2\Delta_0$, the singularity of $\mathbf{\Gamma}_\mu$ must arise from a singularity of $[1 - \lambda\boldsymbol{\phi}(q)]^{-1}$. For this matrix to be singular, the determinant of its inverse must be zero. Therefore, the dispersion law for the collective mode (or modes) is

$$\det [1 - \lambda\boldsymbol{\phi}(\mathbf{q}, \Omega_\mathbf{q})] = 0 \quad (8\text{-}123)$$

where $\Omega_\mathbf{q}$ is the frequency of the mode in question. If we assume s-state pairing and as in Chapter 7 take $w(\mathbf{k})$ to be

$$w(k) = \begin{cases} 1 & |\epsilon_k| < \omega_c \\ 0 & \text{otherwise} \end{cases} \quad (8\text{-}124)$$

there is a root of (8-123) which for $|\mathbf{q}|\xi_0 \ll 1$, satisfies the dispersion law

$$\Omega_\mathbf{q} = \frac{v_F}{(3)^{1/2}} |\mathbf{q}| \quad (8\text{-}125)$$

This sound-wave mode was first discovered by Bogoliubov.[52] Physically, it corresponds to long wavelength density fluctuations of the electron system as a whole. Since the pairing correlations are not expected to be changed appreciably by slowly varying the electron density in space and time, one might expect such a collective mode on physical grounds. In fact (8-125) follows if one uses the standard hydrodynamic expression for the speed of sound s:

$$s^2 = \frac{dP}{d\rho} \quad (8\text{-}126)$$

where ρ and P are the mass density and pressure of the *free*-electron gas. Therefore within this approximation pairing correlations play no direct role in determining the velocity of this mode, their main function being to remove low-lying single-particle states which would otherwise lead to damping of the wave.

Returning to the solution (8-120) for the vertex function, if the two-body potential is purely s-wave, the transverse part of χ_μ vanishes, as one can easily see on symmetry grounds from (8-122). Therefore vertex corrections *do not affect* the Meissner kernel in this case. If there is a finite attractive d-wave part of the potential, d-state excitons exist and will contribute to the vertex function. Calculations by Rickayzen show that these collective corrections to the Meissner kernel are in general small.

If there is a strongly attractive d-wave potential one should see a precursor for infrared absorption below the gap edge, owing to creation of d-state excitons.[138, 139] Such anomalies were first observed by Ginsberg, Richards, and Tinkham,[71, 72] although as Tsuneto[138] has shown, the predicted absorption is an order of

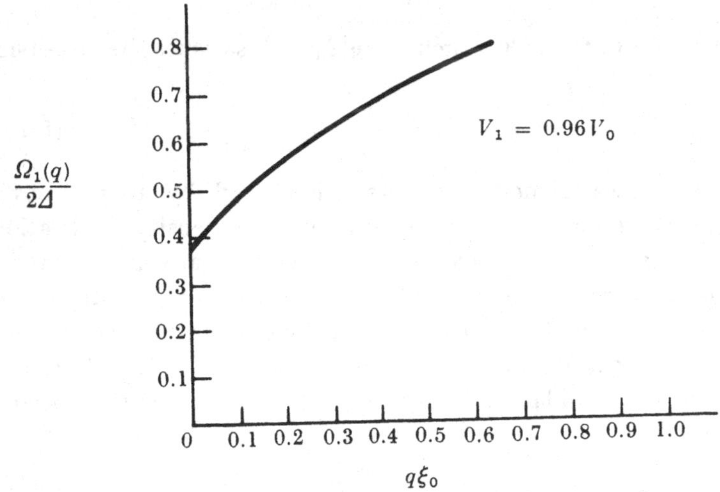

FIGURE 8-6 Energy of a p-state exciton as a function of momentum.

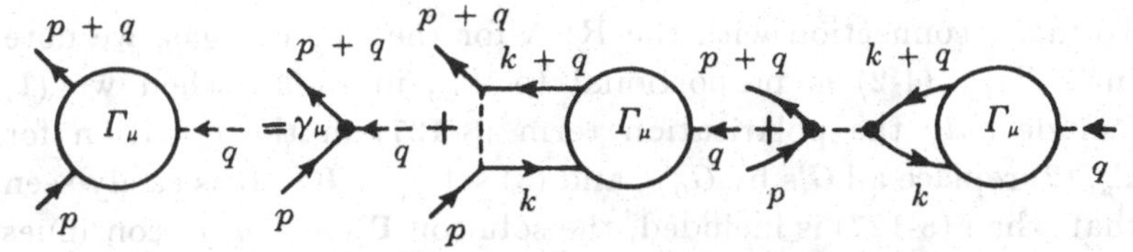

FIGURE 8-7 Vacuum-polarization correction to the equation shown in Figure 8-5. This correction is all important in the longitudinal part of $P_{\mu\nu}$ and leads to plasma oscillations.

magnitude weaker than that observed experimentally. Since the calculations were carried out on a continuum model, momentum was conserved in the absorption process. Owing to the fact that the excitons pass into the single-particle continuum for $q \gtrsim 1/\xi_0$, as shown in Figure 8-6, only the long wavelength components of the penetrating field contribute to exciton creation, and these small q-components are small compared to those for $q \sim 1/\lambda \gg 1/\xi_0$, thereby giving a small absorption. Experimentally, the precursor is essentially unaffected by impurities.[130b] Since the exciton state is destroyed by impurities,[130c] it appears **that the precursor is due to another mechanism. [Subsequent experiments have shown the precursor absorption to be spurious].**

Thus far we have neglected vacuum-polarization processes in the vertex function $\mathbf{\Gamma}_\mu$. As we saw in Chapter 6, these processes dominate the long wavelength polarizability of the electron gas because of the long-range Coulomb potential and therefore *must* be included in $\mathbf{\Gamma}$. The vertex equation including vacuum polarization processes is illustrated in Figure 8-7. The only change in (8-109) is to add the term

$$-iV_B(q)\int \mathrm{Tr}\,\{\tau_3\mathbf{G}(k+q)\mathbf{\Gamma}_\mu(k+q,k)\mathbf{G}(k)\}\frac{d^4k}{(2\pi)^4} \quad (8\text{-}127)$$

to the right-hand side. Here $V_B(q)$ is the sum of the bare Coulomb and bare longitudinal phonon interactions:

$$V_B(q) = \frac{4\pi e^2}{q^2} + |g_{ql}|^2 D_{0l}(q) \quad (8\text{-}128)$$

To make connection with the RPA for the electron gas, we note that P_{RPA} (4-2) is proportional to P_{00} in (8-99) when we (1) include *only* the polarization term (8-127) in the equation for $\mathbf{\Gamma}_\mu$, (2) replace all G's by G_0's, and (3) set $g_q = 0$. It is easily seen that when (8-127) is included, the solution $\mathbf{\Gamma}_\mu(p + q, p)$ continues to satisfy the GWI and $K_{\mu\nu}$ is still manifestly gauge-invariant. By explicitly solving this improved vertex equation, one can see that the Bogoliubov sound-wave mode continues to exist if $V_B(q)$ approaches a finite value as $\mathbf{q} \to 0$, as Anderson[47] first showed. In the presence of the Coulomb potential, which of course always exists in real metals, the Bogoliubov-Anderson mode is pushed up to high energy and becomes the plasma oscillation of the electron system. Therefore, the $q \equiv 0$ singularity of $\mathbf{\Gamma}_\mu(p + q, p)$ required by (8-115) *does not* imply a low-lying boson mode in physical metals due to the long-range Coulomb interaction between electrons.

8-7 FLUX QUANTIZATION

A qualitatively new effect arises when we investigate the electromagnetic behavior of a multiply connected superconducting system, e.g., a long, hollow cylinder. In this case, flux can be trapped in the hole and persist in the absence of an externally applied field. On the basis of London's "rigidity" concept he concluded that for a cylinder with walls thick compared to the penetration depth λ, flux could be trapped only in multiples of $hc/e = 4 \times 10^{-7}$ gauss cm^2.[1] This value of the flux quantum follows if one assumes that the *only* low-lying current-carrying states of the superfluid are those given by multiplying the superfluid ground state by a single-valued phase factor, as we saw in Chapter 1. We shall see below that there are *two* distinct sets of low-lying states, one being the set considered by London, the other arising from phase factors multiplying a basic state which is not included in London's set. Owing to these two sets of states, the flux quantum in superconductors is actually $hc/2e$, i.e., one-half the London unit. The even multiples are associated with London-type states, while the odd multiples are due to the other series of states, as Byers and Yang[19] first pointed out. The value $hc/2e$ was observed experimentally by Deaver and Fairbank[20a] and by Doll and Näbauer[20b] prior to Byers and Yang's work.

Perhaps the simplest way of understanding flux quantization is by considering a long, hollow cylinder of inner and outer radii a and b, respectively. Suppose the cylinder is initially in the normal state in the presence of a magnetic field and the temperature is lowered so that the cylinder becomes superconducting. Owing to the Meissner effect, the magnetic field will be expelled from the material and in general there will be a finite magnetic flux Φ trapped in the hole. If we assume the wall thickness $b - a$ is much greater than the penetration depth, the magnetic field in the penetration layer will be a small perturbation on the system as a whole and cannot affect our results. If we use cylindrical coordinates (r, θ, z), the vector potential is given by

$$\oint \mathbf{A}(\mathbf{r}') \cdot d\mathbf{r}' = \int \mathbf{B} \cdot d\mathbf{S} = \Phi(r) \tag{8-129}$$

where the line integral is taken around a circle of radius r and $\Phi(r)$ is the flux enclosed by the path. Since $\Phi(r)$ goes to a constant Φ (the total trapped flux) for $r - a \gg \lambda$, we write

$$A_\theta(r) = \frac{\Phi}{2\pi r} + \frac{\Phi(r) - \Phi}{2\pi r} \equiv A_\theta^{(0)}(r) + A_\theta^{(1)}(r) \tag{8-130}$$

and only include $A_\theta^{(0)}(r)$ in zero order, $A^{(1)}$ being treated as a perturbation. We first consider the single-particle states defined in the presence of $A^{(0)}$ and then pair up these states to form the superconducting phase. The azimuthal part of the single-particle eigenfunctions satisfy

$$\frac{\left[p_\theta + \dfrac{eA_\theta^{(0)}}{c}\right]^2}{2m_e}\psi_M(\theta) = \frac{\hbar^2}{2m_e r^2}(M + \varphi)^2\psi_M(\theta) \tag{8-131a}$$

where

$$\varphi \equiv \frac{e\Phi}{hc} \tag{8-131b}$$

is the flux measured in units of London's flux quantum and

$$\psi_M(\theta) = e^{iM\theta} \tag{8-131c}$$

In order that Ψ_M be single-valued, M must be an integer. The dynamics is simplified if we assume the thickness of the cylinder $b - a$ is small compared to the radius a. The angular kinetic energy is then $\hbar^2(M + \varphi)^2/2m_e a^2$, i.e., a parabolic function of M, centered about $M = -\varphi$. If we are to obtain a low-energy state of the system we must pair single-particle states which are (a) degenerate with each other and (b) coupled to other paired states by the two-body potential. Condition (a) means that the paired states M and $\bar{M}$ must satisfy

$$|M + \varphi| = |\bar{M} + \varphi| \tag{8-132}$$

This condition can be satisfied if $M = \bar{M}$ but conservation of angular momentum forbids the two-body potential from connecting states paired in this manner. The other choice is $M + \varphi = -(\bar{M} + \varphi)$, that is, $M \equiv m - \varphi$ and $\bar{M} = -m - \varphi$ are paired, so that the pairing is symmetric in M-space about the value $M = -\varphi$. Since M and $\bar{M}$ are required to be integers, it follows that m and φ are both integers or both half-odd integers. Therefore, we conclude from this result and (8-131b) that one obtains a large pairing energy and therefore a low-energy state of the system only if the trapped flux is given by

$$\Phi = n\left(\frac{hc}{2e}\right) \tag{8-133}$$

where n is an integer. In Figure 8-8a, b, c, and d we illustrate the pairing for $n = 0$, 1, 2, and 3, respectively. Geometrically, one simply pairs symmetrically about the value $M = -n/2$. Notice that the states being paired for $n = 0$ and $n = 2$ (Figure 8-8a and c) differ only by a shift of all the angular momentum quantum numbers by the fixed amount -1. Thus, the system wave functions for the $n = 0$ and $n = 2$ states are related by the phase factor

$$\psi_2(\mathbf{r}_1, \mathbf{r}_2 \cdots \mathbf{r}_N) = e^{-i\sum_j \theta_j}\psi_0(\mathbf{r}_1, \mathbf{r}_2 \cdots \mathbf{r}_N) \tag{8-134}$$

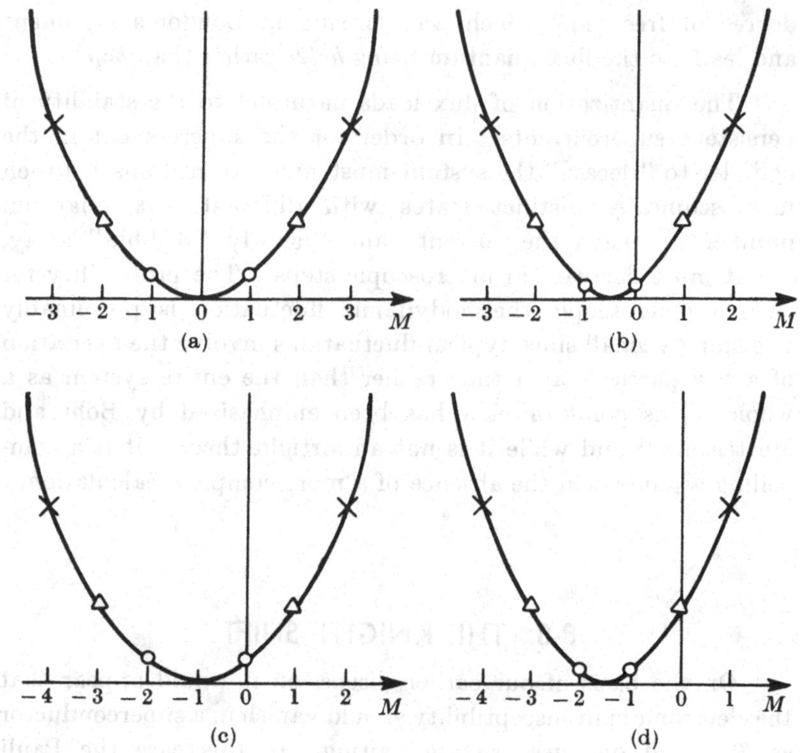

FIGURE 8-8 Pairings of azimuthal quantum numbers for flux quantum numbers $n = 0, 1, 2, 3$ are shown in (a), (b), (c), and (d), respectively [see Eq. (8-133)].

in agreement with London's argument. The states for $n = 1$ and $n = 3$ are related in the same manner:

$$\psi_3(\mathbf{r}_1, \mathbf{r}_2 \cdots \mathbf{r}_N) = e^{-i \sum_j \theta_j} \psi_1(\mathbf{r}_1, \mathbf{r}_2 \cdots \mathbf{r}_N) \qquad (8\text{-}135)$$

The essential point is that the even n- and odd n-states are *not* related by a phase factor. For example, the $n = 0$ and $n = 1$ states differ by sliding *only* the mates to the right of $M = 0$ left one notch in forming the $n = 1$ state from $n = 0$. This cannot be done by a phase factor, yet the states differ little in energy, contrary to London's assumption. It is this "extra degree of freedom" which was missing in London's argument and leads to the flux quantum being $hc/2e$ rather than hc/e.

The quantization of flux leads naturally to the stability of persistent supercurrents. In order for the supercurrent in the cylinder to "decay," the system must make transitions between macroscopically distinct states with different flux quantum number n. Thus the current cannot slowly "dribble" away, but it must decrease in macroscopic steps. The probability for such a macroscopic thermodynamic fluctuation is presumably vanishingly small since typical fluctuations involve the excitation of a few particles at a time rather than the entire system as a whole. This point of view has been emphasized by Bohr and Mottelson,[141] and while it is not an airtight theory, it is a compelling argument in the absence of a more complete calculation.

8-8 THE KNIGHT SHIFT

On the basis of our earlier discussion it would appear that the electronic spin susceptibility should vanish in a superconductor as $T \to 0$ if one uses s-state pairing. In this case the Pauli principle ensures that the spins are paired in the singlet state so that a finite spin magnetization can result only if the spin Zeeman energy $2\mu_B H$ is greater than $2\Delta_0$, the minimum energy to break up a pair. A means of checking this prediction is the Knight shift[142] (the change in nuclear magnetic resonance frequency due to the coupling of the nuclear spins with the polarization of the electrons). If only the electronic *spin* polarization (as opposed to orbital effects) is important in the shift, the Knight shift gives a measure of the electronic spin susceptibility. Reif[143] found that the shift in superconducting mercury, when extrapolated to $0°K$ was about two-thirds of the value in the normal state, contrary to the simple pairing theory. Androes and Knight[144] found similar results in tin, while the shift in vanadium is nearly the same in the N- and S-phases.

There have been several attempts to explain this discrepancy, none of which is widely accepted at present as providing the essential mechanism. It may well be that a fraction of the observed shift comes from several of the following mechanisms.

Parallel Spin Pairing

If one uses p-state (or any odd l-state) pairing, the antisymmetry of the wave-function forces the spins to be paired in triplet states. Fisher[59a] studied the model in which one pairs states with equal z-components of spin (i.e., up paired with up, down with down). The reduced Hamiltonian is then the sum of two noninteracting parts. In this case there is no energy gap for creating spin polarization since a down-spin pair can be changed into an up-spin pair with no change of energy. Unfortunately, this type of pairing leads to an anisotropic energy gap which vanishes in certain directions and a nonexponential electronic specific heat at low temperature. A more general triplet pairing has been treated by Balian and Werthamer[59c], who include all three components of the triplet state. In a pure material they obtain an isotropic energy gap; however, a small amount of disorder or impurity (which was certainly present in the above experiments) destroys their state as well as that of Fisher.

Spin-Orbit Coupling at the Surface

Since the Meissner effect screens out the magnetic field within a distance $\lambda \sim 5 \times 10^{-6}$ cm of the surface, the Knight shift experiments are carried out on specimens whose dimensions are small compared to λ in order to eliminate line broadening due to the field inhomogeniety. The smallness of the particles led Ferrell[145] to suggest that the spin-orbit coupling near the surface might be sufficiently strong to mix the one-electron spin states during scattering and thereby lead to a finite spin susceptibility. A semiquantitative theory of this effect was worked out by Ferrell and by Anderson,[146] however, we shall not discuss the calculations here. Unfortunately, as mentioned above, recent measurements on vanadium and aluminum,[147] which are light metals and should have appreciably weaker spin-orbit coupling than mercury and tin, give essentially the *same* Knight shift in the normal and superconducting states. This rules out the spin-orbit mechanism, at least in this case.

Collective Magnetization

In the original BCS paper[8] it was suggested that there might be low-lying collective spin wave states in a superconductor which give rise to the observed Knight shift. Bardasis and Schrieffer[139] found that by retaining the p-wave part of the two-body potential, spin wave states can exist in the energy gap. However, to obtain a nonzero long wavelength susceptibility the spectrum must go down to zero energy in this limit. These authors found that two situations can exist. If the p-wave part of the two-body potential is weaker than the s-wave part, the spin waves possess a finite energy as their momentum goes to zero. If the p-wave potential is stronger than the s-wave potential, the spin wave states are unstable and the ground state is then formed by p-state pairing. One is then led back to the difficulties of the first proposal.

Modified Antiparallel Spin Pairing

Soon after the BCS theory was proposed, Heine and Pippard[148] suggested that one might relax the strict BCS rule of pairing a given state $\mathbf{k}$ with only one other state $\bar{\mathbf{k}}$. They argued that if $\mathbf{k}$ is paired with a *group* of states centered about $\bar{\mathbf{k}}$, a finite spin susceptibility might result. Their argument was based on an assumed form of the two-particle density matrix which is not consistent with that given by the most general form of the pairing theory. Since it has not been possible to construct a wavefunction which gives their density matrix, it appears that their basic assumption cannot be justified and we are forced to reject this point of view.

Subsequent to Heine and Pippard's proposal, Schrieffer[149] argued that a finite spin susceptibility would be obtained if the pairing condition were modified in the magnetized state. In particular, he suggested pairing states $\mathbf{k}\uparrow$ and $-\mathbf{k}'\downarrow$, which are degenerate when the spin Zeeman energy is included, rather than the BCS prescription of pairing degenerate orbital states $\mathbf{k}\uparrow$ and $-\mathbf{k}\downarrow$. This scheme then leads to a net spin magnetization. In a pure unbounded specimen the modified pairing gives essentially no pairing energy since momentum conservation forbids a pair with center-of-mass momentum $\mathbf{k} - \mathbf{k}'$ from scattering into a pair state with center-of-mass momentum $\tilde{\mathbf{k}} - \tilde{\mathbf{k}}'$, owing to these momenta being different in general. Under experimental conditions, impurity and surface scattering is sufficiently strong to spread the single-particle states "$\mathbf{k}$" in momentum space by an amount large compared to the "center-of-mass" momentum $\hbar|\mathbf{k} - \mathbf{k}'| \sim 2\mu_B H/v_F$. Therefore, the actual single-particle eigenstates (including these one-particle scattering effects) can be formed into pairs in the above manner and coupled by the two-body potential. The pairing still occurs between *two definite* single-particle states. It is reasonable to assume that the reduction in the pairing energy will be a smoothly varying function of the net spin magnetic moment and therefore a finite spin susceptibility will result. This scheme is very different in point of view from that proposed by Heine and Pippard, who argue that a *group* of single-particle states are strongly correlated in occupancy (even in a pure, unbounded specimen).

Recently Cooper[150] has reinvestigated Schrieffer's idea by introducing a phenomenological two-body potential whose matrix elements are taken to be a slowly varying function of the center-of-mass momentum of *each* pair. This leads to a pairing energy which varies slowly as a function of the spin magnetization and gives a finite spin paramagnetism. Cooper stressed the added possibility of momentum nonconservation being due to a nontranslationally invariant two-body potential.

Orbital Paramagnetism

Clogston, Gossard, Jaccarino, and Yafet[151] have argued convincingly that in vanadium essentially *all* of the Knight shift is due to Kubo–Obata temperature-independent orbital paramagnetism.[152] Since this orbital paramagnetism should be the same in the N- and S-states, the lack of change in the Knight shift in vanadium is presumably explained without modifying the pairing theory. It appears unlikely that this mechanism can account for the observed shift in all cases.

8-9 THE GINSBURG–LANDAU–GOR'KOV THEORY

Thus far we have concentrated on the response of a superconductor to weak electromagnetic fields. There are many important problems, e.g., N-S phase boundary, the intermediate and mixed states, etc., in which the magnetic field enters in a nonperturbative manner. These problems typically involve the energy-gap parameter Δ varying as a function of position in the materials. As we discussed in Chapter 1, the phenomenological theory of Ginsburg and Landau[36] (proposed in 1950) gives in many instances a good account of these strong field situations. An important advance in the microscopic theory was made by Gor'kov,[37] who showed how the GL equations follow from the pairing theory when T is near T_c and the magnetic field varies slowly in space over a coherence length. Gor'kov found that the GL effective wave function $\Psi(\mathbf{r})$ is proportional to the local value of the gap parameter $\Delta(\mathbf{r})$ and the effective charge e^* of the GL theory is equal to $2e$, the charge of a pair of electrons. It is interesting to note that these results were guessed prior to Gor'kov's work, the identification of $\Psi(\mathbf{r})$ and $\Delta(\mathbf{r})$ being suggested by Bardeen[8] and the effective charge $e^* = 2e$ being suggested by Ginsburg[153] prior to the BCS theory in order to fit the GL theory with experiment.

We give a brief summary of Gor'kov's derivation below. To familiarize the reader with Gor'kov's scheme, we use his notation. For simplicity Gor'kov used a nonretarded zero-range attractive potential to describe the pairing interactions. Since this singular potential leads to divergences, it is cut off in momentum space at the appropriate point in the derivation. The vector potential $\mathbf{A}(\mathbf{r})$ is treated self-consistently as we did above in the weak-field case. The system Hamiltonian is then

$$H = -\sum_s \int \psi_s^+(\mathbf{r})\left\{\frac{1}{2m}\left[\nabla - \frac{ie}{c}\mathbf{A}(\mathbf{r})\right]^2 + \mu\right\}\psi_s(\mathbf{r})\, d^3r - v\int \psi_\uparrow^+(\mathbf{r})\psi_\uparrow(\mathbf{r})\psi_\downarrow^+(\mathbf{r})\psi_\downarrow(\mathbf{r})\, d^3r \quad (8\text{-}136)$$

where $e = -|e|$ is the charge of an electron, and we measure single-particle energies relative to the chemical potential μ. The thermodynamic Green's function

$$G(x, x') = -\frac{\operatorname{Tr}\left[e^{-\beta H}T\{\psi_\uparrow(x)\psi_\uparrow^+(x')\}\right]}{\operatorname{Tr} e^{-\beta H}} \equiv -\langle T\{\psi_\uparrow(x)\psi_\uparrow^+(x')\}\rangle \quad (8\text{-}137\text{a})$$

is defined for pure imaginary time

$$\psi(x) \equiv \psi(r, \tau) = e^{H\tau}\psi(\mathbf{r}, 0)e^{-H\tau} \quad (8\text{-}137\text{b})$$

and satisfies the equation of motion

$$\left\{-\frac{\partial}{\partial\tau} + \frac{1}{2m}\left[\nabla - \frac{ie}{c}\mathbf{A}(\mathbf{r})\right]^2 + \mu\right\}G(x, x') + V\langle T\{\psi_\uparrow^+(x')\psi_\downarrow^+(x)\psi_\downarrow(x)\psi_\uparrow(x)\}\rangle = \delta(x - x') \quad (8\text{-}138)$$

This result follows from (8-136) and (8-137) since

$$\frac{\partial\psi_\uparrow(x)}{\partial\tau} = [H, \psi(x)] = \left\{\frac{1}{2m}\left[\nabla - \frac{ie}{c}\mathbf{A}(\mathbf{r})\right]^2 + \mu\right\}\psi_\uparrow(x) + V\psi_\downarrow^+(x)\psi_\downarrow(x)\psi_\uparrow(x) \quad (8\text{-}139)$$

In Gor'kov's scheme, the pairing approximation for this zero-range potential corresponds to factorizing the four-point function in (8-138),

$$\langle T\{\psi_\uparrow^+(x')\psi_\downarrow^+(x)\psi_\downarrow(x)\psi_\uparrow(x)\}\rangle \Rightarrow \langle T\{\psi_\uparrow^+(x')\psi_\downarrow^+(x)\}\rangle\langle\psi_\downarrow(x)\psi_\uparrow(x)\rangle \quad (8\text{-}140)$$

Therefore the equation for G becomes

$$\left\{-\frac{\partial}{\partial\tau} + \frac{1}{2m}\left[\nabla - \frac{ie}{c}\mathbf{A}(\mathbf{r})\right]^2 + \mu\right\}G(x, x') + \Delta(\mathbf{r})F^+(x, x') = \delta(x - x') \quad (8\text{-}141)$$

where the "anomalous" Green's function $F^+(x, x')$ is defined by

$$F^+(x, x') = -\langle T\{\psi_\downarrow^+(x)\psi_\uparrow^+(x')\}\rangle \qquad (8\text{-}142)$$

and the energy-gap parameter $\Delta(\mathbf{r})$ is given by

$$\Delta^*(\mathbf{r}) = V\langle\psi_\downarrow(\mathbf{r})\psi_\uparrow(\mathbf{r})\rangle^* = VF^+(x, x') \qquad (8\text{-}143)$$

The functions G and F^+ correspond to Nambu's G_{11} and G_{21}. Since F^+ is an unknown function, it must be determined from its equation of motion. By making a factorization similar to that in (8-140), (except that the four-point function now contains three ψ^+'s and one ψ), one finds

$$\left\{\frac{\partial}{\partial\tau} + \frac{1}{2m}\left[\nabla + \frac{ie}{c}\mathbf{A}(\mathbf{r})\right]^2 + \mu\right\}F^+(x, x') - \Delta^*(\mathbf{r})G(x, x') = 0 \qquad (8\text{-}144)$$

As we discussed in Chapter 7, the pure imaginary time Green's functions can be expressed in the Fourier series variable $\omega_n = (2n + 1)\pi/\beta$ (n = integer) and one finds the Fourier components $\mathscr{G}_\omega(\mathbf{r}, \mathbf{r}')$ and $\mathscr{F}_\omega(\mathbf{r}, \mathbf{r}')$ satisfy

$$\left\{i\omega_n + \frac{1}{2m}\left[\nabla - \frac{ie}{c}\mathbf{A}(\mathbf{r})\right]^2 + \mu\right\}\mathscr{G}_\omega(\mathbf{r}, \mathbf{r}') + \Delta(\mathbf{r})\mathscr{F}_\omega(\mathbf{r}, \mathbf{r}') = \delta(\mathbf{r} - \mathbf{r}')$$

$$\left\{-i\omega_n + \frac{1}{2m}\left[\nabla + \frac{ie}{c}\mathbf{A}(\mathbf{r})\right]^2 + \mu\right\}\mathscr{F}_\omega(\mathbf{r}, \mathbf{r}') - \Delta^*(\mathbf{r})\mathscr{G}_\omega(\mathbf{r}, \mathbf{r}') = 0 \qquad (8\text{-}145)$$

These equations together with the condition (8-143)

$$\Delta^*(\mathbf{r}) = \frac{V}{\beta}\sum_n \mathscr{F}_\omega^+(\mathbf{r}, \mathbf{r}') \qquad (8\text{-}146)$$

in principle determine the behavior of the superconductor in the presence of an arbitrarily strong potential A at any temperature $k_BT = 1/\beta$.

The nonlinearity of the coupled equations makes them difficult to handle. Gor'kov restricted his attention to the region T near T_c, where the gap parameter is small so that a perturbation expansion in powers of Δ can be carried out (in the spirit of the GL theory). Furthermore, the penetration depth λ becomes large compared to Pippard's coherence length ξ_0 for $T \sim T_c$ and therefore **A** will vary slowly over a coherence length. In this limit the linear relation between the current density and the vector potential reduces to London's equation. To carry out the series solution of (8-145) in powers of Δ, Gor'kov wrote these equations in integral form

$$\mathscr{G}_\omega(\mathbf{r}, \mathbf{r}') = \tilde{\mathscr{G}}_\omega(\mathbf{r}, \mathbf{r}') - \int \tilde{\mathscr{G}}_\omega(\mathbf{r}, \mathbf{s})\, \Delta(\mathbf{s}) \mathscr{F}_\omega{}^+(\mathbf{s}, \mathbf{r}')\, d^3s \quad (8\text{-}147a)$$

$$\mathscr{F}_\omega{}^+(\mathbf{r}, \mathbf{r}') = \int \mathscr{G}_\omega(\mathbf{s}, \mathbf{r}')\, \Delta^*(\mathbf{s}) \tilde{\mathscr{G}}_{-\omega}(\mathbf{s}, \mathbf{r})\, d^3s \quad (8\text{-}147b)$$

where $\tilde{\mathscr{G}}$ is the Green's function for an electron in the normal metal in the presence of the magnetic field

$$\left\{i\omega_n + \frac{1}{2m}\left[\nabla - \frac{ie}{c}\mathbf{A}(\mathbf{r})\right]^2 + \mu\right\}\tilde{\mathscr{G}}_\omega(\mathbf{r}, \mathbf{r}') = \delta(\mathbf{r} - \mathbf{r}') \quad (8\text{-}148)$$

If one solves the equations for F^+, accurate to terms of order Δ^4, one finds that the equation (8-146) determining Δ is

$$\begin{aligned}\Delta^*(\mathbf{r}) = \frac{V}{\beta}\sum_n \int \tilde{\mathscr{G}}_\omega(\mathbf{r}, \mathbf{r}')\tilde{\mathscr{G}}_{-\omega}(\mathbf{r}, \mathbf{r}')\, \Delta^*(\mathbf{r}')\, d^3r' \\ - \frac{V}{\beta}\sum_n \int \tilde{\mathscr{G}}_\omega(\mathbf{s}, \mathbf{r})\tilde{\mathscr{G}}_{-\omega}(\mathbf{s}, \mathbf{l})\tilde{\mathscr{G}}_\omega(\mathbf{m}, \mathbf{l})\tilde{\mathscr{G}}_{-\omega}(\mathbf{m}, \mathbf{r})\, \Delta(\mathbf{s}) \\ \times \Delta^*(\mathbf{l})\, \Delta^*(\mathbf{m})\, d^3s\, d^3l\, d^3m \quad (8\text{-}149)\end{aligned}$$

The first term on the right-hand side of the form

$$\int K(\mathbf{r}, \mathbf{r}')\, \Delta^*(\mathbf{r}')\, d^3r \quad (8\text{-}150a)$$

where the kernel $K(\mathbf{r}, \mathbf{r}')$

$$K(\mathbf{r}, \mathbf{r}') = \frac{1}{\beta}\sum_n \tilde{\mathscr{G}}_\omega(\mathbf{r}, \mathbf{r}')\tilde{\mathscr{G}}_{-\omega}(\mathbf{r}, \mathbf{r}') \quad (8\text{-}150b)$$

is given by

$$K_0(\mathbf{r} - \mathbf{r}') = K_0(R) = \left[\frac{m}{2\pi R}\right]^2 \frac{1}{\beta} \sinh\left(\frac{2\pi R}{\beta v_F}\right) \quad (8\text{-}150c)$$

and

$$\tilde{\mathscr{G}}_\omega{}^0(\mathbf{r} - \mathbf{r}') = -\frac{m}{2\pi R}\exp\left[ip_F R \operatorname{sgn}\omega_n - \frac{|\omega_n|}{v_F}R\right] \quad (8\text{-}150d)$$

if $\mathbf{A} = 0$. Since $\mathbf{A}$ is assumed to vary slowly over a coherence length and $\tilde{\mathscr{G}}_\omega(\mathbf{r}, \mathbf{r}')$ decreases exponentially for $|\mathbf{r} - \mathbf{r}'| > v_F/\omega \simeq \xi_0$ ($\omega \simeq \Delta$), a WKB-like approximation can be used to include $\mathbf{A}$ and one finds

$$K(\mathbf{r}, \mathbf{r}') = K_0(\mathbf{r} - \mathbf{r}') \exp\left[2\frac{ie}{c}(\mathbf{r} - \mathbf{r}')\cdot\mathbf{A}(\mathbf{r})\right] \tag{8-150e}$$

and

$$\tilde{\mathscr{G}}_\omega(\mathbf{r}, \mathbf{r}') = \tilde{\mathscr{G}}_\omega{}^0(\mathbf{r} - \mathbf{r}') \exp\left[\frac{ie}{c}(\mathbf{r} - \mathbf{r}')\cdot\mathbf{A}(\mathbf{r})\right] \tag{8-150f}$$

The singularity of K_0 as $\mathbf{r} \to \mathbf{r}'$ arises from the zero-range two-body potential. If one cuts off the potential $V_{kk'}$ outside the energy range $-\omega_0 \to \omega_0$ (centered about the Fermi surface), one has

$$\begin{aligned}\int K_0(R)\,dR &= N(0)\int_0^{\omega_0}\frac{1}{\epsilon}\tanh\left(\frac{\beta\epsilon}{2}\right)d\epsilon \\ &= N(0)\left[\int_0^{\omega}\frac{1}{\epsilon}\tanh\left(\frac{\beta_c\epsilon}{2}\right)d\epsilon + \int_{\beta_c\omega_0}^{\beta\omega_0}\frac{\tanh x}{x}\,dx\right] \\ &= N(0)\left[\frac{1}{N(0)V} + \ln\left(\frac{T_c}{T}\right)\right]\end{aligned} \tag{8-151}$$

In the reduction we have used the equation determining $k_BT_c = 1/\beta_c$. By expanding the normal metal Green's functions in powers of the small quantity $(e/c)(\mathbf{r} - \mathbf{r}')\cdot\mathbf{A}(\mathbf{r})$ and assuming that $\Delta(\mathbf{r})$ varies slowly over a coherence length, Gor'kov obtains the equation

$$\left\{\frac{1}{2m}\left[\nabla + i\frac{2e}{c}\mathbf{A}(\mathbf{r})\right]^2 + \frac{1}{\lambda_G}\left[\left(1 - \frac{T}{T_c}\right) - \frac{7\zeta(3)}{8(\pi k_BT_c)^2}|\Delta(\mathbf{r})|^2\right]\right\}\Delta^*(\mathbf{r}) = 0 \tag{8-152}$$

where Gor'kov's parameter λ_G is

$$\lambda_G = \frac{7\zeta(3)E_F}{12(\pi k_BT_c)^2} \tag{8-153}$$

and $\zeta(x)$ is the Riemann zeta function.

By introducing the "wave function"

$$\psi(\mathbf{r}) = \frac{\Delta(\mathbf{r})[7\zeta(3)n]^{1/2}}{4\pi T_c} \tag{8-154}$$

one obtains the Ginsburg–Landau-like equation

$$\left\{\frac{1}{2m}\left[\nabla - \frac{ie^*}{c}\mathbf{A}(\mathbf{r})\right]^2 + \frac{1}{\lambda_G}\left[\left(1 - \frac{T}{T_c}\right) - \frac{2}{N}|\psi(\mathbf{r})|^2\right]\right\}\psi(\mathbf{r}) = 0 \tag{8-155}$$

where $e^* = 2e$. One can also calculate the current density

$$\mathbf{J}(\mathbf{r}) = \left[\frac{ie}{m}(\nabla_{r'} - \nabla_r)G(x, x') - \frac{2e^2}{mc}\mathbf{A}(\mathbf{r})G(x, x')\right]_{t'=t^+, \mathbf{r}=\mathbf{r}'} \tag{8-156}$$

to second order in Δ by the perturbation expansion used above and one finds on using the relation between ψ and Δ:

$$\mathbf{J}(\mathbf{r}) = -\frac{ie^*}{2m}(\psi^*\nabla\psi + \psi\nabla\psi^*) - \frac{e^{*2}}{mc}\mathbf{A}(\mathbf{r})|\psi(\mathbf{r})|^2 \tag{8-157}$$

as in the GL theory.

Recently, the derivation of Gor'kov has been extended to all temperatures by Werthamer[154] and by Tewordt,[155] who continue to assume the system is such that **A** and Δ vary slowly over a coherence length. Their equations are somewhat more complicated than the GLG form, as one might expect. Gor'kov has extended his treatment to include finite mean-free-path effects. He finds the equations have the same form as above, except that the "mass" m is increased relative to that of the pure material.

CONCLUSION

In this book we have mainly discussed the microscopic aspects of the theory of superconductivity. The microscopic theory has of course many macroscopic consequences, the discussion of which is beyond the scope of our present treatment. Of particular interest in this respect is the theory of type II or "hard" superconductors. As we mentioned in the introduction these materials do not show a sharp drop of the magnetization curve at the critical field, which is characteristic of ideal type I or "soft" superconductors. Rather, they exhibit a perfect Meissner up to a lower critical field H_{c1}, after which the magnetization drops continuously to zero at the upper critical field H_{c2}, owing to flux penetration. Since the Ginsburg–Landau parameter κ, which determines whether a material falls in class I or II, can be influenced by the concentration of impurities, cold work, etc., one can investigate the transition region between these two distinctly different types of magnetic behavior.[157, 158]

There is also a large body of literature dealing with the effect of magnetic and nonmagnetic impurities on the properties of superconductors.[159, 160] An important advance in this area was made by Anderson,[146] who argued that in the presence of nonmagnetic (time-reversal-invariant) impurities one should pair single-particle states which already include the effects of scattering from the impurities. On the basis of this idea he showed that the dominant effect of a small concentration of nonmagnetic impurities is to remove the crystalline anisotropy of the energy gap and thereby reduce the transition temperature.[161] The gap edge is not broadened, however. For larger impurity concentrations, numerous other effects such as shifts of valence electron concentration, electron and phonon band structure, electron-phonon interaction, etc., become important and the problem of determining T_c is quite involved.

Magnetic impurities generally have a tendency to lower the superconducting transition temperature, because the antiparallel spin correlations in the superconducting state prevents the valence electron-magnetic ion interaction from entering as favorably as in the normal state.[162, 163] There are examples,[160] however, where no localized moment occurs and T_c is increased. In addition, magnetic impurities can, within a limited range of concentration lead to "gapless" superconductivity.[172] This important effect was predicted by Abrikosov and Gor'kov, and observed by Reif and Woolf. The problem of superconductivity in transition metals and the relation between ferromagnetism and superconductivity will no doubt receive considerable attention in the future.

Another problem which has received a good deal of attention is the behavior of small specimens and thin films in the presence of strong magnetic fields.[164] A related problem is the structure of the superconductor normal-metal phase boundary. In addition there are many interesting problems associated with superimposed films of normal and superconducting metals.[165] It is well known that a superconductor can make a nearby normal metal become superconducting by allowing the electrons of the two metals to intermingle. The range over which this effect can take place, the effect of magnetic fields and magnetic impurities on the interaction between the metals, etc., are worthy of further study.

Another area in which the pairing theory has met with considerable success is in the theory of nuclear structure. Following the original suggestion of Bohr, Mottelson, and Pines,[166] numerous workers have used the pairing theory to calculate the single-particle and collective excitation spectra in heavy nuclei. The theory has been remarkably successful in predicting the difference of the low-lying single-particle spectra between even and odd nuclei. In even-even nuclei there is an energy gap of order 1 mev for exciting a neutron or a proton, because of the existence of the pairing interactions. In even-odd or odd-odd nuclei the corresponding excitation energy is smaller by a factor of four or more due to the unpaired particles being present in these nuclei. In addition, the rotational and vibrational spectra are generally brought into close agreement with experiment once the pairing correlations are included. Pick-up and stripping reactions measure the energy distribution of the bare particles and reasonable agreement is obtained between the smeared Fermi surface characteristic of the pairing theory and these experiments. Although the effect of the pairing correlations in nuclei is not as striking as in superconductors, it is clear that these correlations play an important role in determining the properties of nuclei.[167]

Another area where the pairing concept has been applied is in the mass spectrum of elementary particles. It is difficult to resist drawing the analogy between the quasi-particle energy in a superconductor $E_p = (\epsilon_p{}^2 + \Delta^2)^{1/2}$ and the relativistic form $E_p = (p^2 + m^2)^{1/2}$. Nambu[168] and co-workers have based a model of elementary particles on the pairing scheme, as has Fisher.[169] Whether these attempts will play a role in the ultimate resolution of the mass spectrum problem is unclear at present.

The pairing theory has also been applied[170] to the possibility of a superfluid phase of He^3. Thus far no such transition has been observed down to temperatures below 0.01°K.

A problem that has received little attention to date is whether there are systems having strong correlations involving clusters of more than two particles. One knows that alpha-particle correlations are important in light nuclei and Little[171] has experimental data which might be interpreted in terms of large clusters. Nevertheless, we must conclude that the pairing correlations, upon which the pairing theory is based, are the essential correlations required to explain the basic phenomena observed in the superconducting state.

APPENDIX

SECOND-QUANTIZATION FORMALISM

In this appendix we shall give a brief summary of second quantization.

A-1 OCCUPATION-NUMBER REPRESENTATION

Let us consider a system of n identical particles described in the Schrödinger representation by the Hamiltonian

$$H(x_1 \cdots x_n) = \sum_i \frac{p_i^2}{2m} + \sum_i V_1(x_i) + \frac{1}{2} \sum_{i \neq j} V_2(x_i, x_j) \qquad \text{(A-1)}$$

The coordinate x_i labels the position and spin of particle i. Three-body potentials and higher interactions can be included in a straightforward manner, but for the moment we shall confine ourselves to two-body interactions.

The many-body Schrödinger equation is

$$H(x_1 \cdots x_n)\Psi(x_1 \cdots x_n, t) = i\hbar \frac{\partial \Psi(x_1 \cdots x_n, t)}{\partial t} \qquad \text{(A-2)}$$

We introduce a complete set of n-particle wave functions Φ. These are constructed as a properly symmetrized product of one-particle wave functions $u_k(x)$, which form a complete orthonormal set

$$\int u_{k'}^*(x)u_k(x)\,dx = \delta_{kk'} \qquad \text{(orthonormality)} \qquad \text{(A-3)}$$

$$\sum_k u_k^*(x')u_k(x) = \delta(x - x') \qquad \text{(completeness)} \qquad \text{(A-4)}$$

The function Φ is then given by

$$\Phi = \mathscr{S} u_{k_1}(x_1)u_{k_2}(x_2)\cdots u_{k_n}(x_n) \qquad \text{(A-5)}$$

where $\mathscr{S} = (1/n!)\sum P$ in Bose statistics and $\mathscr{S} = (1/n!)\sum(-1)^p P$ in Fermi statistics and the summation is over all $n!$ possible permutations of the coordinates $x_1\cdots x_n$ and p is the order of the permutation. Rather than labeling Φ by the quantum numbers $k_1, k_2\cdots k_n$, we may specify the state by stating how many times each single-particle state enters the product. Let this occupation number be n_k for state k. Then the set of numbers $n_1, n_2\cdots n_k$ uniquely specifies the symmetrized state $\Phi_{n_1,n_2\cdots n_k}$. If we describe a system of n-particles, we have clearly $\sum_k n_k = n$. For Fermi statistics the occupation numbers n_k are restricted to the values 0 and 1, whereas for Bose statistics they can have all possible positive integer values (as well as 0). The functions $\Phi_{n_1\cdots n_k\cdots}(x_1\cdots x_n)$ form a complete orthonormal set of n-particle functions for fermions when they are multiplied by the factor $(n!)^{1/2}$, whereas a multiplication factor $(n!/n_1!n_2!)^{1/2}$ must be included to obtain a complete orthonormal set of Bose functions. (*Note:* 0! is defined as unity.) The orthonormality condition is

$$\int \Phi^*_{n_1', n_2'\cdots}(x_1\cdots x_n)\Phi_{n_1,n_2\cdots}(x_1\cdots x_n)\,dx_1\cdots dx_n = \delta_{n_1',n_1}\delta_{n_2',n_2}\cdots \qquad \text{(A-6)}$$

In general, the total Schrödinger wave function may be expanded in the complete set of the $\Phi_{n_1\cdots n_k\cdots}$:

$$\Psi(x_1\cdots x_n, t) = \sum A(n_1\cdots n_k\cdots, t)\Phi_{n_1\cdots n_k\cdots}(x_1\cdots x_n) \qquad \text{(A-7)}$$

The coefficients $A(n_1 \cdots n_k \cdots, t)$ are now to be interpreted as the wave functions in the occupation-number representation. Their norm gives the probability of finding n_k-particles in state k.

A-2 SECOND QUANTIZATION FOR BOSONS

For Bose statistics we introduce a set of operators a_k and a_k^+ defined by

$$a_k^+ \Phi_{n_1 \cdots n_k \cdots}(x_1 \cdots x_n) = (n_k + 1)^{1/2} \Phi_{n_1 \cdots n_k + 1 \cdots}(x_1 \cdots x_{n+1})$$

$$\text{(A-8)}$$

$$a_k \Phi_{n_1 \cdots n_k \cdot} \cdot (x_1 \cdots x_n) = (n_k)^{1/2} \Phi_{n_1 \cdots n_k - 1 \cdots}(x_1 \cdots x_{n-1})$$

The operator a_k^+ (creation operator) creates an additional particle in state k, and a_k (annihilation operator) destroys a particle in state k. If $n_k = 0$ in Φ, the operator a_k gives 0.

We can see that a_k^+ is the Hermitian conjugate of a_k by noting that the only nonvanishing matrix element of a_k is

$$(a_k)\begin{matrix} n_1 \cdots n_k - 1 \cdots \\ n_1 \cdots n_k \cdots \end{matrix}$$

which is equal to $(n_k)^{1/2}$, and its Hermitian conjugate operator will have as its only nonvanishing matrix element

$$\left[\left(a_k\right)\begin{matrix} n_1 \cdots n_k \\ n_1 \cdots n_k - 1 \end{matrix}\right]^* = (n_k)^{1/2}$$

The operator which has only this nonvanishing matrix element is indeed a_k^+ by definition. If we define a new operator $N_k = a_k^+ a_k$, it follows from definition (A-8) that its eigenvalue equation is

$$N_k \Phi_{n_1 \cdots n_k \cdots}(x_1 \cdots x_n) = n_k \Phi_{n_1 \cdots n_k \cdots}(x_1 \cdots x_n) \qquad \text{(A-9)}$$

Therefore, N_k may be interpreted as the operator which measures the number of particles of state k (the number operator). We may now construct the operator N which measures the total number of particles in the system:

$$N = \sum_k N_k = \sum_k a_k^+ a_k \qquad \text{(A-10)}$$

The commutation relation between the creation and destruction operators is easily established to be

$$[a_k, a_{k'}^+] = \delta_{kk'} \qquad [a_k, a_{k'}] = [a_k^+, a_{k'}^+] = 0 \qquad \text{(A-11)}$$

For instance,

$$(a_k a_k^+ - a_k^+ a_k)\Phi_{n_1 \cdots n_k \cdots} = [(n_k + 1) - n_k]\Phi_{n_1 \cdots n_k \cdots} = 1\Phi_{n_1 \cdots n_k}$$

The Hamiltonian can be expressed in the occupation number representation, and it can be seen that it becomes

$$H = \sum_{kk'} \langle k'|H_1|k\rangle a_{k'}^+ a_k + \tfrac{1}{2} \sum_{k_1'k_2'k_1k_2} \langle k_1', k_2'|V_2|k_1, k_2\rangle a_{k_1'}^+ a_{k_2'}^+ a_{k_1} a_{k_2} \qquad \text{(A-12)}$$

where

$$\langle k'|H_1|k\rangle = \int u_{k'}^*(x)\left\{\frac{p^2}{2m} + V_1(x)\right\}u_k(x)\,dx$$

and

$$\langle k_1', k_2'|V_2|k_1, k_2\rangle = \int u_{k_1'}^*(x_1)u_{k_2'}^*(x_2)V_2(x_1, x_2)u_{k_1}(x_1)u_{k_2}(x)_2 dx_1\,dx_2$$

This can be proved by noticing that all matrix elements of Hamiltonian (A-12) with the complete set $\Phi_{n_1}..$ are equal to the matrix elements as evaluated from the original Hamiltonian (A-1) in configuration space. We shall not give the complete proof but will outline the way in which it can be obtained.

First, we evaluate all matrix elements of the Hamiltonian (A-12). The basic rule is to use the orthonormality of the set Φ and the definition (A-8) for creation and destruction operators. The wave function obtained by operating with (A-12) on the wave function to the right of the matrix element must be equal to the wave function on the left of the matrix element.

1. In diagonal elements only terms with $k = k'$ and $k_1 = k_1'$, $k_2 = k_2'$ or $k_1 = k_2'$, $k_2 = k_1'$ or $k_1 = k_2 = k_1' = k_2'$ give nonzero

contribution because only these terms leave the wave function to the right unchanged. We have, then,

$$\langle \Phi_{n_1 n_2 \cdots} | H | \Phi_{n_1 n_2 \cdots} \rangle = \sum_k \langle k | H_1 | k \rangle n_k + \tfrac{1}{2} \sum_{k_1 \neq k_2} n_{k_2} n_{k_1} \{ \langle k_1 k_2 | V_2 | k_1 k_2 \rangle + \langle k_2 k_1 | V_2 | k_1 k_2 \rangle \} + \tfrac{1}{2} \sum_{k_1} n_{k_1} (n_{k_1} - 1) \langle k_1 k_1 | V_2 | k_1 k_1 \rangle \tag{A-13}$$

2. We now illustrate off-diagonal elements between wave functions which differ in the occupation numbers of two states i and j. Let the wave functions be

$$\Phi_{\cdot\cdot n_i n_j \cdots} \quad \text{and} \quad \Phi_{\cdot\cdot m_i m_j \cdots}$$

with

$$n_i + n_j = m_i + m_j$$

because the number of particles is fixed. The only matrix elements different from zero are those for which either

$$(1) \quad n_j = m_j \mp 1 \qquad n_i = m_i \pm 1$$

or

$$(2) \quad n_j = m_j \mp 2 \qquad n_i = m_i \pm 2$$

On choosing the upper signs, the matrix elements for the first case are

$$\langle k_i | H_1 | k_j \rangle [(m_i + 1) m_j]^{1/2} + \sum_l m_l [(m_i + 1) m_j]^{1/2} \times (\langle k_l k_i | V | k_l k_j \rangle + \langle k_l k_i | V | k_j k_l \rangle) \tag{A-14}$$

and for the second case

$$\tfrac{1}{2} [m_i (m_i - 1)(m_j + 1)(m_j + 2)]^{1/2} \langle k_j k_j | V_2 | k_i k_i \rangle$$

On choosing the lower signs, we have the same expressions with i and j interchanged.

3. We show here off-diagonal elements between wave functions which differ in the occupation numbers of only three states i, j, and l. Those different from zero obey conditions of the type

$$n_i = m_i \pm 1 \qquad n_j = m_j \pm 1 \qquad n_l = m_l \mp 2$$

The matrix element with the upper choice of signs is

$$\tfrac{1}{2}[(m_l + 2)(m_l + 1)m_j m_i]^{1/2}\{\langle k_l k_l | V_2 | k_j k_i\rangle + \langle k_l k_l | V_2 | k_i k_j\rangle\}$$

With the lower choice of signs, we have

$$\tfrac{1}{2}[(m_j + 1)(m_i + 1)(m_l - 1)m_l]^{1/2} \times \{\langle k_i k_j | V_2 | k_l k_l\rangle + \langle k_j k_i | V_2 | k_l k_l\rangle\} \quad \text{(A-15)}$$

4. The following are off-diagonal elements which differ in the occupational numbers of four states k_i, k_j, k_l, and k_s; those different from zero obey the condition

$$n_i = m_i \mp 1 \qquad n_j = m_j \mp 1 \qquad n_l = m_l \pm 1 \qquad n_s = m_s \pm 1$$

The matrix element is

$$[m_i m_j (m_l + 1)(m_s + 1)]^{1/2}\{\langle k_l k_s | V_2 | k_i k_j\rangle + \langle k_l k_s | V_2 | k_j k_i\rangle\} \quad \text{(A-16)}$$

with the upper choice of signs. A matrix element of the same type, but with l, s, and i, j interchanged is obtained with the lower choice of signs.

The same matrix elements as in 1, 2, 3, and 4 are obtained from Hamiltonian (A-1). Let us consider as an example the last case; the same procedure will apply to the other cases. We want to evaluate

$$\langle \Phi \cdots m_i - 1, m_j - 1, m_l + 1, m_s + 1 | \tfrac{1}{2} \sum_{pq} V_2(pq) | \Phi \cdots m_i, m_j, m_l, m_s\rangle$$

First, we can simplify the normalization constants by integrating over all variables other than p and q whatever p and q may be. We have

$$\begin{aligned}
&[(m_i - 1)!(m_j - 1)!(m_l + 1)!(m_s + 1)! m_i! m_j! m_l! m_s!]^{-1/2} \\
&\quad \times \langle u_i(1) \cdots u_i(m_i - 1) u_l(m_i) \cdots u_l(m_i + m_l) u_j(m_i + m_l + 1) \cdots \\
&\quad \times u_j(m_i + m_j + m_l - 1) u_s(m_i + m_j + m_l) \cdots \\
&\quad \times u_s(m_i + m_j + m_l + m_s) | \tfrac{1}{2} \sum_{pq} V_2(p, q) | P u_i(1) \cdots \\
&\quad \times u_i(m_i) u_l(m_i + 1) \cdots u_l(m_i + m_l) u_j(m_i + m_l + 1) \cdots \\
&\quad \times u_j(m_i + m_j + m_l) u_s(m_i + m_j + m_l + 1) \cdots \\
&\qquad\qquad \times u_s(m_i + m_j + m_l + m_s)\rangle
\end{aligned}$$

where P indicates all possible permutations of particles. When $p = m_i$ and $q = m_i + m_j + m_l$, or vice versa, we obtain from the integral the quantity

$$\langle k_l k_s | V_2 | k_i k_j \rangle + \langle k_l k_s | V_2 | k_j k_i \rangle$$

multiplied by a factor $m_i! m_j! m_l! m_s!$, because of the permutations which interchange particles on the same state on the wave function to the right. Because of the summation over p and q in the operator, we must consider the sum of all terms with

$$m_i \leqslant p \leqslant (m_i + m_l)$$

and

$$(m_i + m_j + m_l) \leqslant q \leqslant (m_i + m_j + m_l + m_s)$$

this introduces another multiplicative factor $(m_l + 1)(m_s + 1)$. By multiplication of all factors with the normalization constants, we obtain

$$[m_i m_j (m_l + 1)(m_s + 1)]^{1/2} \{\langle k_l k_s | V_2 | k_i k_j \rangle + \langle k_l k_s | V_2 | k_j k_i \rangle\}$$

which is the same as Eq. (A-16).

Let us define some new operators which do not depend on state k but depend on the variable x as

$$\psi(x) = \sum_k u_k(x) a_k$$

and

$$\psi^+(x) = \sum_k u_k(x)^* a_k^+ \tag{A-17}$$

They are called "wave field" operators and satisfy the commutation relations

$$[\psi(x), \psi^+(x')] = \sum_{kk'} u_k(x) u_{k'}^*(x') [a_k, a_{k'}^+] = \delta(x - x')$$
$$[\psi(x), \psi(x')] = [\psi(x), \psi(x')]^+ = 0 \tag{A-18}$$

To clarify the usefulness of the wave field operators, we may notice that $\rho(x) = \psi^+(x)\psi(x)$ represents the density of particles in x-space and the number operator is

$$N = \int \rho(x)\, dx = \sum_{kk'} a_{k'}^+ a_k \int u_{k'}^+(x) u_k(x)\, dx = \sum_k a_k^+ a_k \tag{A-19}$$

In many instances it is useful to have the Fourier transform of ρ, given by

$$\rho_q = \int e^{i\mathbf{q}\cdot\mathbf{x}}\rho(x)\,dx = \sum_{kk'} a_{k'}{}^+ a_k \int e^{i\mathbf{q}\cdot\mathbf{x}} u_{k'}{}^*(x) u_k(x)\,dx \qquad \text{(A-20)}$$

When the $u_k(x)$ are given simply by plane waves $e^{i\mathbf{k}\cdot\mathbf{x}}$ and we normalize in a box of unit volume, we have $\mathbf{k}' = \mathbf{k} - \mathbf{q}$ and, consequently,

$$\rho = \sum_{\mathbf{k}'} a_{\mathbf{k}'}{}^+ a_{\mathbf{k}'+\mathbf{q}} \qquad \text{(A-21)}$$

We may express the Hamiltonian operator in terms of the field variable $\psi(x)$; from Eq. (A-12) and the definition of the ψ we obtain

$$H = \int \psi^+(x) H_1(x) \psi(x)\,dx + \tfrac{1}{2}\int \psi^+(x)\psi^+(x') V_2(x, x')\psi(x')\psi(x)\,dx\,dx' \qquad \text{(A-22)}$$

The order of the operators ensures that the term $i = j$ has been omitted in the two-body potential. If $V_2(x, x') = V_2(x - x')$, that is to say, if our two-body operator is translationally invariant, the Hamiltonian can be written as

$$H = \int \psi^+(x) H_1(x) \psi(x)\,dx + \tfrac{1}{2}\sum_q V_2(q)\eta(\rho_q{}^+ \rho_q) \qquad \text{(A-23)}$$

where η is the normally ordered product such that all the ψ^+ are placed to the left and all ψ to the right in the product. The proof is obtained by expanding $V_2(x - x')$ as its Fourier transform and using the definition of ρ_q.

The prescription for expressing an n-body interaction in the occupation-number representation is now clear:

$$V_\nu = \frac{1}{\nu!}\int \psi^+(x_1)\cdots\psi^+(x_\nu) V_\nu(x_1\cdots x_\nu)\psi(x_\nu)\cdots\psi(x_1)\,dx_1\cdots dx_\nu \qquad \text{(A-24)}$$

From this point on we shall suppress the variables $x_1\cdots x_n$ in Φ and represent Φ by

$$\Phi_{n_1\cdots n_i}(x_1\cdots x_n) = |n_1\cdots n_i\cdots\rangle \qquad \text{(A-25)}$$

Equally well, $|n_1 \cdots n_i \cdots\rangle$ can be considered to be a vector in a Hilbert space. This vector has components $\langle x_1 \cdots x_n | n_1 \cdots n_i \cdots\rangle$ along a complete set of position eigenvectors $|x_1 \cdots x_n\rangle$. These components are indentified with the function

$$\Phi_{n_1} \cdots n_i \cdots (x_1 \cdots x_n)$$

The n-body Schrödinger equation may now be expressed in the a_k-language as

$$H\Psi(t) = i\hbar \frac{\partial \Psi(t)}{\partial t}$$

where

$$\Psi(t) = \sum_{n_1, n_2 \cdots} A(n_1, n_2 \cdots t)|n_1, n_2 \cdots\rangle \tag{A-26}$$

and

$$H = H_1 + V_2 + \cdots + V_\nu + \cdots$$

as given in second-quantization formalism.

A-3 SECOND QUANTIZATION FOR FERMIONS

For Fermi statistics we introduce creation and destruction operators $c_k{}^+$ and c_k, formally given by Eq. (A-8), referred to the antisymmetric wave function. These operators satisfy an anti-commutation relation

$$\{c_k{}^+, c_{k'}\} = \delta_{kk'}$$

and

$$\{c_k, c_{k'}\} = 0 = \{c_k{}^+, c_{k'}{}^+\} \tag{A-27}$$

where

$$\{A, B\} = AB + BA$$

It can be seen that this choice of commutation relations restricts the occupation number of the states k to 0 or 1 as required by Fermi statistics. In fact, $c_k c_k \Phi_{n_k} = 0 = (n_k - 1)^{1/2}(n_k)^{1/2}\Phi_{n_k - 2}$. With this restriction in mind, the definition of the operator c_k is

$$\begin{aligned} c_k \Phi_{\cdots n_k} &= \Phi_{\cdots n_k - 1} & \{n_k &= 1 \\ &= 0 & \{n_k &= 0 \\ c_k{}^+ \Phi_{\cdots n_k} &= \Phi_{\cdots n_k + 1 \cdots} & \{n_k &= 0 \\ &= 0 & \{n_k &= 1 \end{aligned} \tag{A-28}$$

The Hamiltonian operator in the second-quantization formalism is given by

$$H = \sum_{k'ks} \langle k's|H_1|ks\rangle c_{k's}{}^+ c_{ks} + \frac{1}{2} \sum_{\substack{ijkl \\ ss'}} \langle ij|V|lk\rangle c_{is}{}^+ c_{js'}{}^+ c_{ks'} c_{ls} \tag{A-29}$$

where s and s' label the spin of the particle.

The wave fields are defined as

$$\psi(x) = \sum_k u_k(x) c_k$$

and they satisfy the anticommutation equations

$$\{\psi(x), \psi^+(x')\} = \delta(x - x')$$
$$\{\psi(x), \psi(x')\} = 0 = \{\psi^+(x), \psi^+(x')\}$$

Note that in (A-29) the order of the destruction operators with regard to the matrix-element indices is the inverse of that of the creation operators. The ν-body operator of Eq. (A-24) is

$$V_\nu = \frac{1}{\nu!} \sum_{k_\nu' s_\nu' \cdots k_1' s_1' k_\nu s_\nu \cdots k_1 s_1} \langle k_\nu' \cdots k_1'|V_\nu|k_\nu \cdots k_1\rangle \times c_{k_\nu' s_\nu'}{}^+ \cdots c_{k_1' s_1'}{}^+ c_{k_1 s_1} \cdots c_{k_\nu s_\nu} \tag{A-24'}$$

The ordering is not entirely arbitrary because of sign changes arising from anticommutation of fermion operators; that of (A-29), (A-24′), (A-24), and (A-22) gives the correct sign for either fermion or boson operators.

If one is dealing with particles of nonzero spin, the coordinate x represents both space and spin variables, as does the variable k. Integrals over x represent integrals over space and sums over spin variables.

MACROSCOPIC QUANTUM PHENOMENA FROM PAIRING IN SUPERCONDUCTORS

Nobel Lecture, December 11, 1972
by

J. R. Schrieffer

University of Pennsylvania, Philadelphia, Pa.

I. Introduction

It gives me great pleasure to have the opportunity to join my colleagues John Bardeen and Leon Cooper in discussing with you the theory of superconductivity. Since the discovery of superconductivity by H. Kamerlingh Onnes in 1911, an enormous effort has been devoted by a spectrum of outstanding scientists to understanding this phenomenon. As in most developments in our branch of science, the accomplishments honored by this Nobel prize were made possible by a large number of developments preceding them. A general understanding of these developments is important as a backdrop for our own contribution.

On December 11, 1913, Kamerlingh Onnes discussed in his Nobel lecture (1) his striking discovery that on cooling mercury to near the absolute zero of temperature, the electrical resistance became vanishingly small, but this disappearance "did not take place gradually but *abruptly*." His Fig. 17 is reproduced as Fig. 1. He said, "Thus, mercury at 4.2 K has entered a new state

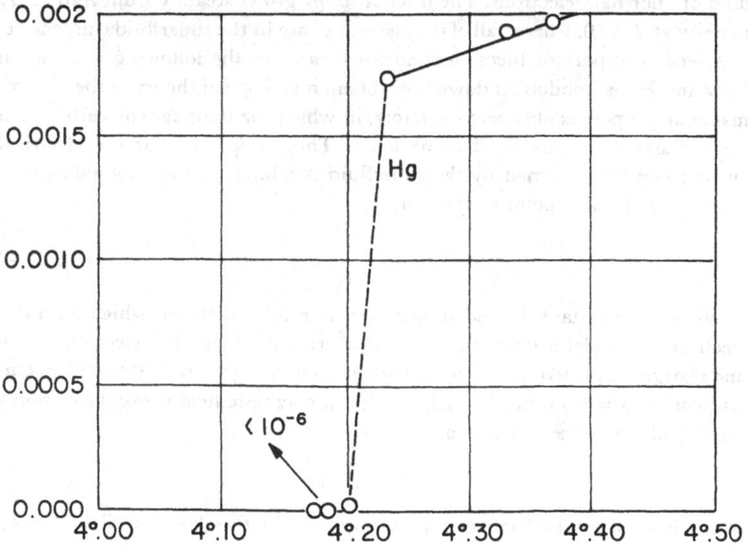

Fig. 1

which owing to its particular electrical properties can be called the state of superconductivity." He found this state could be destroyed by applying a sufficiently strong magnetic field, now called the critical field H_c. In April–June, 1914, Onnes discovered that a current, once induced in a closed loop of superconducting wire, persists for long periods without decay, as he later graphically demonstrated by carrying a loop of superconducting wire containing a persistent current from Leiden to Cambridge.

In 1933, W. Meissner and R. Ochsenfeld (2) discovered that a superconductor is a perfect diamagnet as well as a perfect conductor. The magnetic field vanishes in the interior of a bulk specimen, even when cooled down below the transition temperature in the presence of a magnetic field. The diamagnetic currents which flow in a thin penetration layer near the surface of a simply connected body to shield the interior from an externally applied field are stable rather than metastable. On the other hand, persistent currents flowing in a multiply connected body, e.g., a loop, are metastable.

An important advance in the understanding of superconductivity occurred in 1934, when C. J. Gorter and H. B. G. Casimir (3) advanced a two fluid model to account for the observed second order phase transition at T_c and other thermodynamic properties. They proposed that the total density of electrons ϱ could be divided into two components

$$\varrho = \varrho_s + \varrho_n \tag{1}$$

where a fraction ϱ_s/ϱ_n of the electrons can be viewed as being condensed into a "superfluid," which is primarily responsible for the remarkable properties of superconductors, while the remaining electrons form an interpenetrating fluid of "normal" electrons. The fraction ϱ_s/ϱ_n grows steadily from zero at T_c to unity at $T = 0$, where "all of the electrons" are in the superfluid condensate.

A second important theoretical advance came in the following year, when Fritz and Hans London set down their phenomenological theory of the electromagnetic properties of superconductors, in which the diamagnetic rather than electric aspects are assumed to be basic. They proposed that the electrical current density $\mathbf{j}_s$ carried by the superfluid is related to the magnetic vector potential $\mathbf{A}$ at each point in space by

$$\mathbf{j}_s = -\frac{1}{\Lambda c}\mathbf{A} \tag{2}$$

where Λ is a constant dependent on the material in question, which for a free electron gas model is given by $\Lambda = m/\varrho_s e^2$, m and e being the electronic mass and charge, respectively. A is to be chosen such that $\nabla \cdot \mathbf{A} = 0$ to ensure current conservation. From (2) it follows that a magnetic field is excluded from a superconductor except within a distance

$$\lambda_L = \sqrt{\Lambda c^2/4\pi}$$

which is of order 10^{-6} cm in typical superconductors for T well below T_c. Observed values of λ are generally several times the London value.

In the same year (1935) Fritz London (4) suggested how the diamagnetic property (2) might follow from quantum mechanics, if there was a "rigidity" or stiffness of the wavefunction ψ of the superconducting state such that ψ was essentially unchanged by the presence of an externally applied magnetic field. This concept is basic to much of the theoretical development since that time, in that it sets the stage for the gap in the excitation spectrum of a superconductor which separates the energy of superfluid electrons from the energy of electrons in the normal fluid. As Leon Cooper will discuss, this gap plays a central role in the properties of superconductors.

In his book published in 1950, F. London extended his theoretical conjectures by suggesting that a superconductor is a "quantum structure on a macroscopic scale [which is a] kind of solidification or condensation of the average momentum distribution" of the electrons. This momentum space condensation locks the average momentum of each electron to a common value which extends over appreciable distance in space. A specific type of condensation in momentum space is central to the work Bardeen, Cooper and I did together. It is a great tribute to the insight of the early workers in this field that many of the important general concepts were correctly conceived before the microscopic theory was developed. Their insight was of significant aid in our own work.

The phenomenological London theory was extended in 1950 by Ginzburg and Landau (5) to include a spatial variation of ϱ_s. They suggested that ϱ_s/ϱ be written in terms of a phenomenological condensate wavefunction $\psi(r)$ as $\varrho_s(r)/\varrho = |\psi(r)|^2$ and that the free energy difference ΔF between the superconducting and normal states at temperature T be given by

$$\Delta F = \int \left\{ \frac{\hbar^2}{2\bar{m}} \left| \left(\nabla + \frac{\bar{e}}{c} A(r) \right) \psi(r) \right|^2 - a(T) |\psi(r)|^2 + \frac{b(T)}{2} \left| \psi(r) \right|^4 \right\} d^3r \qquad (3)$$

where $\bar{e}$, $\bar{m}$, a and b are phenomenological constants, with $a(T_c) = 0$.

They applied this approach to the calculation of boundary energies between normal and superconducting phases and to other problems.

As John Bardeen will discuss, a significant step in understanding which forces cause the condensation into the superfluid came with the experimental discovery of the isotope effect by E. Maxwell and, independently, by Reynolds, et al. (6). Their work indicated that superconductivity arises from the interaction of electrons with lattice vibrations, or phonons. Quite independently, Herbert Fröhlich (7) developed a theory based on electron-phonon interactions which yielded the isotope effect but failed to predict other superconducting properties. A somewhat similar approach by Bardeen (8) stimulated by the isotope effect experiments also ran into difficulties. N. Bohr, W. Heisenberg and other distinguished theorists had continuing interest in the general problem, but met with similar difficulties.

An important concept was introduced by A. B. Pippard (9) in 1953. On the basis of a broad range of experimental facts he concluded that a coherence length ξ is associated with the superconducting state such that a perturbation of the superconductor at a point necessarily influences the superfluid within a distance ξ of that point. For pure metals, $\xi \sim 10^{-4}$ cm. for $T \ll T_c$. He generalized the London equation (3) to a non-local form and accounted for the fact that the experimental value of the penetration depth is several times larger than the London value. Subsequently, Bardeen (10) showed that Pippard's non-local relation would likely follow from an energy gap model.

A major problem in constructing a first principles theory was the fact that the physically important condensation energy ΔF amounts typically to only 10^{-8} electron volts (e.V.) per electron, while the uncertainty in calculating the total energy of the electron-phonon system in even the normal state amounted to of order 1 e.V. per electron. Clearly, one had to isolate those correlations peculiar to the superconducting phase and treat them accurately, the remaining large effects presumably being the same in the two phases and therefore cancelling. Landau's theory of a Fermi liquid (11), developed to account for the properties of liquid He^3, formed a good starting point for such a scheme. Landau argued that as long as the interactions between the particles (He^3 atoms in his case, electrons in our case) do not lead to discontinuous changes in the microscopic properties of the system, a "quasi-particle" description of the low energy excitations is legitimate; that is, excitations of the fully interacting normal phase are in one-to-one correspondence with the excitations of a non-interacting fermi gas. The effective mass m and the Fermi velocity v_F of the quasi-particles differ from their free electron values, but aside from a weak decay rate which vanishes for states at the Fermi surface there is no essential change. It is the residual interaction between the quasi-particles which is responsible for the special correlations characterizing superconductivity. The ground state wavefunction of the superconductor ψ_0 is then represented by a particular superposition of these normal state configurations, Φ_n.

A clue to the nature of the states Φ_n entering strongly in ψ_0 is given by combining Pippard's coherence length ξ with Heisenberg's uncertainty principle

$$\Delta p \sim \hbar/\xi \sim 10^{-4} p_F \qquad (4)$$

where p_F is the Fermi momentum. Thus, Ψ_0 is made up of states with quasi-particles (electrons) being excited above the normal ground state by a momentum of order Δp. Since electrons can only be excited to states which are initially empty, it is plausible that only electronic states within a momentum $10^{-4}\ p_F$ of the Fermi surface are involved significantly in the condensation, i.e., about 10^{-4} of the electrons are significantly affected. This view fits nicely with the fact that the condensation energy is observed to be of order $10^{-4}\varrho \cdot k_B T_c$. Thus, electrons within an energy $\sim v_F \Delta p \simeq kT_c$ of the Fermi surface have their energies lowered by of order kT_c in the condensation. In summary, the problem was how to account for the phase transition in which a condensation of electrons occurs in momentum space for electrons very near the Fermi surface. A proper theory should automatically account for the perfect conductivity and diamagnetism, as well as for the energy gap in the excitation spectrum.

II. The Pairing Concept

In 1955, stimulated by writing a review article on the status of the theory of superconductivity, John Bardeen decided to renew the attack on the problem.

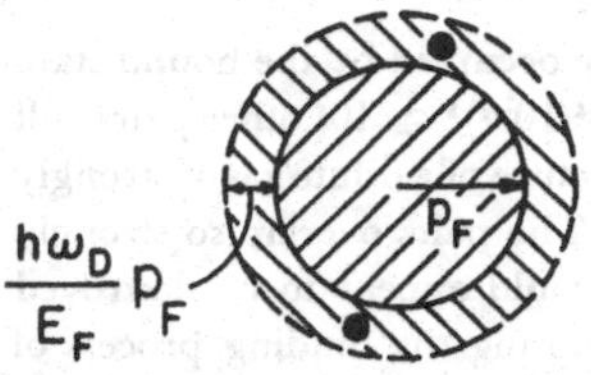

He invited Leon Cooper, whose background was in elementary particle physics and who was at that time working with C. N. Yang at the Institute for Advanced Study to join in the effort starting in the fall of 1955. I had the good fortune to be a graduate student of Bardeen at that time, and, having finished my graduate preliminary work, I was delighted to accept an invitation to join them.

We focused on trying to understand how to construct a ground state Ψ_0 formed as a coherent superposition of normal state configurations Φ_n,

$$\Psi_0 = \sum_n a_n \Phi_n \tag{5}$$

such that the energy would be as low as possible. Since the energy is given in terms of the Hamiltonian H by

$$E_0 = (\Psi_0, H\psi_0) = \sum_{n,n'} a_{n'}^* \, a_n \, (\Phi_{n'}, H\Phi_n) \tag{6}$$

we attempted to make E_0 minimum by restricting the coefficients a_n so that only states which gave negative off-diagonal matrix elements would enter (6). In this case all terms would add in phase and E_0 would be low.

By studying the eigenvalue spectrum of a class of matrices with off-diagonal elements all of one sign (negative), Cooper discovered that frequently a single eigenvalue is split off from the bottom of the spectrum. He worked out the problem of two electrons interacting via an attractive potential-V above a quiescent Fermi sea, i.e., the electrons in the sea were not influenced by V and the extra pair was restricted to states within an energy $\hbar\omega_D$ above the Fermi surface, as illustrated in Fig. 2. As a consequence of the non-zero density of quasi-particle states $N(0)$ at the Fermi surface, he found the energy eigenvalue spectrum for two electrons having zero total momentum had a bound state split off from the continuum of scattering states, the binding energy being

$$E_B \cong \hbar\omega_D e^{-\frac{2}{N(0)V}} \tag{7}$$

if the matrix elements of the potential are constant equal to V in the region of interaction. This important result, published in 1956 (12), showed that, regardless of how weak the residual interaction between quasi-particles is, if the interaction is attractive the system is unstable with respect to the formation of bound pairs of electrons. Further, if E_B is taken to be of order $k_B T_c$, the uncertainty principle shows the average separation between electrons in the bound state is of order 10^{-4} cm.

While Cooper's result was highly suggestive, a major problem arose. If, as we discussed above, a fraction 10^{-4} of the electrons is significantly involved in the condensation, the average spacing between these condensed electrons is roughly 10^{-6} cm. Therefore, within the volume occupied by the bound state of a given pair, the centers of approximately $(10^{-4}/10^{-6})^3 \cong 10^6$ other pairs will be found, on the average. Thus, rather than a picture of a dilute gas of strongly bound pairs, quite the opposite picture is true. The pairs overlap so strongly in space that the mechanism of condensation would appear to be destroyed due to the numerous pair-pair collisions interrupting the binding process of a given pair.

Returning to the variational approach, we noted that the matrix elements $(\Phi_{n'}, H\Phi_n)$ in (6) alternate randomly in sign as one randomly varies n and n' over the normal state configurations. Clearly this cannot be corrected to obtain a low value of E_0 by adjusting the sign of the a_n's since there are $\mathcal{N}^2$ matrix elements to be corrected with only $\mathcal{N}$ parameters a_n. We noticed that if the sum in (6) is restricted to include only configurations in which, if any quasi-particle state, say k, s, is occupied ($s = \uparrow$ or $\downarrow$ is the spin index), its "mate" state $\bar{k}$, $\bar{s}$ is also occupied, then the matrix elements of H between such states would have a unique sign and a coherent lowering of the energy would be obtained. This correlated occupancy of pairs of states in momentum space is consonant with London's concept of a condensation in momentum.

In choosing the state $\bar{k}$, $\bar{s}$ to be paired with a given state k, s, it is important to note that in a perfect crystal lattice, the interaction between quasi-particles conserves total (crystal) momentum. Thus, as a given pair of quasi-particles interact, their center of mass momentum is conserved. To obtain the largest number of non-zero matrix elements, and hence the lowest energy, one must choose the total momentum of each pair to be the same, that is

$$k + \bar{k} = q. \tag{8}$$

States with $q \neq 0$ represent states with net current flow. The lowest energy state is for $q = 0$, that is, the pairing is such that if any state $k\uparrow$ is occupied in an admissible Φn, so is $-k\downarrow$ occupied. The choice of $\downarrow\uparrow$ spin pairing is not restrictive since it encompasses triplet and singlet paired states.

Through this reasoning, the problem was reduced to finding the ground state of the reduced Hamiltonian

$$H_{\text{red}} = \sum_{ks} \epsilon_k n_{ks} - \sum_{kk'} V_{k'k}\, b_{k'}^{+} b_k. \tag{9}$$

The first term in this equation gives the unperturbed energy of the quasi-particles forming the pairs, while the second term is the pairing interaction in which a pair of quasi-particles in $(k\uparrow, -k\downarrow)$ scatter to $(k'\uparrow, -k'\uparrow)$. The operators $b_k^{+} = c_{k\uparrow}^{+}\, c_{-k\downarrow}^{+}$, being a product of two fermion (quasi-particle) creation operators, do not satisfy Bose statistics, since $b_k^{+2} = 0$. This point is essential to the theory and leads to the energy gap being present not only for dissociating a pair but also for making a pair move with a total momentum different from the common momentum of the rest of the pairs. It is this feature which enforces long range order in the superfluid over macroscopic distances.

III. The Ground State

In constructing the ground state wavefunction, it seemed clear that the average occupancy of a pair state $(k\uparrow, -k\downarrow)$ should be unity for k far below the Fermi surface and 0 for k far above it, the fall off occurring symmetrically about k_F over a range of momenta of order

$$\Delta k \sim \frac{1}{\xi} \sim 10^4 \text{ cm}^{-1}.$$

One could not use a trial Ψ_0 as one in which each pair state is definitely occupied or definitely empty since the pairs could not scatter and lower the energy in this case. Rather there had to be an amplitude, say v_k, that $(k\uparrow, -k\downarrow)$ is occupied in Ψ_0 and consequently an amplitude $u_k = \sqrt{1-v_k^2}$ that the pair state is empty. After we had made a number of unsuccessful attempts to construct a wavefunction sufficiently simple to allow calculations to be carried out, it occurred to me that since an enormous number ($\sim 10^{19}$) of pair states $(k'\uparrow, -k'\downarrow)$ are involved in scattering into and out of a given pair state $(k\uparrow, -k\downarrow)$, the "instantaneous" occupancy of this pair state should be essentially uncorrelated with the occupancy of the other pair states at that "instant". Rather, only the *average* occupancies of these pair states are related.

On this basis, I wrote down the trial ground state as a product of operators —one for each pair state—acting on the vacuum (state of no electrons),

$$\Psi_0 = \pi_k \, (u_k + v_k b_k) \, |0\rangle, \tag{10}$$

where $u_k = \sqrt{1-v_k^2}$. Since the pair creation operators b_k^+ commute for different k's, it is clear that Ψ_0 represents uncorrelated occupancy of the various pair states. I recall being quite concerned at the time that Ψ_0 was an admixture of states with different numbers of electrons, a wholly new concept to me, and as I later learned to others as well. Since by varying v_k the mean number of electrons varied, I used a Lagrange multiplier μ (the chemical potential) to make sure that the mean number of electrons (N_{op}) represented by Ψ_0 was the desired number N. Thus by minimizing

$$E_0 - \mu N = (\Psi_0, [H_{red} - \mu N_{op}]\Psi_0)$$

with respect to v_k, I found that v_k was given by

$$v_k^2 = \tfrac{1}{2}\left[1 - \frac{(\epsilon_k - \mu)}{E_k}\right] \tag{11}$$

where

$$E_k = \sqrt{(\epsilon_k - \mu)^2 + \Delta_k^2} \tag{12}$$

and the parameter Δ_k satisfied what is now called the energy gap equation:

$$\Delta_k = -\sum V_{k'k} \frac{\Delta_{k'}}{2E_{k'}} \tag{13}$$

From this expression, it followed that for the simple model

$$V_{k'k} = \begin{cases} V, & |\epsilon_k - \mu| < \hbar\omega_D \text{ and } |\epsilon_{k'} - \mu| < \hbar\omega_D \\ 0, & \text{otherwise} \end{cases}$$

$$\Delta = \hbar \, \omega_D e - \frac{1}{N(0)\,V} \tag{14}$$

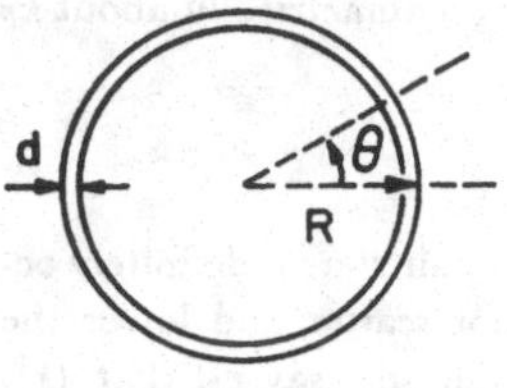

and the condensation energy at zero temperature is

$$\Delta F = \tfrac{1}{2}\mathcal{N}(0)\Delta^2 \tag{15}$$

The idea occurred to me while I was in New York at the end of January, 1957, and I returned to Urbana a few days later where John Bardeen quickly recognized what he believed to be the essential validity of the scheme, much to my pleasure and amazement. Leon Cooper will pick up the story from here to describe our excitement in the weeks that followed, and our pleasure in unfolding the properties of the excited states.

IV. Quantum Phenomena on a Macroscopic Scale

Superconductors are remarkable in that they exhibit quantum effects on a broad range of scales. The persistence of current flow in a loop of wire many meters in diameter illustrates that the pairing condensation makes the superfluid wavefunction coherent over macroscopic distances. On the other hand, the absorption of short wavelength sound and light by a superconductor is sharply reduced from the normal state value, as Leon Cooper will discuss. I will concentrate on the large scale quantum effects here.

The stability of persistent currents is best understood by considering a circular loop of superconducting wire as shown in Fig. 3. For an ideal small diameter wire, one would use the eigenstates $e^{im\theta}$, ($m = 0, \pm 1, \pm 2, \ldots$), of the angular momentum L_z about the symmetry axis to form the pairing. In the ground state no net current flows and one pairs m$\uparrow$ with $-m\downarrow$, instead of $k\uparrow$ with $-k\downarrow$ as in a bulk superconductor. In both cases, the paired states are time reversed conjugates, a general feature of the ground state. In a current carrying state, one pairs $(m+\nu)\uparrow$ with $(-m+\nu)\downarrow$, ($\nu = 0, \pm 1, \pm 2 \ldots$), so that the total angular momentum of each pair is identical, $2\hbar\,\nu$. It is this commonality of the center of mass angular momentum of each pair which preserves the condensation energy and long range order even in states with current flow. Another set of flow states which interweave with these states is formed by pairing $(m+\nu)\uparrow$ with $(-m+\nu+1)\downarrow$, ($\nu = 0, \pm 1, \pm 2 \ldots$), with the pair angular momentum being $(2\nu+1)\hbar$. The totality of states forms a set with all integer multiples n of $\hbar$ for allowed total angular momentum of pairs. Thus, even though the pairs greatly overlap in space, the system exhibits quantization effects as if the pairs were well defined.

There are two important consequences of the above discussion. First, the fact that the coherent condensate continues to exist in flow states shows that to scatter a pair out of the (rotating) condensate requires an increase of energy.

Crudely speaking, slowing down a given pair requires it ot give up its binding energy and hence this process will occur only as a fluctuation. These fluctuations average out to zero. The only way in which the flow can stop is if all pairs simultaneously change their pairing condition from, say, ν to $\nu-1$. In this process the system must fluctate to the normal state, at least in a section of the wire, in order to change the pairing. This requires an energy of order the condensation energy ΔF. A thermal fluctuation of this size is an exceedingly rare event and therefore the current persists.

The second striking consequence of the pair angular momentum quantization is that the magnetic flux Φ trapped within the loop is also quantized,

$$\Phi_n = n \cdot \frac{hc}{2e} \qquad (n = 0, \pm 1, \pm 2 \ldots). \tag{16}$$

This result follows from the fact that if the wire diameter d is large compared to the penetration depth λ, the electric current in the center of the wire is essentially zero, so that the canonical angular momentum of a pair is

$$L_{\text{pair}} = \frac{2e}{c} r_{\text{pair}} \times A \tag{17}$$

where r_{pair} is the center of mass coordinate of a pair and A is the magnetic vector potential. If one integrates L_{pair}, around the loop along a path in the center of the wire, the integral is nh, while the integral of the right hand side of (17) is $\frac{2e}{c}\Phi$.

A similar argument was given by F. London (4b) except that he considered only states in which the superfluid flows as a whole without a change in its internal strucutre, i.e., states analogous to the $(m+\nu)\uparrow$, $(-m+\nu)\downarrow$ set. He found $\Phi z = n \cdot hc/e$. The pairing $(m+\nu)\uparrow$, $(m+\nu+1)\downarrow$ cannot be obtained by adding ν to each state, yet this type of pairing gives an energy as low as the more conventional flow states and these states enter experimentally on the same basis as those considered by London. Experiments by Deaver and Fairbank (13), and independently by Doll and Näbauer (14) confirmed the flux quantization phenomenon and provided support for the pairing concept by showing that $2e$ rather than e enters the flux quantum. Following these experiments a clear discussion of flux quantization in the pairing scheme was given by Beyers and Yang (15).

The idea that electron pairs were somehow important in superconductivity has been considered for some time (16, 17). Since the superfluidity of liquid He^4 is qualitatively accounted for by Bose condensation, and since pairs of electrons behave in some respects as a boson, the idea is attractive. The essential point is that while a dilute gas of tightly bound pairs of electrons might behave like a Bose gas (18) this is not the case when the mean spacing between pairs is very small compared to the size of a given pair. In this case the inner structure of the pair, i.e., the fact that it is made of fermions, is essential; it is this which distinguishes the pairing condensation, with its energy gap for single pair translation as well as dissociation, from the spectrum of a Bose con-

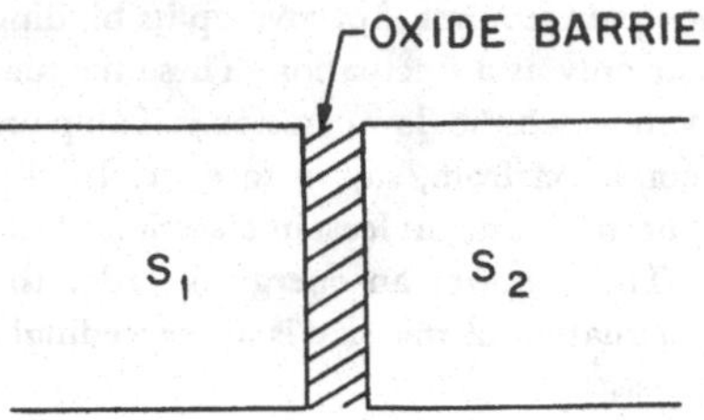

densate, in which the low energy exictations are Bose-like rather than Fermi-like as occurs in acutal superconductors. As London emphasized, the condensation is an ordering in occupying momentum space, and not a space-like condensation of clusters which then undergo Bose condensation.

In 1960, Ivar Giaever (19) carried out pioneering experiments in which electrons in one superconductor (S_1) tunneled through a thin oxide layer (~ 20—30 Å) to a second superconductor (S_2) as shown in Fig. 4. Giaever's experiments were dramatic evidence of the energy gap for quasi-particle excitations. Subsequently, Brian Josephson made a highly significant contribution by showing theoretically that a superfluid current could flow between S_1 and S_2 with zero applied bias. Thus, the superfluid wavefunction is coherent not only in S_1 and S_2 separately, but throughout the entire system, $S_1 - 0 - S_2$, under suitable circumstances. While the condensate amplitude is small in the oxide, it is sufficient to lock the phases of S_1 and S_2 together, as has been discussed in detail by Josephson (20) and by P. W. Anderson (21).

To understand the meaning of phase in this context, it is useful to go back to the ground state wavefunction Ψ_0, (10). Suppose we write the parameter v_k as $|v_k|$ exp $i\varphi$ and choose u_k to be real. If we expand out the k-product in Ψ_0, we note that the terms containing $\mathcal{N}$ pairs will have a phase factor exp $(i\,\mathcal{N}\varphi)$, that is, each occupied pair state contributes a phase φ to Ψ_0. Let this wavefunction, say $\Psi_0^{(1)}$ represent S_1, and have phase φ_1. Similarly, let $\Psi_0^{(2)}$ represent S_2 and have phase angle φ_2. If we write the state of the combined system as a product

$$\Psi_0^{(1,2)} = \Psi_0^{(1)}\,\Psi_0^{(2)} \tag{18}$$

then by expanding out the double product we see that the phase of that part of $\Psi_0^{(1,2)}$ which has $\mathcal{N}_1$ pairs in S_1 and $\mathcal{N}_2$ pairs in S_2 is $\mathcal{N}_1\,\varphi_1 + \mathcal{N}_2\varphi_2$. For a truly isolated system, $2(\mathcal{N}_1+\mathcal{N}_2) = 2\mathcal{N}$ is a fixed number of electrons; however $\mathcal{N}_1$ and $\mathcal{N}_2$ are not separately fixed and, as Josephson showed, the energy of the combined system is minimized when $\varphi_1 = \varphi_2$ due to tunneling of electrons between the superconductors. Furthermore, if $\varphi_1 = \varphi_2$, a current flows between S_1 and S_2

$$j = j_1 \sin(\varphi_1 - \varphi_2) \tag{19}$$

If $\varphi_1 - \varphi_2 = \varphi$ is constant in time, a constant current flows with no voltage applied across the junction. If a bias voltage is V applied between S_1 and S_2, then, according to quantum mechanics, the phase changes as

$$\frac{2eV}{\hbar} = \frac{d\varphi}{dt} \tag{20}$$

Hence a constant voltage applied across such a junction produces an alternating current of frequency

$$\nu = \frac{2eV}{\hbar} = 483 \text{ THz/V}. \tag{21}$$

These effects predicted by Josephson were observed experimentally in a series of beautiful experiments (22) by many scientists, which I cannot discuss in detail here for lack of time. I would mention, as an example, the work of Langenberg and his collaborators (23) at the University of Pennsylvania on the precision determination of the fundamental constant e/h using the frequency-voltage relation obeyed by the alternating Josephson supercurrent. These experiments have decreased the uncertainty in our experimental knowledge of this constant by several orders of magnitude and provide, in combination with other experiments, the most accurate available value of the Sommerfeld fine structure constant. They have resulted in the resolution of several discrepancies between theory and experiment in quantum electrodynamics and in the development of an "atomic" voltage standard which is now being used by the United States National Bureau of Standards to maintain the U.S. legal volt.

V. Conclusion

As I have attempted to sketch, the development of the theory of superconductivity was truly a collaborative effort, involving not only John Bardeen, Leon Cooper and myself, but also a host of outstanding scientists working over a period of half a century. As my colleagues will discuss, the theory opened up the field for many exciting new developments, both scientific and technological, many of which no doubt lie in the future. I feel highly honored to have played a role in this work and I deeply appreciate the honor you have bestowed on me in awarding us the Nobel prize.

References

1. Kamerlingh Onnes, H., Nobel Lectures, Vol. 1, pp. 306–336.
2. Meissner, W. and Ochsenfeld, R., Naturwiss. *21*, 787 (1933).
3. Gorter, C. J. and Casimir, H. B. G., Phys. Z. *35*, 963 (1934); Z. Techn. Phys. *15*, 539 (1934).
4. London, F., [a] Phys. Rev. *24*, 562 (1948); [b] Superfluids, Vol. 1 (John Wiley & Sons, New York, 1950).
5. Ginzburg, V. L. and Landau, L. D., J. Exp. Theor. Phys. (U.S.S.R.) *20*, 1064 (1950).
6. Maxwell, E., Phys. Rev. *78*, 477 (1950); Reynolds, C. A., Serin, B., Wright W. H. and Nesbitt, L. B., Phys. Rev. *78*, 487 (1950).
7. Fröhlich, H., Phys. Rev. *79*, 845 (1950).

8. Bardeen, J., Rev. Mod. Phys. *23*, 261 (1951).
9. Pippard, A. B., Proc. Royal Soc. (London) A*216*, 547 (1953).
10. Bardeen, J., [a] Phys. Rev. *97*, 1724 (1955); [b] Encyclopedia of Physics, Vol. 15 (Springer-Verlag, Berlin, 1956), p. 274.
11. Landau, L. D., J. Exp. Theor. Phys. (U.S.S.R.) *30* (*3*), 1058 (920) (1956); *32* (*5*), 59 (101) (1957).
12. Cooper, L. N., Phys. Rev. *104*, 1189 (1956).
13. Deaver, B. S. Jr., and Fairbank, W. M., Phys. Rev. Letters *7*, 43 (1961).
14. Doll, R. and Näbauer, M., Phys. Rev. Letters *7*, 51 (1961).
15. Beyers, N. and Yang, C. N., Phys. Rev. Letters *7*, 46 (1961).
16. Ginzburg, V. L., Usp. Fiz. Nauk *48*, 25 (1952); transl. Fortsch. d. Phys. *1*, 101 (1953).
17. Schafroth, M. R., Phys. Rev. *96*, 1442 (1954); *100*, 463 (1955).
18. Schafroth, M. R., Blatt, J. M. and Butler, S. T., Helv. Phys. Acta *30*, 93 (1957).
19. Giaever, I., Phys. Rev. Letters *5*, 147 (1960).
20. Josephson, B. D., Phys. Letters *1*, 251 (1962); Advan. Phys. *14*, 419 (1965).
21. Anderson, P. W., in Lectures on the Many-body Problem, edited by E. R. Caianiello (Academic Press, Inc. New York, 1964), Vol. II.
22. See Superconductivity, Parks, R. D., ed. (Dekker New York, 1969).
23. See, for example, Parker, W. H. Taylor B. N. and Langenberg, D. N. Phys. Rev. Letters *18*, 287 (1967); Finnegan, T. F. Denenstein A. and Langenberg, D. N. Phys. Rev. B*4*, 1487 (1971).

MICROSCOPIC QUANTUM INTERFERENCE EFFECTS IN THE THEORY OF SUPERCONDUCTIVITY

Nobel Lecture, December 11, 1972

by

LEON N COOPER

Physics Department, Brown University, Providence, Rhode Island

It is an honor and a pleasure to speak to you today about the theory of superconductivity. In a short lecture one can no more than touch on the long history of experimental and theoretical work on this subject before 1957. Nor can one hope to give an adequate account of how our understanding of superconductivity has evolved since that time. The theory (1) we presented in 1957, applied to uniform materials in the weak coupling limit so defining an ideal superconductor, has been extended in almost every imaginable direction. To these developments so many authors have contributed (2) that we can make no pretense of doing them justice. I will confine myself here to an outline of some of the main features of our 1957 theory, an indication of directions taken since and a discussion of quantum interference effects due to the singlet-spin pairing in superconductors which might be considered the microscopic analogue of the effects discussed by Professor Schrieffer.

NORMAL METAL

Although attempts to construct an electron theory of electrical conductivity date from the time of Drude and Lorentz, an understanding of normal metal conduction electrons in modern terms awaited the development of the quantum theory. Soon thereafter Sommerfeld and Bloch introduced what has evolved into the present description of the electron fluid. (3) There the conduction electrons of the normal metal are described by single particle wave functions. In the periodic potential produced by the fixed lattice and the conduction electrons themselves, according to Bloch's theorem, these are modulated plane waves:

$$\O_{\boldsymbol{K}}(\boldsymbol{r}) = \mathrm{u}_{\boldsymbol{K}}(\boldsymbol{r})\, \mathrm{e}^{i\boldsymbol{k}\cdot\boldsymbol{r}},$$

where $u_{\boldsymbol{K}}(\boldsymbol{r})$ is a two component spinor with the lattice periodicity. We use $\boldsymbol{K}$ to designate simultaneously the wave vector $\boldsymbol{k}$, and the spin state σ: $\boldsymbol{K} \equiv \boldsymbol{k}, \uparrow$; $-\boldsymbol{K} \equiv -\boldsymbol{k}, \downarrow$. The single particle Bloch functions satisfy a Schrödinger equation

$$\left[-\frac{\hbar^2}{2m}\nabla^2 + V_0(\boldsymbol{r}) \right] \O_{\boldsymbol{K}} = \varepsilon_{\boldsymbol{K}} \O_{\boldsymbol{K}}$$

where $V_0(\boldsymbol{r})$ is the periodic potential and in general might be a linear operator to include exchange terms.

The Pauli exclusion principle requires that the many electron wave function be antisymmetric in all of its coordinates. As a result no two electrons can be

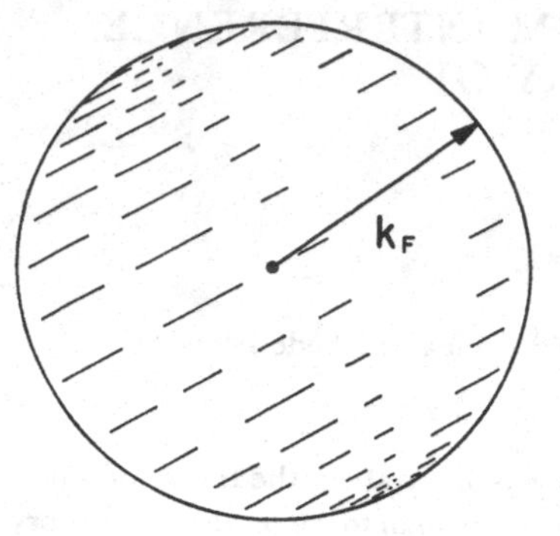

Fig. 1.
The normal ground state wavefunction, Φ_0, is a filled Fermi sphere for both spin directions.

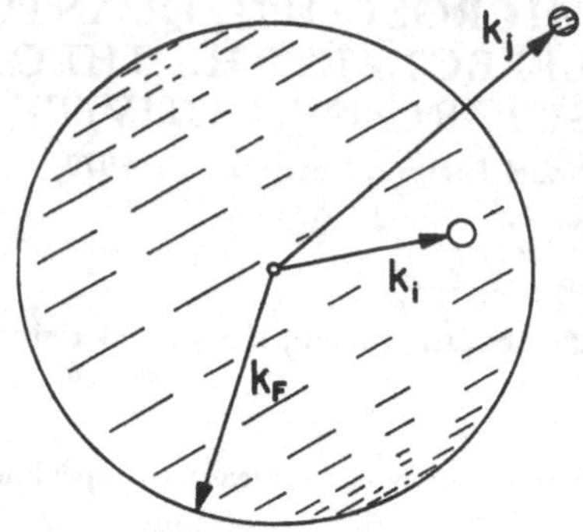

Fig. 2.
An excitation of the normal system.

in the same single particle Bloch state. The energy of the entire system is

$$W = \sum_{i=1}^{2\mathcal{N}} \mathcal{E}_i$$

where $\mathcal{E}_i$ is the Bloch energy of the ith single electron state. The ground state of the system is obtained when the lowest $\mathcal{N}$ Bloch states of each spin are occupied by single electrons; this can be pictured in momentum space as the filling in of a Fermi sphere, Fig. 1. In the ground-state wave function there is no correlation between electrons of opposite spin and only a statistical correlation (through the general anti-symmetry requirement on the total wave function) of electrons of the same spin.

Single particle excitations are given by wave functions identical to the ground state except that one electron states $k_i < k_F$ are replaced by others $k_j < k_F$. This may be pictured in momentum space as opening vacancies below the Fermi surface and placing excited electrons above, Fig. 2. The energy difference between the ground state and the excited state with the particle excitation k_j and the hole excitation k_i is

$$\mathcal{E}_j - \mathcal{E}_i = \mathcal{E}_j - \mathcal{E}_F - (\mathcal{E}_i - \mathcal{E}_F) = \varepsilon_j - \varepsilon_i = |\varepsilon_j| + |\varepsilon_i|$$

where we define ε as the energy measured relative to the Fermi energy

$$\varepsilon_i = \mathcal{E}_i - \mathcal{E}_F.$$

When Coulomb, lattice-electron and other interactions, which have been omitted in constructing the independent particle Bloch model are taken into account, various modifications which have been discussed by Professor Schrieffer are introduced into both the ground state wave function and the excitations. These may be summarized as follows: The normal metal is described by a ground state Φ_0 and by an excitation spectrum which, in addition to the various collective excitations, consists of quasi-fermions which satisfy the usual anticommutation relations. It is defined by the sharpness of the Fermi surface, the finite density of excitations, and the continuous decline of the single particle excitation energy to zero as the Fermi surface is approached.

Electron Correlations that Produce Superconductivity

For a description of the superconducting phase we expect to include correlations that are not present in the normal metal. Professor Schrieffer has discussed the correlations introduced by an attractive electron-electron interaction and Professor Bardeen will discuss the role of the electron-phonon interaction in producing the electron-electron interaction which is responsible for superconductivity. It seems to be the case that any attractive interaction between the fermions in a many-fermion system can produce a superconducting-like state. This is believed at present to be the case in nuclei, in the interior of neutron stars and has possibly been observed (4) very recently in He^3. We will therefore develop the consequences of an attractive two-body interaction in a degenerate many-fermion system without enquiring further about its source.

The fundamental qualitative difference between the superconducting and normal ground state wave function is produced when the large degeneracy of the single particle electron levels in the normal state is removed. If we visualize the Hamiltonian matrix which results from an attractive two-body interaction in the basis of normal metal configurations, we find in this enormous matrix, sub-matrices in which all single-particle states except for one pair of electrons remain unchanged. These two electrons can scatter via the electron-electron interaction to all states of the same total momentum. We may envisage the pair wending its way (so to speak) over all states unoccupied by other electrons. [The electron-electron interaction in which we are interested is both weak and slowly varying over the Fermi surface. This and the fact that the energy involved in the transition into the superconducting state is small leads us to guess that only single particle excitations in a small shell near the Fermi surface play a role. It turns out, further, that due to exchange terms in the electron-electron matrix element, the effective interaction in metals between electrons of singlet spin is much stronger than that between electrons of triplet spin—thus our preoccupation with singlet spin correlations near the Fermi surface.] Since every such state is connected to every other, if the interaction is attractive and does not vary rapidly, we are presented with submatrices of the entire Hamiltonian of the form shown in Fig. 3. For purposes of illustration we have set all off diagonal matrix elements equal to the constant $-V$ and the diagonal terms equal to zero (the single particle excitation energy at the Fermi surface) as though all the initial electron levels were completely degenerate. Needless to say, these simplifications are not essential to the qualitative result.

Diagonalizing this matrix results in an energy level structure with $M-1$ levels raised in energy to $E = +V$ while one level (which is a superposition of all of the original levels and quite different in character) is lowered in energy to

$$E = -(M-1)V.$$

Since M, the number of unoccupied levels, is proportional to the volume of the container while V, the scattering matrix element, is proportional to 1/volume, the product is independent of the volume. Thus the removal of

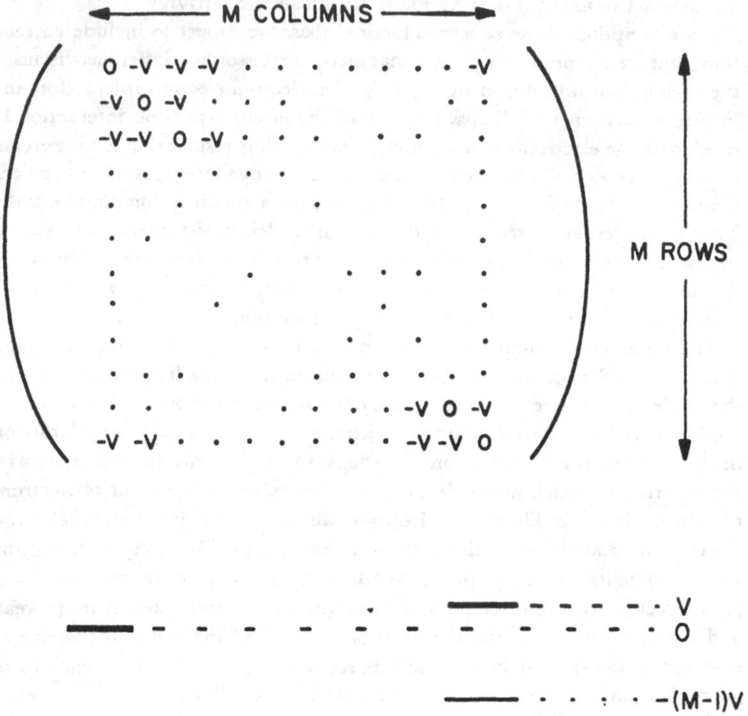

For V = 0,
M levels
at E = 0

For V > 0,M —1 levels at
E = V and one level
at E = —(M—1) V

Fig. 3.

the degeneracy produces a single level separated from the others by a volume independent energy gap.

To incorporate this into a solution of the full Hamiltonian, one must devise a technique by which all of the electrons pairs can scatter while obeying the exclusion principle. The wave function which accomplishes this has been discussed by Professor Schrieffer. Each pair gains an energy due to the removal of the degeneracy as above and one obtains the maximum correlation of the entire wave function if the pairs all have the same total momentum. This gives a coherence to the wave function in which for a combination of dynamical and statistical reasons there is a strong preference for momentum zero, singlet spin correlations, while for statistical reasons alone there is an equally strong preference that all of the correlations have the same total momentum.

In what follows I shall present an outline of our 1957 theory modified by introducing the quasi-particles of Bogoliubov and Valatin. (5) This leads to a formulation which is generally applicable to a wide range of calculations

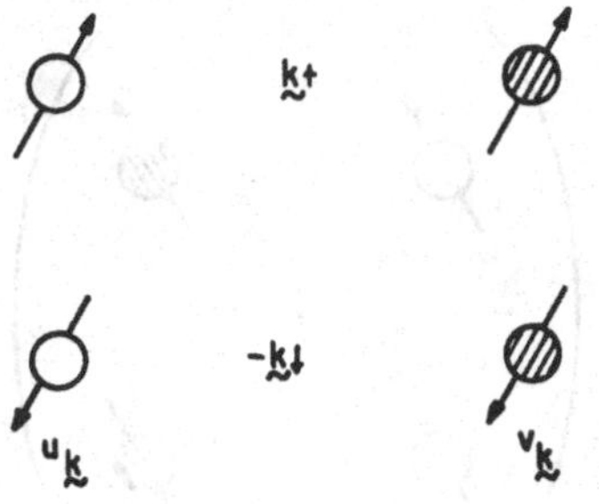

Fig. 4.
The ground state of the superconductor is a linear superposition of states in which pairs ($k\uparrow -k\downarrow$) are occupied or unoccupied.

in a manner analogous to similar calculations in the theory of normal metals.

We limit the interactions to terms which scatter (and thus correlate) singlet zero-momentum pairs. To do this, it is convenient to introduce the pair operators:

$$b_k = c_{-K}c_K$$
$$b_k^* = c_K^* c_{-K}^*$$

and using these we extract from the full Hamiltonian the so-called reduced Hamiltonian

$$H_{\text{reduced}} = \sum_{k<k_f} 2|\varepsilon| b_k b_k^* + \sum_{k>k_f} 2\varepsilon b_k^* b_k + \sum_{kk'} V_{k'k} b_{k'}^* b_k$$

where $V_{k'k}$ is the scattering matrix element between the pair states k and k'.

Ground State

As Professor Schrieffer has explained, the ground state of the superconductor is a linear superposition of pair states in which the pairs ($k\uparrow$, $-k\downarrow$) are occupied or unoccupied as indicated in Fig. 4. It can be decomposed into two disjoint vectors—one in which the pair state k is occupied, $\emptyset_k$ and one in which it is unoccupied, $\emptyset_{(k)}$:

$$\psi_0 = u_k\emptyset_{(k)} + v_k\emptyset_k.$$

The probability amplitude that the pair state k is (is not) occupied in the ground state is then $v_k(u_k)$. Normalization requires that $|u|^2+|v|^2 = 1$. The phase of the ground state wave function may be chosen so that with no loss o generality u_k is real. We can then write

$$u = (1-h)^{1/2}$$
$$v = h^{1/2}\, e^{i\varphi}$$

where

$$0 \leqslant h \leqslant 1.$$

A further decomposition of the ground state wave function of the superconductor in which the pair states k and k' are either occupied or unoccupied Fig. 5 is:

$$\psi_0 = u_k u_{k'}\emptyset_{(k),(k')} + u_k v_{k'}\emptyset_{(k),k'} + v_k u_{k'}\emptyset_{k,(k')} + v_k v_{k'}\emptyset_{k,k'}\ .$$

This is a Hartree-like approximation in the probability amplitudes for the occupation of pair states. It can be shown that for a fermion system the wave

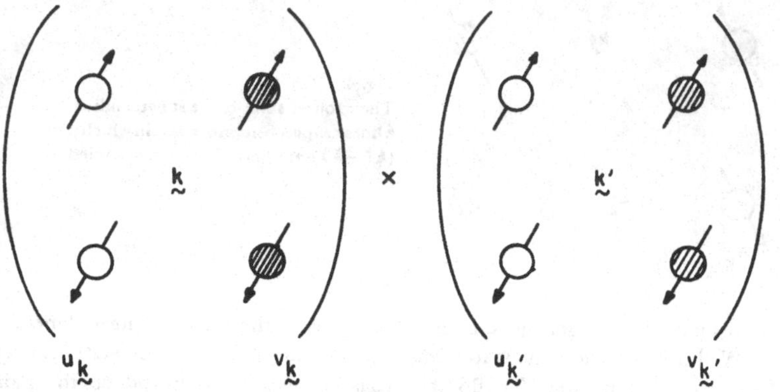

Fig. 5.
A decomposition of the ground state of the superconductor into states in which the pair states *k* and *k'* are either occupied or unoccupied.

function cannot have this property unless there are a variable number of particles. To terms of order $1/N$, however, this decomposition is possible for a fixed number of particles; the errors introduced go to zero as the number of particles become infinite. (6)

The correlation energy, W_c, is the expectation value of H_{red} for the state ψ_0

$$W_c = (\psi_0, H_{red}\psi_0) = W_c\,[h,\varphi].$$

Setting the variation of W_c with respect to h and φ equal to zero in order to minimize the energy gives

$$h = 1/2\ (1-\varepsilon/E)$$
$$E = (\varepsilon^2+|\Delta|^2)^{1/2}$$

where

$$\Delta = |\Delta| e^{i\varphi}$$

satisfies the integral equation

$$\Delta(\boldsymbol{k}) = -1/2 \sum_{\boldsymbol{k}'} V_{\boldsymbol{kk}'} \frac{\Delta(\boldsymbol{k}')}{E(\boldsymbol{k}')}.$$

If a non-zero solution of this integral equation exists, $W_c < 0$ and the "normal" Fermi sea is unstable under the formation of correlated pairs.

In the wave function that results there are strong correlations between pairs of electrons with opposite spin and zero total momentum. These correlations are built from normal excitations near the Fermi surface and extend over spatial distances typically of the order of 10^{-4} cm. They can be constructed due to the large wave numbers available because of the exclusion principle. Thus with a small additional expenditure of kinetic energy there can be a greater gain in the potential energy term. Professor Schrieffer has discussed some of the properties of this state and the condensation energy associated with it.

Single-Particle Excitations

In considering the excited states of the superconductor it is useful, as for the

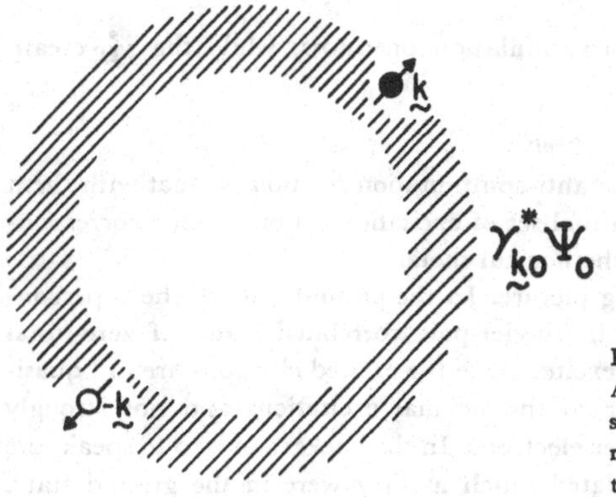

Fig. 6.
A single par icle excitation of the superconduc or in one-to-one correspondenc with an excitation of the normal fermion system.

normal metal, to make a distinction between single-particle and collective excitations; it is the single-particle excitation spectrum whose alteration is responsible for superfluid properties. For the superconductor excited (quasi-particle) states can be defined in one-to-one correspondence with the excitations of the normal metal. One finds, for example, that the expectation value of H_{red} for the excitation Fig. 6 is given by

$$E_k = \sqrt{\varepsilon_k^2 + |\Delta|^2}.$$

In contrast to the normal system, for the superconductor even as ε goes to zero E remains larger than zero, its lowest possible value being $E = |\Delta|$. One can therefore produce single particle excitations from the superconducting ground state only with the expenditure of a small but finite amount of energy. This is called the energy gap; its existence severely inhibits single particle processes and is in general responsible for the superfluid behavior of the electron gas. [In a gapless superconductor it is the finite value of $\Delta(\boldsymbol{r})$, the order parameter, rather than the energy gap as such that becomes responsible for the superfluid properties.] In the ideal superconductor, the energy gap appears because not a single pair can be broken nor can a single element of phase space be removed without a finite expenditure of energy. If a single pair is broken, one loses its correlation energy; if one removes an element of phase space from the system, the number of possible transitions of all the pairs is reduced resulting in both cases in an increase in the energy which does not go to zero as the volume of the system increases.

The ground state of the superconductor and the excitation spectrum described above can conveniently be treated by introducing a linear combination of c^* and c, the creation and annihilation operators of normal fermions. This is the transformation of Bogoliubov and Valatin (5):

$$\gamma_{k0}^* = u_k c_K^* - v_k^* c_{-K}$$
$$\gamma_{k1}^* = v_k^* c_K + u_k c_{-K}^*$$

It follows that

$$\gamma_{ki} \psi_0 = 0$$

so that the γ_{ki} play the role of annihilation operators, while the γ^*_{ki} create excitations

$$\gamma^*_{ki} \cdots \gamma^*_{mj}\, \psi_0 = \psi_{ki}, \cdots {}_{mj}.$$

The γ operators satisfy Fermi anti-commutation relations so that with them we obtain a complete orthonormal set of excitations in one-to-one correspondence with the excitations of the normal metal.

We can sketch the following picture. In the ground state of the superconductor all the electrons are in singlet-pair correlated states of zero total momentum. In an m electron excited state the excited electrons are in "quasi-particle" states, very similar to the normal excitations and not strongly correlated with any of the other electrons. In the background, so to speak, the other electrons are still correlated much as they were in the ground state. The excited electrons behave in a manner similar to normal electrons; they can be easily scattered or excited further. But the background electrons—those which remain correlated—retain their special behavior; they are difficult to scatter or to excite.

Thus, one can identify two almost independent fluids. The correlated portion of the wave function shows the resistance to change and the very small specific heat characteristic of the superfluid, while the excitations behave very much like normal electrons, displaying an almost normal specific heat and resistance. When a steady electric field is applied to the metal, the superfluid electrons short out the normal ones, but with higher frequency fields the resistive properties of the excited electrons can be observed. [7]

Thermodynamic Properties, the Ideal Superconductor

We can obtain the thermodynamic properties of the superconductor using the ground state and excitation spectrum just described. The free energy of the system is given by

$$F[h, \varphi, f] = W_c(T) - TS,$$

where T is the absolute temperature and S is the entropy; f is the superconducting Fermi function which gives the probability of single-particle excitations. The entropy of the system comes entirely from the excitations as the correlated portion of the wave function is non-degenerate. The free energy becomes a function of $f(\boldsymbol{k})$ and $h(\boldsymbol{k})$, where $f(\boldsymbol{k})$ is the probability that the state $\boldsymbol{k}$ is occupied by an excitation or a quasi-particle, and $h(\boldsymbol{k})$ is the relative probability that the state $\boldsymbol{k}$ is occupied by a pair given that is not occupied by a quasi-particle. Thus some states are occupied by quasi-particles and the unoccupied phase space is available for the formation of the coherent background of the remaining electrons. Since a portion of phase space is occupied by excitations at finite temperatures, making it unavailable for the transitions of bound pairs, the correlation energy is a function of the temperature, $W_c(T)$. As T increases, $W_c(T)$ and at the same time Δ decrease until the critical temperature is reached and the system reverts to the normal phase.

Since the excitations of the superconductor are independent and in a one-to-one correspondence with those of the normal metal, the entropy of an excited configuration is given by an expression identical with that for the normal metal except that the Fermi function, $f(\boldsymbol{k})$, refers to quasi-particle excitations. The correlation energy at finite temperature is given by an expression similar to that at $T = 0$ with the available phase space modified by the occupation functions $f(\boldsymbol{k})$. Setting the variation of F with respect to h, φ, and f equal to zero gives:

$$h = 1/2\ (1-\varepsilon/E)$$

$$E = \sqrt{\varepsilon^2 + |\Delta|^2}$$

and

$$f = \frac{1}{1 + \exp(E/k_B T)}$$

where

$$\Delta = |\Delta| e^{i\varphi}$$

is now temperature-dependent and satisfies the fundamental integral equation of the theory

$$\Delta_k(T) = -1/2 \sum_{kk'} V_{kk'} \frac{\Delta_{k'}(T)}{E_{k'}(T)} \tanh\left(\frac{E_{k'}(T)}{2k_B T}\right).$$

The form of these equations is the same as that at $T = 0$ except that the energy gap varies with the temperature. The equation for the energy gap can be satisfied with non-zero values of Δ only in a restricted temperature range. The upper bound of this temperature range is defined as T_c, the critical temperature. For $T < T_c$, singlet spin zero momentum electrons are strongly correlated, there is an energy gap associated with exciting electrons from the correlated part of the wave function and $E(\boldsymbol{k})$ is bounded below by $|\Delta|$. In this region the system has properties qualitatively different from the normal metal.

In the region $T > T_c$, $\Delta = 0$ and we have in every respect the normal solution. In particular f, the distribution function for excitations, becomes just the Fermi function for excited electrons $k > k_F$, and for holes $k < k_F$

$$f = \frac{1}{1 + \exp(|\varepsilon|/k_B T)}.$$

If we make our simplifications of 1957, (defining in this way an 'ideal' superconductor)

$$V_{k'k} = -V \qquad |\varepsilon| < \hbar\omega_{av}$$
$$= 0 \qquad \text{otherwise}$$

and replace the energy dependent density of states by its value at the Fermi surface, $\mathcal{N}(0)$, the integral equation for Δ becomes

$$1 = \mathcal{N}(0)\, V \int_0^{\hbar\omega_{av}} \frac{d\varepsilon}{\sqrt{\varepsilon^2 + |\Delta|^2}} \tanh\left(\frac{\sqrt{\varepsilon^2 + |\Delta|^2}}{2k_B T}\right).$$

The solution of this equation, Fig. 7, gives $\Delta(T)$ and with this f and h. We can then calculate the free energy of the superconducting state and obtain the thermodynamic properties of the system.

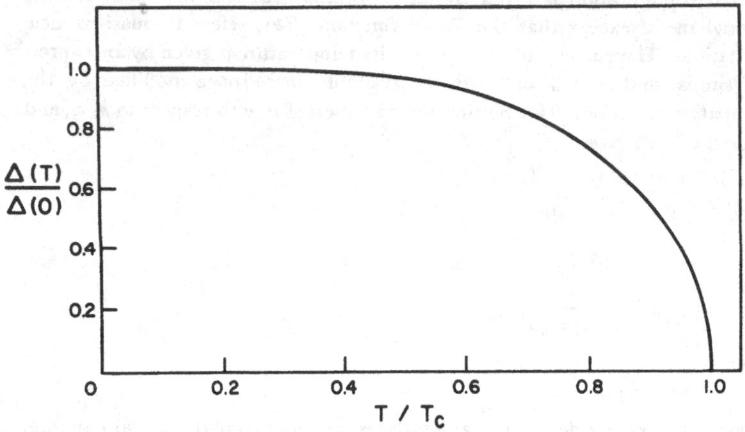

Fig. 7.
Variation of the energy gap with temperature for the ideal superconductor.

In particular one finds that at T_c (in the absence of a magnetic field) there is a second-order transition (no latent heat: $W_c = 0$ at T_c) and a discontinuity in the specific heat. At very low temperatures the specific heat goes to zero exponentially. For this ideal superconductor one also obtains a law of corresponding states in which the ratio

$$\frac{\gamma T_c^2}{H_0^2} = 0.170,$$

where

$$\gamma = 2/3\pi^2 \mathcal{N}(0) k_B^2.$$

The experimental data scatter about the number 0.170. The ratio of Δ to $k_B T_c$ is given as a universal constant

$$\Delta / k_B T_c = 1.75.$$

There are no arbitrary parameters in the idealized theory. In the region of empirical interest all thermodynamic properties are determined by the quantities γ and $\hbar w_{av}\ e^{-1/N(0)V}$. The first, γ, is found by observation of the normal specific heat, while the second is found from the critical temperature, given by

$$k_B T_c = 1.14\ \hbar\omega_{av} e^{-1/N(0)V}.$$

At the absolute zero

$$\Delta = \hbar\omega_{av}/\sinh\left(\frac{1}{\mathcal{N}(0)V}\right).$$

Further, defining a weak coupling limit $[\mathcal{N}(0)V \ll 1]$ which is one region of interest empirically, we obtain

$$\Delta \simeq 2\hbar\omega_{av} e^{-1/N(0)V}.$$

The energy difference between the normal and superconducting states becomes (again in the weak coupling limit)

$$W_s - W_n = W_c = -2\mathcal{N}(0)(\hbar\omega_{av})^2\ e^{-2/N(0)V}.$$

The dependence of the correlation energy on $(\hbar\omega_{av})^2$ gives the isotope effect, while the exponential factor reduces the correlation energy from the dimensionally expected $\mathcal{N}(0)(\hbar\omega_{av})^2$ to the much smaller observed value. This, however, is more a demonstration that the isotope effect is consistent with our model rather than a consequence of it, as will be discussed further by Professor Bardeen.

The thermodynamic properties calculated for the ideal superconductor are in qualitative agreement with experiment for weakly coupled superconductors. Very detailed comparison between experiment and theory has been made by many authors. A summary of the recent status may be found in reference (2). When one considers that in the theory of the ideal superconductor the existence of an actual metal is no more than hinted at (We have in fact done all the calculations considering weakly interacting fermions in a container.) so that in principle (with appropriate modifications) the calculations apply to neutron stars as well as metals, we must regard detailed quantitative agreement as a gift from above. We should be content if there is a single metal for which such agreement exists. [Pure single crystals of tin or vanadium are possible candidates.]

To make comparison between theory and experiments on actual metals, a plethora of detailed considerations must be made. Professor Bardeen will discuss developments in the theory of the electron-phonon interaction and the resulting dependence of the electron-electron interaction and superconducting properties on the phonon spectrum and the range of the Coulomb repulsion. Crystal symmetry, Brillouin zone structure and the actual wave function (*S*, *P* or *D* states) of the conduction electrons all play a role in determining real metal behavior. There is a fundamental distinction between superconductors w ich always show a Meissner effect and those (type II) which allow magnetic field penetration in units of the flux quantum.

When one considers, in addition, specimens with impurities (magnetic and otherwise) superimposed films, small samples, and so on, one obtains a variety of situations, developed in the years since 1957 by many authors, whose richness and detail takes volumes to discuss. The theory of the ideal superconductor has so far allowed the addition of those extensions and modifications necessary to describe, in what must be considered remarkable detail, all of the experience actually encountered.

Microscopic Interference Effects

In its interaction with external perturbations the superconductor displays remarkable interference effects which result from the paired nature of the wave function and are not at all present in similar normal metal interactions. Neither would they be present in any ordinary two-fluid model. These "coherence effects" are in a sense manifestations of interference in spin and momentum space on a microscopic scale, analogous to the macroscopic quantum effects due to interference in ordinary space which Professor Schrieffer discussed. They depend on the behavior under time reversal of the perturbing fields. (8) It is intriguing to speculate that if one could somehow amplify them

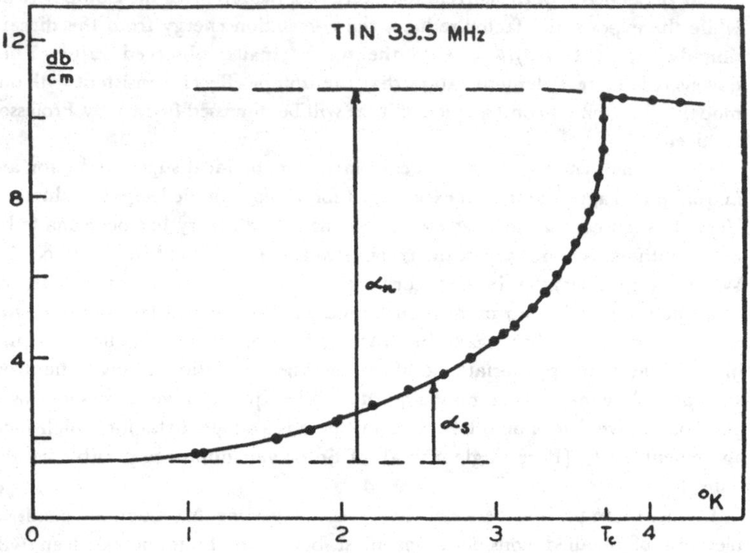

Fig. 8.
Ultrasonic attenuation as a function of temperature across the superconducting transition as measured by Morse and Bohm.

properly, the time reversal symmetry of a fundamental interaction might be tested. Further, if helium 3 does in fact display a phase transition analogous to the superconducting transition in metals as may be indicated by recent experiments (4) and this is a spin triplet state, the coherences effects would be greatly altered.

Near the transition temperature these coherence effects produce quite dramatic contrasts in the behavior of coefficients which measure interactions with the conduction electrons. Historically, the comparison with theory of the behavior of the relaxation rate of nuclear spins (9) and the attenuation of longitudinal ultrasonic waves in clean samples (10) as the temperature is decreased through T_c provided an early test of the detailed structure of the theory.

The attenuation of longitudinal acoustic waves due to their interaction with the conduction electrons in a metal undergoes a very rapid drop (10a) as the temperature drops below T_c. Since the scattering of phonons from "normal" electrons is responsible for most of the acoustic attenuation, a drop was to be expected; but the rapidity of the decrease measured by Morse and Bohm (10b) Fig. 8 was difficult to reconcile with estimates of the decrease in the normal electron component of a two-fluid model.

The rate of relaxation of nuclear spins was measured by Hebel and Slichter (9a) in zero magnetic field in superconducting aluminum from 0.94 K to 4.2 K just at the time of the development of our 1957 theory. Redfield and Anderson (9b) confirmed and extended their results. The dominant relaxation mechanism is provided by interaction with the conduction electrons so that one would expect, on the basis of a two-fluid model, that this rate should decrease below the transition temperature due to the diminishing density of "normal" electrons. The experimental results however show just the reverse. The relaxation rate does not drop but increases by a factor of more than two just below the transition temperature. Fig 13. This observed increase in the nuclear spin relaxation rate and the very sharp drop in the acoustic attenuation coefficient as the temperature is decreased through T_c impose contradictory requirements on a conventional two-fluid model.

To illustrate how such effects come about in our theory, we consider the transition probability per unit time of a process involving electronic transitions from the excited state $\boldsymbol{k}$ to the state $\boldsymbol{k}'$ with the emission to or absorption of energy from the interacting field. What is to be calculated is the rate of transition between an initial state $|i>$ and a final state $|f>$ with the absorption or emission of the energy $\hbar\omega_{|\boldsymbol{k}'-\boldsymbol{k}|}$ (a phonon for example in the interaction of sound waves with the superconductor). All of this properly summed over final states and averaged with statistical factors over initial states may be written:

$$\omega = \frac{2\pi}{\hbar} \frac{\sum_{i,f} \exp(-W_i/k_B T) \left| <f\left| H_{\text{int}} \right| i> \right|^2 \delta(W_f - W_i)}{\sum_i \exp(-W_i/k_B T)}$$

We focus our attention on the matrix element $<f|H_{\text{int}}|i>$. This typically contains as one of its factors matrix elements between excited states of the superconductor of the operator

$$B = \sum_{\boldsymbol{K}\boldsymbol{K}'} B_{\boldsymbol{K}'\boldsymbol{K}} c^*_{\boldsymbol{K}'} c_{\boldsymbol{K}}$$

where $c^*_{\boldsymbol{K}'}$ and $c_{\boldsymbol{K}}$ are the creation and annihilation operators for electrons in the states $\boldsymbol{K}'$ and $\boldsymbol{K}$, and $B_{\boldsymbol{K}'\boldsymbol{K}}$ is the matrix element between the states $\boldsymbol{K}'$ and $\boldsymbol{K}$ of the configuration space operator $B(\boldsymbol{r})$

$$B_{\boldsymbol{K}'\boldsymbol{K}} = <\boldsymbol{K}' \left| B(\boldsymbol{r}) \right| \boldsymbol{K}>.$$

The operator B is the electronic part of the matrix element between the full final and initial state

$$<f|H_{\text{int}}|i> = m_{fi}<f|B|i>.$$

In the normal system scattering from single-particle electron states $\boldsymbol{K}$ to $\boldsymbol{K}'$ is independent of scattering from $-\boldsymbol{K}'$ to $-\boldsymbol{K}$. But the superconducting states are linear superpositions of $(\boldsymbol{K}, -\boldsymbol{K})$ occupied and unoccupied. Because of this states with excitations $\boldsymbol{k}\uparrow$ and $\boldsymbol{k}'\uparrow$ are connected not only by $c^*_{\boldsymbol{k}'\uparrow} c_{\boldsymbol{k}\uparrow}$ but also by $c^*_{-\boldsymbol{k}\downarrow} c_{-\boldsymbol{k}'\downarrow}$; if the state $|f>$ contains the single-particle excitation $\boldsymbol{k}'\uparrow$ while the state $|i>$ contains $\boldsymbol{k}\uparrow$, as a result of the superposition of occupied and unoccupied pair states in the coherent part of the wave function, these are connected not only by $B_{\boldsymbol{K}'\boldsymbol{K}}\, c^*_{\boldsymbol{K}'} c_{\boldsymbol{K}}$ but also by $B_{-\boldsymbol{K}-\boldsymbol{K}'}\, c^*_{-\boldsymbol{K}} c_{-\boldsymbol{K}'}$.

For operators which do not flip spins we therefore write:

$$B = \sum_{\boldsymbol{k}\boldsymbol{k}'} (B_{\boldsymbol{K}'\boldsymbol{K}}\, c^*_{\boldsymbol{K}'} c_{\boldsymbol{K}} + B_{-\boldsymbol{K}-\boldsymbol{K}'}\, c^*_{-\boldsymbol{K}} c_{-\boldsymbol{K}'}).$$

Many of the operators, B, we encounter (e.g., the electric current, or the charge density operator) have a well-defined behavior under the operation of time reversal so that

$$B_{\boldsymbol{K}'\boldsymbol{K}} = \pm B_{-\boldsymbol{K}-\boldsymbol{K}'} \equiv B_{\boldsymbol{k}'\boldsymbol{k}}.$$

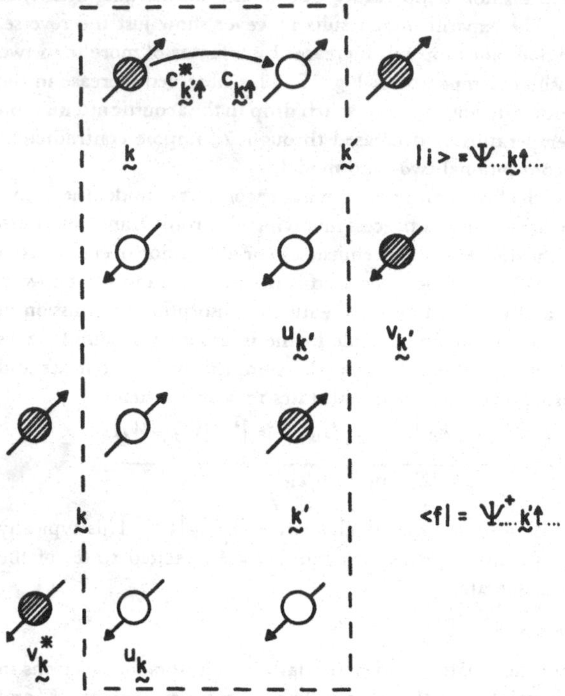

Fig. 9.
The two states $|i\rangle$ and $\langle f|$ shown are connected by $c^*_{k'\uparrow}c_{k\uparrow}$ with the amplitude $u_{k'}u_k$.

Then B becomes

$$B = \sum_{kk'} B_{k'k} (c^*_{k'\uparrow}c_{k\uparrow} \pm c^*_{-k\downarrow}c_{-k'\downarrow})$$

where the upper (lower) sign results for operators even (odd) under time reversal.

The matrix element of B between the initial state, $\psi \ldots {}_{k\uparrow} \ldots$, and the final state $\psi \ldots {}_{k'\uparrow} \ldots$ contains contributions from $c^*_{k'\uparrow}c_{k\uparrow}$ Fig. 9 and unexpectedly from $c^*_{-k\downarrow}c_{-k'\downarrow}$ Fig. 10. As a result the matrix element squared $|\langle f|B|i\rangle|^2$ contains terms of the form

$$|B_{k'k}|^2 \; |(u_{k'}u_k \mp v_{k'}v^*_k)|^2,$$

where the sign is determined by the behavior of B under time reversal:

upper sign B even under time reversal
lower sign B odd under time reversal.

Applied to processes involving the emission or absorption of boson quanta such as phonons or photons, the squared matrix element above is averaged with the appropriate statistical factors over initial and summed over final states; substracting emission from absorption probability per unit time, we obtain typically

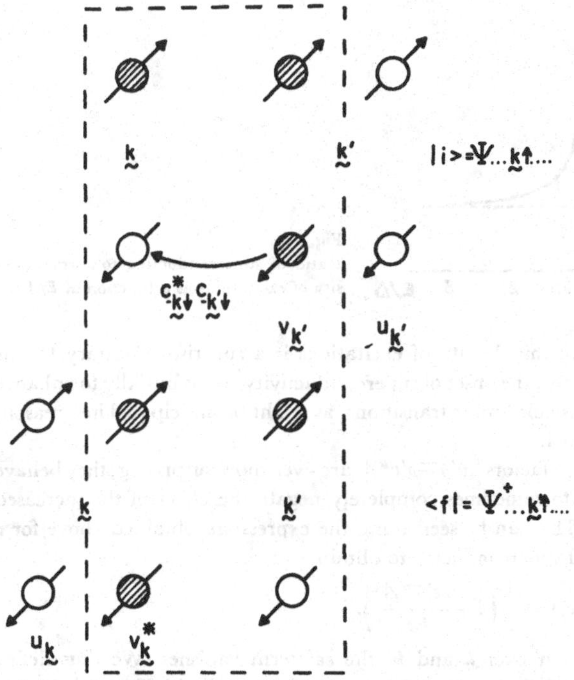

Fig. 10.
The two states $|i>$ and $<f|$ are also connected by $c^*_{k\downarrow}c_{-k'\downarrow}$ with the amplitude $-v_{k'}v_k^*$.

$$a = \frac{4\pi}{\hbar} |m|^2 \sum_{kk'} |(u_{k'}u_k \mp v_{k'}v_k^*)|^2 \; (f_{k'}-f_k) \; \delta(E_{k'}-E_k-\hbar\omega_{|k-k'|})$$

where f_k is the occupation probability in the superconductor for the excitation $k\uparrow$ or $k\downarrow$. [In the expression above we have considered only quasiparticle or quasi-hole scattering processes (not including processes in which a pair of excitations is created or annihilated from the coherent part of the wave function) since $\hbar\omega_{|k'-k|} < \Delta$, is the usual region of interest for the ultrasonic attenuation and nuclear spin relaxation we shall contrast.]

For the ideal superconductor, there is isotropy around the Fermi surface and symmetry between particles and holes; therefore sums of the form $\sum_k$ can be converted to integrals over the superconducting excitation energy, E:

$$\sum_k \rightarrow 2\mathcal{N}(0) \int_\Delta^\infty \frac{E}{\sqrt{E^2-\Delta^2}} \, dE$$

where $\mathcal{N}(0)\dfrac{E}{\sqrt{E^2-\Delta^2}} = \mathcal{N}(0)\dfrac{E}{\varepsilon}$ is the density of excitations in the superconductor, Fig. 11.

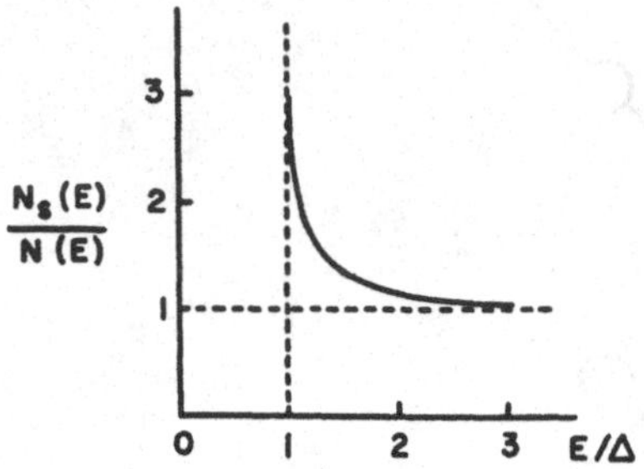

Fig. 11.
Ratio of superconducting to normal density of excitations as a function of E/Δ.

The appearance of this density of excitations is a surprise. Contrary to our intuitive expectations, the onset of superconductivity seems initially to enhance rather than diminish electronic transitions, as might be anticipated in a reasonable two-fluid model.

But the coherence factors $|(u'u \mp v'v^*)|^2$ are even more surprising; they behave in such a way as to sometimes completely negate the effect of the increased density of states. This can be seen using the expressions obtained above for u and v for the ideal superconductor to obtain

$$(u'u \mp v'v)^2 = \frac{1}{2}\left(1+\frac{\varepsilon\varepsilon' \mp \Delta^2}{EE'}\right).$$

In the integration over $\boldsymbol{k}$ and $\boldsymbol{k}'$ the $\varepsilon\varepsilon'$ term vanishes. We thus define $(u'u \mp v'v)_s^2$; in usual limit where $\hbar\omega_{|\boldsymbol{k}'-\boldsymbol{k}|} \ll \Delta$, $\varepsilon \simeq \varepsilon'$ and $E \simeq E'$, this becomes

$$(u^2-v^2)_s^2 \to \frac{1}{2}\left(\frac{\varepsilon^2}{E^2}\right) \qquad \text{operators even under time reversal}$$

$$(u^2+v^2)_s^2 \to \frac{1}{2}\left(1+\frac{E^2}{\Delta^2}\right) \qquad \text{operators odd under time reversal.}$$

For operators even under time reversal, therefore, the decrease of the coherence factors near $\varepsilon = 0$ just cancels the increase due to the density of states. For the operators odd under time reversal the effect of the increase of the density of states is not cancelled and should be observed as an increase in the rate of the corresponding process.

In general the interaction Hamiltonian for a field interacting with the superconductor (being basically an electromagnetic interaction) is invariant under the operation of time reversal. However, the operator B might be the electric current $\boldsymbol{j}(\boldsymbol{r})$ (for electromagnetic interactions) the electric charge density $\varrho(\boldsymbol{r})$ (for the electron-phonon interaction) or the z component of the electron spin operator, σ_z (for the nuclear spin relaxation interaction). Since under time-reversal

$\boldsymbol{j}(\boldsymbol{r}, \mathrm{t}) \to -\boldsymbol{j}(\boldsymbol{r}, -t)$ (electromagnetic interaction)
$\varrho(\boldsymbol{r}, \mathrm{t}) \to +\varrho(\boldsymbol{r}, -t)$ (electron-phonon interaction)
$\sigma_z(t) \to -\sigma_z(-t)$ (nuclear spin relaxation interaction)

these show strikingly different interference effects.

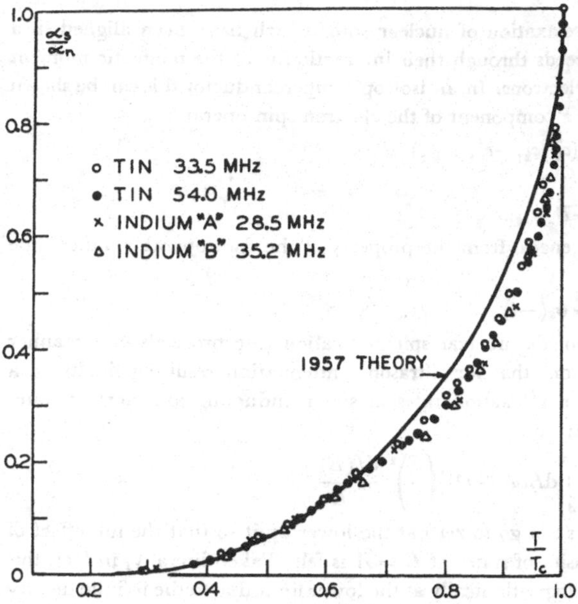

Fig. 12.
Comparison of observed ultrasonic attenuation with the ideal theory. The data are due to Morse and Bohm.

Ultrasonic attenuation in the ideal pure superconductor for $ql \gg 1$ (the product of the phonon wave number and the electron mean free path) depends in a fundamental way on the absorption and emission of phonons. Since the matrix elements have a very weak dependence on changes near the Fermi surface in occupation of states other than $\boldsymbol{k}$ or $\boldsymbol{k}'$ that occur in the normal to superconducting transition, calculations within the quasi-particle model can be compared in a very direct manner with similar calculations for the normal metal, as $B_{k'k}$ is the same in both states. The ratio of the attenuation in the normal and superconducting states becomes:

$$\frac{a_s}{a_n} = -4\int_{\Delta}^{\infty} dE(u^2-v^2)_s^2 \left(\frac{E}{\varepsilon}\right)^2 \frac{df(E)}{dE}.$$

Since $(u^2-v^2)_s^2 \to \frac{1}{2}\left(\frac{\varepsilon}{E}\right)^2$, the coherence factors cancel the density of states giving

$$\frac{a_s}{a_n} = 2f(\Delta(T)) = \frac{2}{1+\exp\left(\dfrac{\Delta(T)}{k_B T}\right)}.$$

Morse and Bohm (10b) used this result to obtain a direct experimental determination of the variation of Δ with T. Comparison of their attenuation data with the theoretical curve is shown in Figure 12.

In contrast the relaxation of nuclear spins which have been aligned in a magnetic field proceeds through their interaction with the magnetic moment of the conduction electrons. In an isotropic superconductor this can be shown to depend upon the z component of the electron spin operator

$$B_{K'K} = B(c^*_{k'\uparrow}c_{k\uparrow} - c^*_{-k\downarrow}c_{-k'\downarrow})$$

so that

$$B_{K'K} = -B_{-K-K'}.$$

This follows in general from the property of the spin operator under time reversal

$$\sigma_z(t) = -\sigma_z(-t).$$

The calculation of the nuclear spin relaxation rate proceeds in a manner not too different from that for ultrasonic attenuation resulting finally in a ratio of nuclear spin relaxation rates in superconducting and normal states in the same sample:

$$\frac{R_s}{R_n} = -4\int_{\Delta}^{\infty} dE(u^2+v^2)^2_s\left(\frac{E}{\varepsilon}\right)^2 \frac{df(E)}{dE}.$$

But $(u^2+v^2)^2_s$ does not go to zero at the lower limit so that the full effect of the increase in density of states at $E=\Delta$ is felt. Taken literally, in fact, this expression diverges logarithmically at the lower limit due to the infinite density of states. When the Zeeman energy difference between the spin up and spin down states is included, the integral is no longer divergent but the integrand is much too large. Hebel and Slichter, by putting in a broadening of levels phenomenologically, could produce agreement between theory and experiment. More recently Fibich (11) by including the effect of thermal phonons has obtained the agreement between theory and experiment shown in Fig. 13.

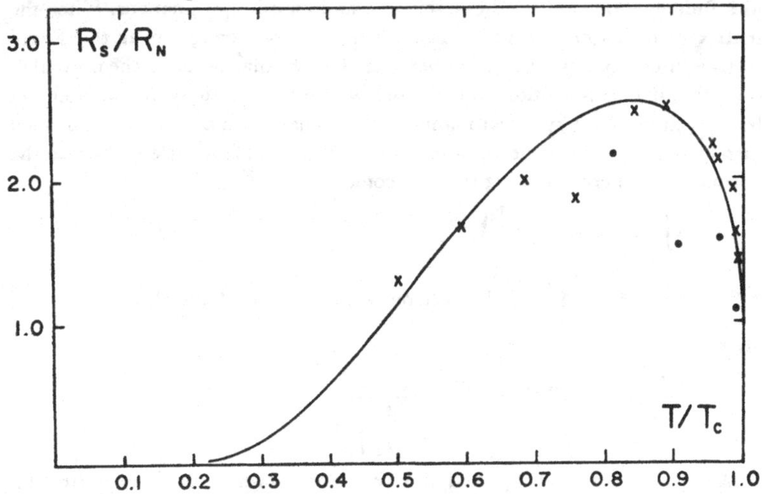

Fig. 13.
Comparison of observed nuclear spin relaxation rate with theory. The circles represent experimental data of Hebel and Slichter, the crosses data by Redfield and Anderson.

Interference effects manifest themselves in a similar manner in the interaction of electromagnetic radiation with the superconductor. Near T_c the absorption is dominated by quasi-particle scattering matrix elements of the type we have described. Near $T = 0$, the number of quasi-particle excitations goes to zero and the matrix elements that contribute are those in which quasi-particle pairs are created from ψ_0. For absorption these latter occur only when $\hbar\omega > 2\Delta$. For the linear response of the superconductor to a static magnetic field, the interference occurs in such a manner that the paramagnetic contribution goes to zero leaving the diamagnetic part which gives the Meissner effect.

The theory developed in 1957 and applied to the equilibrium properties of uniform materials in the weak coupling region has been extended in numerous directions by many authors. Professor Schrieffer has spoken of Josephson junctions and macroscopic quantum interference effects; Professor Bardeen will discuss the modifications of the theory when the electron-phonon interactions are strong. The treatment of ultrasonic attenuation, generalized to include situations in uniform superconductors in which $ql < 1$, gives a surprisingly similar result to that above. (12) There have been extensive developments using Green's function methods (13) appropriate for type II superconductors, materials with magnetic impurities and non-uniform materials or boundary regions where the order parameter is a function of the spatial coordinates. (14) With these methods formal problems of gauge invariance and/or current conservation have been resolved in a very elegant manner. (15) In addition, many calculations (16) of great complexity and detail for type II superconductors have treated ultrasonic attenuation, nuclear spin relaxation and other phenomena in the clean and dirty limits (few or large numbers of impurities). The results cited above are modified in various ways. For example, the average density of excitation levels is less sharply peaked at T_c in a type II superconductor; the coherence effects also change somewhat in these altered circumstances but nevertheless play an important role. Overall one can say that the theory has been amenable to these generalizations and that agreement with experiment is good.

It is now believed that the finite many-nucleon system that is the atomic nucleus enters a correlated state analogous to that of a superconductor. (17) Similar considerations have been applied to many-fermion systems as diverse as neutron stars, (18) liquid He^3, (19) and to elementary fermions. (20) In addition the idea of spontaneously broken symmetry of a degenerate vacuum has been applied widely in elementary particle theory and recently in the theory of weak interactions. (21) What the electron-phonon interaction has produced between electrons in metals may be produced by the van der Waals interaction between atoms in He^3, the nuclear interaction in nuclei and neutron stars, and the fundamental interactions in elementary fermions. Whatever the success of these attempts, for the theoretician the possible existence of this correlated paired state must in the future be considered for any degenerate many-fermion system where there is some kind of effective attraction between fermions for transitions near the Fermi surface.

In the past few weeks my colleagues and I have been asked many times: "What are the practical uses of your theory?" Although even a summary inspection of the proceedings of conferences on superconductivity and its applications would give an immediate sense of the experimental, theoretical and developmental work in this field as well as expectations, hopes and anticipations—from applications in heavy electrical machinery to measuring devices of extraordinary sensitivity and new elements with very rapid switching speeds for computers — I, personally, feel somewhat uneasy responding. The discovery of the phenomena and the development of the theory is a vast work to which many scientists have contributed. In addition there are numerous practical uses of the phenomena for which theory rightly should not take credit. A theory (though it may guide us in reaching them) does not produce the treasures the world holds. And the treasures themselves occasionally dazzle our attention; for we are not so wealthy that we may regard them as irrelevant.

But a theory is more. It is an ordering of experience that both makes experience meaningful and is a pleasure to regard in its own right. Henri Poincaré wrote (22):

> Le savant doit ordonner; on fait la science
> avec des faits comme une maison avec des
> pierres; mais une accumulation de faits
> n'est pas plus une science qu'un tas de
> pierres n'est une maison.

One can build from ordinary stone a humble house or the finest chateau. Either is constructed to enclose a space, to keep out the rain and the cold. They differ in the ambition and resources of their builder and the art by which he has achieved his end. A theory, built of ordinary materials, also may serve many a humble function. But when we enter and regard the relations in the space of ideas, we see columns of remarkable height and arches of daring breadth. They vault the fine structure constant, from the magnetic moment of the electron to the behavior of metallic junctions near the absolute zero; they span the distance from materials at the lowest temperatures to those in the interior of stars, from the properties of operators under time reversal to the behavior of attenuation coefficients just beyond the transition temperature.

I believe that I speak for my colleagues in theoretical science as well as myself when I say that our ultimate, our warmest pleasure in the midst of one of these incredible structures comes with the realization that what we have made is not only useful but is indeed a beautiful way to enclose a space.

References and Notes

1. Bardeen, J., Cooper, L. N. and Schrieffer, J. R., Phys. Rev. *108*, 1175 (1957).
2. An account of the situation as of 1969 may be found in the two volumes: Superconductivity, edited by R. D. Parks, Marcel Dekker, Inc., New York City (1969).
3. Sommerfeld, A., Z. Physik *47*, 1 (1928). Bloch, F., Z. Physik *52* 555 (1928).
4. Osheroff, D. D., Gully, W. J., Richardson R. C. and Lee, D. M., Phys. Rev. Letters *29*, 920 (1972); Leggett, A. J., Phys. Rev. Letters *29*, 1227 (1972).

5. Bogoliubov, N. N., Nuovo Cimento *7*, 794 (1958); Usp. Fiz. Nauk *67*, 549 (1959); Valatin, J. G., Nuovo Cimento *7*, 843 (1958).
6. Bardeen, J. and Rickayzen, G., Phys. Rev. *118*, 936 (1960); Mattis, D. C. and Lieb, E., J. Math. Phys. *2*, 602 (1961); Bogoliubov, N. N., Zubarev D. N. and Tserkovnikov, Yu. A., Zh. Eksperim. i Teor. Fiz. *39*, 120 (1960) translated: Soviet Phys. JETP *12*, 88 (1961).
7. For example, Glover, R. E. III and Tinkham, M., Phys. Rev. *108*, 243 (1957); Biondi, M. A. and Garfunkel, M. P., Phys. Rev. 116, 853 (1959).
8. The importance of the coupling of time reversed states in constructing electron pairs was emphasized by P. W. Anderson,; for example, Anderson, P. W., J. Phys. Chem. Solids *11*, 26 (1959)
9. a) Hebel, L. C. and Slichter, C. P., Phys. Rev. *113*, 1504 (1959).
   b) Redfield, A. G. and Anderson, A. G., Phys. Rev. *116*, 583 (1959).
10. a) Bommel, H. E., Phys. Rev. *96*, 220 (1954).
    b) Morse, R. W. and Bohm, H. V., Phys. Rev. *108*, 1094 (1957).
11. Fibich, M., Phys. Rev. Letters *14*, 561 (1965).
12. For example, Tsuneto T., Phys. Rev. *121*, 402 (1961).
13. Gor'kov, L. P., Zh. Eksperim. i Teor. Fiz. *34*, 735 (1958) translated: Soviet Physics JETP *7*, 505 (1958); also Martin P. C. and Schwinger J., Phys. Rev. *115*, 1342 (1959); Kadanoff L. P. and Martin P. C., Phys. Rev. *124*, 670 (1961).
14. Abrikosov, A. A. and Gor'kov, L. P., Zh. Eksperim. i Teor. Fiz. *39*, 1781 (1960) translated: Soviet Physics JETP *12*, 1243 (1961); de Gennes P. G., Superconductivity of Metals and Alloys, Benjamin, New York (1966).
15. For example, Ambegaokar V. and Kadanoff L. P., Nuovo Cimento *22*, 914 (1961).
16. For example, Caroli C. and Matricon J., Physik Kondensierten Materie *3*, 380 (1965); Maki K., Phys. Rev. *141*, 331 (1966); *156*, 437 (1967); Groupe de Supraconductivité d'Orsay, Physik Kondensierten Materie *5*, 141 (1966); Eppel D., Pesch W. and Tewordt L., Z. Physik *197*; 46 (1966); McLean F. B. and Houghton A., Annals of Physics *48*, 43 (1968).
17. Bohr, A., Mottelson, B. R. and Pines, D., Phys. Rev. *110*, 936 (1958); Migdal, A. B., Nuclear Phys. *13*, 655 (1959).
18. Ginzburg, V. L. and Kirzhnits, D. A., Zh. Eksperim. i Theor. Fiz. *47*, 2006 (1964) translated: Soviet Physics JETP *20*, 1346 (1965); Pines D., Baym G. and Pethick C., Nature *224*, 673 (1969).
19. Many authors have explored the possibility of a superconducting-like transition in He^3. Among the most recent contributions see reference 4.
20. For example. Nambu Y. and Jona-Lasinio G., Phys. Rev. *122*, 345 (1961).
21. Goldstone, J., Nuovo Cimento *19*, 154 (1961); Weinberg, S., Phys. Rev. Letters *19*, 1264 (1967).
22. Poincaré, H., La Science et l'Hypothèse, Flammarion, Paris, pg. 168 (1902). "The scientist must order; science is made with facts as a house with stones; but an accumulation of facts is no more a science than a heap of stones is a house."

ELECTRON-PHONON INTERACTIONS AND SUPERCONDUCTIVITY

Nobel Lecture, December 11, 1972

By John Bardeen

Departments of Physics and of Electrical Engineering
University of Illinois
Urbana, Illinois

1
Introduction

Our present understanding of superconductivity has arisen from a close interplay of theory and experiment. It would have been very difficult to have arrived at the theory by purely deductive reasoning from the basic equations of quantum mechanics. Even if someone had done so, no one would have believed that such remarkable properties would really occur in nature. But, as you well know, that is not the way it happened, a great deal had been learned about the experimental properties of superconductors and phenomenological equations had been given to describe many aspects before the microscopic theory was developed. Some of these have been discussed by Schrieffer and by Cooper in their talks.

My first introduction to superconductivity came in the 1930's and I greatly profited from reading David Shoenberg's little book on superconductivity, [1] which gave an excellent summary of the experimental findings and of the phenomenological theories that had been developed. At that time it was known that superconductivity results from a phase change of the electronic structure and the Meissner effect showed that thermodynamics could be applied successfully to the superconductive equilibrium state. The two fluid Gorter—Casimir model was used to describe the thermal properties and the London brothers had given their famous phenomenological theory of the electrodynamic properties. Most impressive were Fritz London's speculations, given in 1935 at a meeting of the Royal Society in London, [2] that superconductivity is a quantum phenomenon on a macroscopic scale. He also gave what may be the first indication of an energy gap when he stated that "the electrons be coupled by some form of interaction in such a way that the lowest state may be separated by a finite interval from the excited ones." He strongly urged that, based on the Meissner effect, the diamagnetic aspects of superconductivity are the really basic property.

My first abortive attempt to construct a theory, [3] in 1940, was strongly influenced by London's ideas and the key idea was small energy gaps at the Fermi surface arising from small lattice displacements. However, this work was interrupted by several years of wartime research, and then after the war I joined the group at the Bell Telephone Laboratories where my work turned to semiconductors. It was not until 1950, as a result of the discovery of the isotope effect, that I again began to become interested in superconductivity, and shortly after moved to the University of Illinois.

The year 1950 was notable in several respects for superconductivity theory. The experimental discovery of the isotope effect [4, 5] and the independent prediction of H. Fröhlich [6] that superconductivity arises from interaction between the electrons and phonons (the quanta of the lattice vibrations) gave the first clear indication of the directions along which a microscopic theory might be sought. Also in the same year appeared the phenomenological Ginzburg—Landau equations which give an excellent description of superconductivity near T_c in terms of a complex order parameter, as mentioned by Schrieffer in his talk. Finally, it was in 1950 that Fritz London's book [7] on superconductivity appeared. This book included very perceptive comments about the nature of the microscopic theory that have turned out to be remarkably accurate. He suggested that superconductivity requires "a kind of solidification or condensation of the average momentum distribution." He also predicted the phenomenon of flux quantization, which was not observed for another dozen years.

The field of superconductivity is a vast one with many ramifications. Even in a series of three talks, it is possible to touch on only a few highlights. In this talk, I thought that it might be interesting to trace the development of the role of electron-phonon interactions in superconductivity from its beginnings in 1950 up to the present day, both before and after the development of the microscopic theory in 1957. By concentrating on this one area, I hope to give some impression of the great progress that has been made in depth of understanding of the phenomena of superconductivity. Through developments by many people, [8] electron-phonon interactions have grown from a qualitative concept to such an extent that measurements on superconductors are now used to derive detailed quantitative information about the interaction and its energy dependence. Further, for many of the simpler metals and alloys, it is possible to derive the interaction from first principles and calculate the transition temperature and other superconducting properties.

The theoretical methods used make use of the methods of quantum field theory as adopted to the many-body problem, including Green's functions, Feynman diagrams, Dyson equations and renormalization concepts. Following Matsubara, temperature plays the role of an imaginary time. Even if you are not familiar with diagrammatic methods, I hope that you will be able to follow the physical arguments involved.

In 1950, diagrammatic methods were just being introduced into quantum field theory to account for the interaction of electrons with the field of photons. It was several years before they were developed with full power for application to the quantum statistical mechanics of many interacting particles. Following Matsubara, those prominent in the development of the theoretical methods include Kubo, Martin and Schwinger, and particularly the Soviet physicists, Migdal, Galitski, Abrikosov, Dzyaloshinski, and Gor'kov. The methods were first introduced to superconductivity theory by Gor'kov [9] and a little later in a somewhat different form by Kadanoff and Martin. [10] Problems of superconductivity have provided many applications for the powerful Green's function methods of many-body theory and these applications have helped to further develop the theory.

Diagrammatic methods were first applied to discuss electron-phonon interactions in normal metals by Migdal [11] and his method was extended to superconductors by Eliashberg. [12] A similar approach was given by Nambu. [13] The theories are accurate to terms of order $(m/M)^{1/2}$, where m is the mass of the electron and M the mass of the ion, and so give quite accurate quantitative accounts of the properties of both normal metals and superconductors.

We will first give a brief discussion of the electron-phonon interactions as applied to superconductivity theory from 1950 to 1957, when the pairing theory was introduced, then discuss the Migdal theory as applied to normal metals, and finally discuss Eliashberg's extension to superconductors and subsequent developments. We will close by saying a few words about applications of the pairing theory to systems other than those involving electron-phonon interactions in metals.

2
Developments from 1950—1957

The isotope effect was discovered in the spring of 1950 by Reynolds, Serin, et al, [4] at Rutgers University and by E. Maxwell [5] at the U. S. National Bureau of Standards. Both groups measured the transition temperatures of separated mercury isotopes and found a positive result that could be interpreted as $T_c M^{1/2} \simeq$ constant, where M is the isotopic mass. If the mass of the ions is important, their motion and thus the lattice vibrations must be involved.

Independently, Fröhlich, [6] who was then spending the spring term at Purdue University, attempted to develop a theory of superconductivity based on the self-energy of the electrons in the field of phonons. He heard about the isotope effect in mid-May, shortly before he submitted his paper for publication and was delighted to find very strong experimental confirmation of his ideas. He used a Hamiltonian, now called the Fröhlich Hamiltonian, in which interactions between electrons and phonons are included but Coulomb interactions are omitted except as they can be included in the energies of the individual electrons and phonons. Fröhlich used a perturbation theory approach and found an instability of the Fermi surface if the electron-phonon interaction were sufficiently strong.

When I heard about the isotope effect in early May in a telephone call from Serin, I attempted to revive my earlier theory of energy gaps at the Fermi surface, with the gaps now arising from dynamic interactions with the phonons rather than from small static lattice displacements. [14] I used a variational method rather than a perturbation approach but the theory was also based on the electron self-energy in the field of phonons. While we were very hopeful at the time, it soon was found that both theories had grave difficulties, not easy to overcome. [15] It became evident that nearly all of the self-energy is included in the normal state and is little changed in the transition. A theory involving a true many-body interaction between the electrons seemed to be required to account for superconductivity. Schafroth [16] showed that starting with the Fröhlich Hamiltonian, one cannot derive the Meissner effect in any order of perturbation theory. Migdal's theory, [11] supposedly correct to terms of order $(m/M)^{1/2}$, gave no gap or instability at the Fermi surface and no indication of superconductivity.

Of course Coulomb interactions really are present. The effective direct Coulomb interaction between electrons is shielded by the other electrons and the electrons also shield the ions involved in the vibrational motion. Pines and I derived an effective electron-electron interaction starting from a Hamiltonian in which phonon and Coulomb terms are included from the start. [17] As is the case for the Fröhlich Hamiltonian, the matrix element for scattering of a pair of electrons near the Fermi surface from exchange of virtual phonons is negative (attractive) if the energy difference between the electron states involved is less than the phonon energy. As discussed by Schrieffer, the attractive nature of the interaction was a key factor in the development of the microscopic theory. In addition to the phonon induced interaction, there is the repulsive screened Coulomb interaction, and the criterion for superconductivity is that the attractive phonon interaction dominate the Coulomb interaction for states near the Fermi surface. [18]

During the early 1950's there was increasing evidence for an energy gap at the Fermi surface. [19] Also very important was Pippard's proposed non-local modification [20] of the London electrodynamics which introduced a new length the coherence distance, ξ_0, into the theory. In 1955 I wrote a review article [17] on the theory of superconductivity for the Handbuch der Physik, which was published in 1956. The central theme of the article was the energy gap, and it was shown that Pippard's version of the electrodynamics would likely follow from an energy gap model. Also included was a review of electron-phonon interactions. It was pointed out that the evidence suggested that all phonons are involved in the transition, not just the long wave length phonons, and that their frequencies are changed very little in the normal-superconducting transition. Thus one should be able to use the effective interaction between electrons as a basis for a true many-body theory of the superconducting state. Schrieffer and Cooper described in their talks how we were eventually able to accomplish this goal.

3
Green's Function Method for Normal Metals

By use of Green's function methods, Migdal [11] derived a solution of Fröhlich's Hamiltonian, $H = H_{el}+H_{ph}+H_{el\text{-}ph}$, for normal metals valid for abritrarily strong coupling and which involves errors only of order $(m/M)^{1/2}$. The Green's functions are defined by thermal average of time ordered operators for the electrons and phonons, respectively

$$G = -i\langle T\psi(1)\psi^{+}(2)\rangle \tag{1a}$$

$$D = -i\langle T\phi(1)\phi^{+}(2)\rangle \tag{1b}$$

Here $\psi(r,t)$ is the wave field operator for electron quasi-particles and $\emptyset(r,t)$ for the phonons, the symbols 1 and 2 represent the space-time points (r_1,t_1) and (r_2,t_2) and the brackets represent thermal averages over an ensemble.

Fourier transforms of the Green's functions for $H_0 = H_{el}+H_{ph}$ for non-interacting electrons and phonons are

$$G_0(P) = \frac{1}{\omega_n - \varepsilon_0(k) + i\delta_k} \tag{2a}$$

$$D_0(Q) = \left\{\frac{1}{\nu_n - \omega_0(q) + i\delta} - \frac{1}{\nu_n + \omega_0(q) - i\delta}\right\}, \tag{2b}$$

where $P = (k,\omega_n)$ and $Q = (q,\nu_n)$ are four vectors, $\varepsilon_0(k)$ is the bare electron quasiparticle energy referred to the Fermi surface, $\omega_0(q)$ the bare phonon frequency and ω_n and ν_n the Matsubara frequencies

$$\omega_n = (2n+1)\pi i k_B T; \quad \nu_n = 2n\pi i k_B T \tag{3}$$

for Fermi and Bose particles, respectively.

As a result of the electron-phonon interaction, $H_{el\text{-}ph}$, both electron and phonon energies are renormalized. The renormalized propagators, G and D, can be given by a sum over Feynman diagrams, each of which represents a term in the perturbation expansion. We shall use light lines to represent the bare propagators, G_0 and D_0, heavy lines for the renormalized propagators, G and D, straight lines for the electrons and curly lines for the phonons.

The electron-phonon interaction is described by the vertex

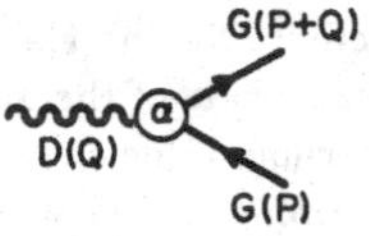

which represents scattering of an electron or hole by emission or absorption of a phonon or creation of an electron and hole by absorption of a phonon by an electron in the Fermi sea. Migdal showed that renormalization of the vertex represents only a small correction, of order $(m/M)^{1/2}$, a result in accord with the Born-Oppenheimer adiabatic-approximation. If terms of this order are neglected, the electron and phonon self-energy corrections are given by the lowest order diagrams provided that fully renormalized propagators are used in these diagrams.

The electron self-energy $\Sigma(P)$ in the Dyson equation:

$$G(P) = G_0(P) + G_0(P)\Sigma(P)G(P) \tag{4}$$

is given by the diagram

(5)

The phonon self-energy, $\pi(Q)$, defined by

(6)

is given by

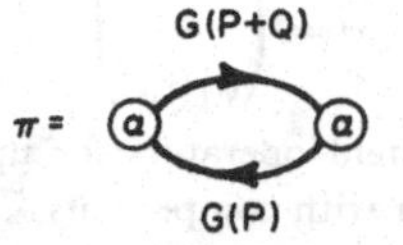

Since to order $(m/M)^{1/2}$ one can use an unrenormalized vertex function $\alpha = \alpha_0$, the Dyson equations form a closed system such that both $\Sigma(P)$ and $\pi(Q)$ can be determined. The phonon self-energy, $\pi(Q)$, gives only a small renormalization of the phonon frequencies. As to the electrons, Migdal noted that we are interested in states k very close to k_F, so that to a close approximation $\Sigma(k,\omega)$ depends only on the frequency. For an isotropic system,

$$\Sigma(k,\omega) \simeq \Sigma(k_F,\omega) \equiv \Sigma(\omega) \tag{7}$$

The renormalized electron quasi-particle energy, ω_k, is then given by a root of

$$\varepsilon(k) = \omega_k = \varepsilon_0(k) + \Sigma(\omega_k) \tag{8}$$

In the thermal Green's function formalism, one may make an analytic continuation from the imaginary frequencies, ω_n, to the real ω axis to determine $\Sigma(\omega)$.

Although $\Sigma(\omega)$ is small compared with the Fermi energy, E_F, it changes rapidly with energy and so can affect the density of states at the Fermi surface and thus the low temperature electronic specific heat. The mass renormalization factor m^*/m, at the Fermi surface may be expressed in terms of a parameter λ:

$$m^*/m = Z(k_F) = 1 + \lambda = (d\varepsilon_0/dk)_F/(d\varepsilon/dk)_F \tag{9}$$

In modern notation, the experession for λ is

$$\lambda = 2 \int_0^\infty d\omega \frac{\alpha^2(\omega)F(\omega)}{\omega}, \tag{10}$$

where $F(\omega)$ is the density of phonon states in energy and $\alpha^2(\omega)$ is the square of the electron-phonon coupling constant averaged over polarization directions of the phonons. Note that λ is always positive so that the Fermi surface is stable if the lattice is stable. Values of λ for various metals range from about 0.5 to 1.5. The parameter λ corresponds roughly to the $N(0)V_{phonon}$ of the BCS theory.

4 Nambu-Eliashberg Theory for Superconductors

Migdal's theory has important consequences that have been verified experimentally for normal metals, but gave no clue as to the origin of superconductivity. Following the introduction of the BCS theory, Gor'kov showed that pairing could be introduced through the anomalous Green's function

$$F(P) = i < T\psi_\uparrow\psi_\downarrow >, \tag{11}$$

Nambu showed that both types of Green's functions can be conveniently included with use of a spinor notation

$$\psi = \begin{pmatrix} \psi_{\uparrow}(r,t) \\ \\ \psi_{\downarrow}^{+}(r,t) \end{pmatrix} \tag{12}$$

where $\psi_{\uparrow}$ and $\psi_{\downarrow}$ are wave field operators for up and down spin electrons and a matrix Green's function with components

$$\tilde{G}_{\alpha\beta} = -\mathrm{i}\langle T\psi_{\alpha}\psi_{\beta}^{+}\rangle \tag{13}$$

Thus G_{11} and G_{22} are the single particle Green's functions for up and down spin particles and $G_{12} = G_{21}^{*} = F(P)$ is the anomalous Green's function of Gor'kov.

There are two self-energies, Σ_1 and Σ_2, defined by the matrix

$$\tilde{\Sigma} = \begin{pmatrix} \Sigma_1 & \Sigma_2 \\ \\ \Sigma_2 & \Sigma_1 \end{pmatrix} \tag{14}$$

Eliashberg noted that one can describe superconductors to the same accuracy as normal metals if one calculates the self-energies with the same diagrams that Migdal used, but with Nambu matrix propagators in place of the usual normal state Green's functions. The matrix equation for $\tilde{G}$ is

$$\tilde{G} = \tilde{G}_0 + \hat{G}_0\tilde{\Sigma}\tilde{G} \tag{15}$$

The matrix equation for $\tilde{\Sigma}$ yields a pair of coupled integral equations for Σ_1 and Σ_2. Again Σ_1 and Σ_2 depend mainly on the frequency and are essentially independent of the momentum variables. Following Nambu, [13] one may define a renormalization factor $Z_s(\omega)$ and a pair potential, $\Delta(\omega)$, for isotropic systems through the equations:

$$\omega Z_s(\omega) = \omega + \Sigma_1(\omega) \tag{16}$$

$$\Delta(\omega) = \Sigma_2(\omega)/Z(\omega). \tag{17}$$

Both Z_s and Δ can be complex and include quasi-particle life-time effects. Eliashberg derived coupled non-linear integral equations for $Z_s(\omega)$ and $\Delta(\omega)$ which involve the electron-phonon interaction in the function $\alpha^2(\omega)F(\omega)$.

The Eliashberg equations have been used with great success to calculate the properties of strongly coupled superconductors for which the frequency dependence of Z and Δ is important. They reduce to the BCS theory and to the nearly equivalent theory of Bogoliubov [21] based on the principle of "compensation of dangerous diagrams" when the coupling is weak. By weak coupling is meant that the significant phonon frequencies are very large compared with $k_B T_c$, so that $\Delta(\omega)$ can be regarded as a constant independent of frequency in the important range of energies extending to at most a few $k_B T_c$. In weak coupling one may also neglect the difference in quasi-particle energy renormalization and assume that $Z_s = Z_n$.

The first solutions of the Eliashberg equations were obtained by Morel and Anderson [22] for an Einstein frequency spectrum. Coulomb interactions were included, following Bogoliubov, by introducing a parameter μ^* which renormalizes the screened Coulomb interaction to the same energy range as the phonon interaction, In weak coupling, $N(0)V = \lambda - \mu^*$. They estimated λ from electronic specific heat data and μ^* from the electron density and thus the transition temperatures, T_c, for a number of metals. Order-of-magnitude agreement with experiment was found. Later work, based in large part on tunneling data, has yielded precise information on the electron-phonon interaction for both weak and strongly-coupled superconductors.

4

ANALYSIS OF TUNNELING DATA

From the voltage dependence of the tunneling current between a normal metal and a superconductor one can derive $\Delta(\omega)$ and thus get direct information about the Green's function for electrons in the superconductor. It is possible to go further and derive empirically from tunneling data the electron-phonon coupling, $\alpha^2(\omega)F(\omega)$, as a function of energy. That electron tunneling should provide a powerful method for investigating the energy gap in superconductors was suggested by I. Giaever, [23] and he first observed the effect in the spring of 1960.

The principle of the method is illustrated in Fig. 1. At very low temperatures, the derivative of the tunneling current with respect to voltage is proportional to the density of states in energy in the superconductor. Thus the ratio of the density of states in the metal in the superconducting phase, $\mathcal{N}_\mathrm{S}$, to that of the same metal in the normal phase, $\mathcal{N}_\mathrm{n}$, at an energy eV above the Fermi surface is given by

$$\frac{\mathcal{N}_\mathrm{s}(eV)}{\mathcal{N}_\mathrm{n}} = \frac{(\mathrm{d}I/\mathrm{d}V)_\mathrm{ns}}{(\mathrm{d}I/\mathrm{d}V)_\mathrm{nn}} \tag{18}$$

$$\left(\frac{\mathrm{d}I}{\mathrm{d}V}\right)_\mathrm{ns} \sim N_\mathrm{s}(\omega) \sim \frac{\omega}{\sqrt{\omega^2 - \Delta^2}}$$

Tunneling from a normal metal into a superconductor

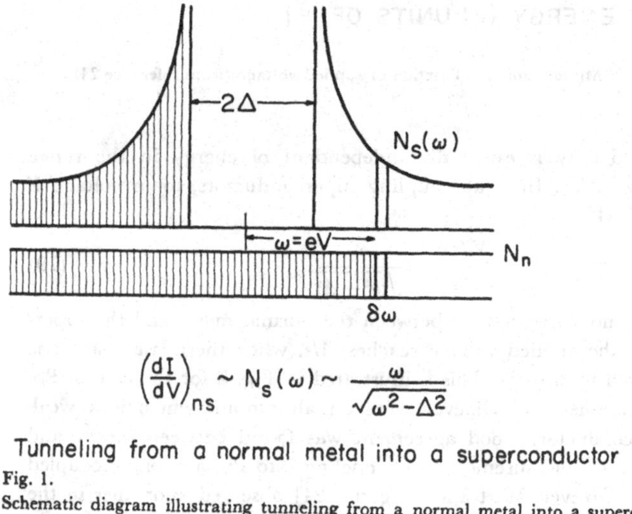

Fig. 1.

Schematic diagram illustrating tunneling from a normal metal into a superconductor near $T = 0°\mathrm{K}$. Shown in the lower part of the diagram is the uniform density of states in energy of electrons in the normal metal, with the occupied states shifted by an energy eV from an applied voltage V across the junction. The upper part of the diagram shows the density of states in energy in the superconductor, with an energy gap $2\triangle$. The effect of an increment of voltage δV giving an energy change $\delta\omega$ is to allow tunneling from states in the range $\delta\omega$. Since the tunneling probability is proportional to density of states $N_\mathrm{s}(\omega)$, the increment in current δI is proportional to $N_\mathrm{s}(\omega)\delta V$.

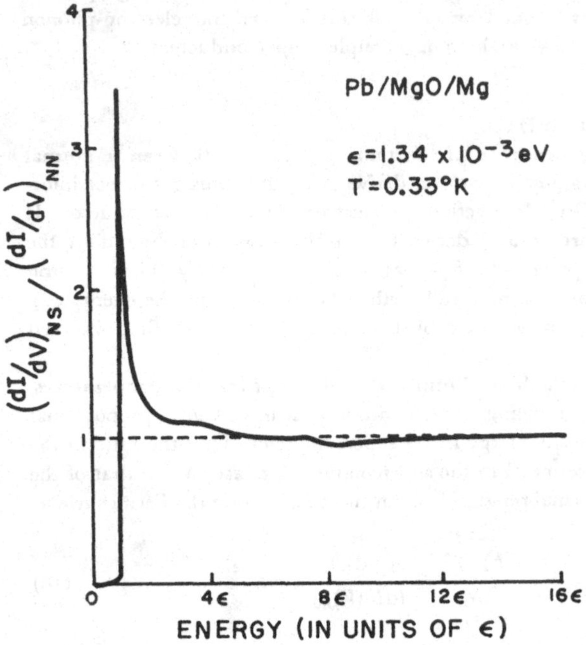

Fig. 2.
Conductance of a Pb-Mg junction as a function of applied voltage (from reference 24).

The normal density is essentially independent of energy in the range involved (a few meV). In weak coupling superconductors, for a voltage V and energy $\omega = eV$,

$$\frac{\mathcal{N}_s(\omega)}{\mathcal{N}_n} = \frac{\omega}{\sqrt{\omega^2 - \Delta^2}}. \tag{19}$$

As $T \to 0$ K, no current flows between the normal metal and the superconductor until the applied voltage reaches Δ/e, when there is a sharp rise in dI/dV followed by a drop. This is illustrated in Fig. 2 for the case of Pb.

The first experiments of Giaever were on aluminum, which is a weak coupling superconductor. Good agreement was found between theory and experiment. In later measurements on tunneling into Pb, a strongly coupled superconductor, Giaever, Hart and Megerle [24] observed anomalies in the density of states that appeared to be associated with phonons, as shown in Fig. 2. These results were confirmed by more complete and accurate tunneling data on Pb by J. M. Rowell et al. [25]

In the meantime, in the summer of 1961, Schrieffer had derived numerical solutions of the Eliashberg equations working with a group engaged in developing methods for computer control using graphical display methods. [26] He and co-workers calculated the complex $\Delta(\omega)$ for a Debye frequency spectrum. Later, at the University of Pennsylvania, he together with J. W. Wilkins and D. J. Scalapino [27] continued work on the problem with a view to explaining the observed anomalies on Pb. They showed that for the general case of a complex $\Delta(\omega)$

$$\frac{(\mathrm{d}I/\mathrm{d}V)_{\mathrm{ns}}}{(\mathrm{d}I/\mathrm{d}V)_{\mathrm{nn}}} = \frac{\mathcal{N}_{\mathrm{s}}(\omega)}{\mathcal{N}_{\mathrm{n}}} = \mathrm{Re}\left\{\frac{\omega}{\sqrt{\omega^2-\Delta^2(\omega)}}\right\} \tag{20}$$

where Re represents the real part. From measurements of the ratio over the complete range of voltages, one can use Kramers-Krönig relations to obtain both the real and imaginary parts of $\Delta(\omega) = \Delta_1(\omega)+\mathrm{i}\Delta_2(\omega)$. From analysis of the data, one can obtain the Green's functions which in turn can be used to calculate the various thermal and transport properties of superconductors. This has been done with great success, even for such strongly-coupled super conductors as lead and mercury.

For lead, Schrieffer et al, used a phonon spectrum consisting of two Lorentzian peaks, one for transverse waves and one for longitudinal and obtained a good fit to the experimental data for $T << T_c$. The calculations were extended up to T_c for Pb, Hg, and Al by Swihart, Wada and Scalapino, [28] again finding good agreement with experiment.

In analysis of tunneling data, one would like to find a phonon interaction spectrum, $\alpha^2(\omega)F(\omega)$, and a Coulomb interaction parameter, μ^*, which when inserted into the Eliashberg equations will yield a solution consistent with the tunneling data. W. L. McMillan devised a computer program such that one could work backwards and derive $\alpha^2(\omega)F(\omega)$ and μ^* directly from the tunneling data. His program has been widely used since then and has been applied to a number of superconducting metals and alloys, including, Al, Pb, Sn, the transition elements Ta and Nb, a rare earth, La, and the compound Nb_3Sn. In all cases it has been found that the phonon mechanism is dominant with reasonable values of μ^*. Peaks in the phonon spectrum agree with peaks in the phonon density of states as found from neutron scattering data, as shown in Fig. 3 for the case of Pb. In Fig. 4 is shown the real and imaginary parts of $\Delta(\omega)$ for Pb as derived from tunneling data.

One can go further and calculate the various thermodynamic and other properties. Good agreement with experiment is found for strongly coupled superconductors even when there are significant deviations from the weak coupling limits. For example, the weak-coupling BCS expression for the condensation energy at $T = 0$ K is

$$E_{\mathrm{BCS}} = \frac{1}{2}\mathcal{N}(0)\mathcal{Z}_{\mathrm{n}}\Delta_{\mathrm{o}}^2 \tag{21}$$

where $\mathcal{N}(0)\mathcal{Z}_{\mathrm{n}}$ is the phonon enhanced density of states and Δ_{o} is the gap parameter at $T = 0$ K. The theoretical expression with $\mathcal{Z}_{\mathrm{S}}(\omega)$ and $\Delta(\omega)$ derived from tunneling data, again for the case of Pb, gives [29, 30, 31]

$$E_{\mathrm{theor}} = 0.78\, E_{\mathrm{BCS}} \tag{22}$$

in excellent agreement with the experimental value

$$E_{\mathrm{exp}} = (0.76+0.02)\, E_{\mathrm{BCS}}. \tag{23}$$

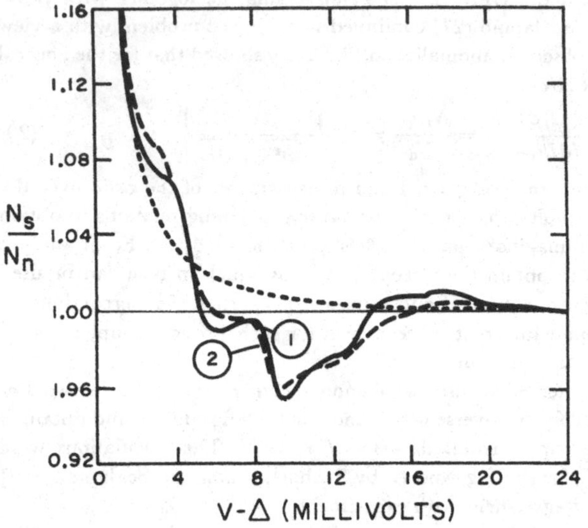

Fig. 3.
Density of states versus energy for Pb. Solid line, calculated by Schrieffer et al; long dashed ine, observed from tunneling; short dashed line, BCS weak coupling theory.

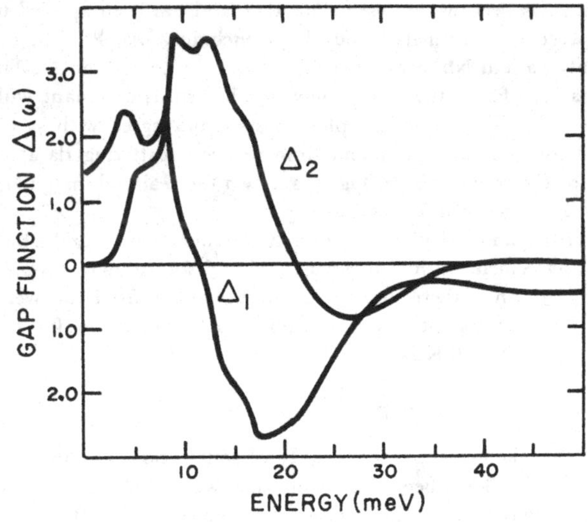

Real and imaginary parts of Δ versus $\omega - \Delta_0$ for Pb.

Fig. 4.
Real and imaginary parts of $\Delta(\omega) = \Delta_1(\omega) + i\Delta_2(\omega)$ versus energy for Pb. (After McMillan & Rowell).

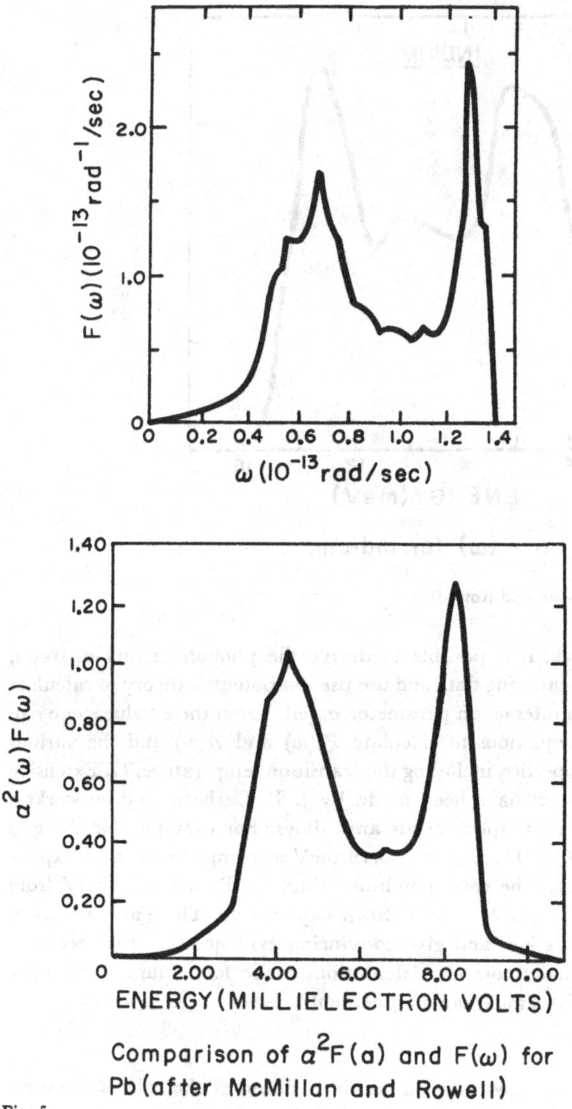

Fig. 5.
Comparison of α^2F for Pb derived from tunneling data with phonon density of states from neutron scattering data of Stedman et al. [8]

In Figs. 5, 6, 7, and 8 are shown other examples of $\alpha^2(\omega)F(\omega)$ derived from tunneling data for Pb, In, [31] La, [32] and Nb_3Sn. [33] In all cases the results are completely consistent with the phonon mechanism. Coulomb interactions play only a minor role, with μ^* varying only slowly from one metal to another, and generally in the range 0.1—02.

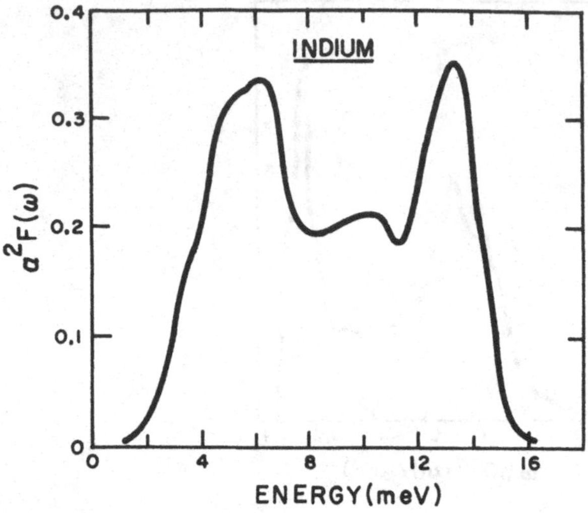

$a^2F(\omega)$ for indium.

Fig. 6.
α^2F for In (after McMillan and Rowell).

As a further check, it is possible to derive the phonon density of states, $F(\omega)$ from neutron scattering data and use pseudo-potential theory to calculate the electron-phonon interaction parameter $\alpha_q(\omega)$. From these values, one can use the Eliashberg equations to calculate $Z_s(\omega)$ and $\Delta(\omega)$ and the various superconducting properties, including the transition temperature, T_c. Extensive calculations of this sort have been made by J. P. Carbotte and co-workers [34] for several of the simpler metals and alloys. For example, for the gap edge, Δ_0, in Al at $T = 0$ K they find 0.19 meV as compared with an experimental value of 0.17. The corresponding values for Pb are 1.49 meV from theory as compared with 1.35 meV from experiment. These are essentially first principles calculations and give convincing evidence that the theory as formulated is essentially correct. Calculations made for a number of other metals and alloys give similar good agreement.

Conclusions

In this talk we have traced how our understanding of the role of electron-phonon interactions in superconductivity has developed from a concept to a precise quantitative theory. The self-energy and pair potential, and thus the Green's functions, can be derived either empirically from tunneling data or directly from microscopic theory with use of the Eliashberg equations. Physicists, both experimental and theoretical, from different parts of the world have contributed importantly to these developments.

All evidence indicates that the electron-phonon interaction is the dominant mechanism in the cases studied so far, which include many simple metals,

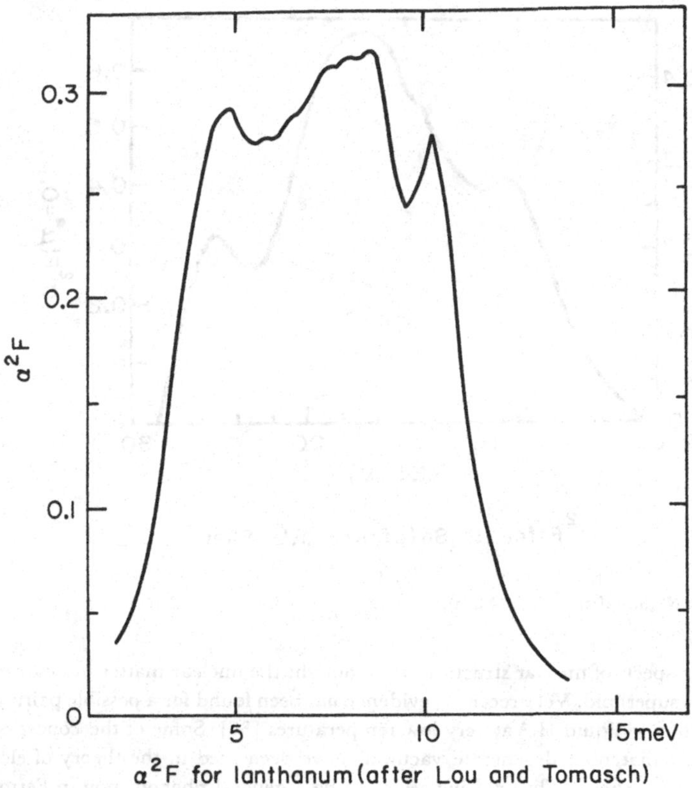

Fig. 7.
α^2F for La (after Lou and Tomasch).

transition metals, a rare earth, and various alloys and compounds. Except possibly for the metallic form of hydrogen, [35] which is presumed to exist at very high pressures, it is unlikely that the phonon mechanism will yield substantially higher transition temperatures than the present maximum of about 21 K for a compound of Nb, Al and Ge.

Other mechanisms have been suggested for obtaining higher transition temperatures. One of these is to get an effective attractive interaction between electrons from exchange of virtual excitons, or electron-hole pairs. This requires a semiconductor in close proximity to the metal in a layer or sandwich structure. At present, one can not say whether or not such structures are feasible and in no case has the exciton mechanism been shown to exist. As Ginzburg has emphasized, this problem (as well as other proposed mechanisms) deserves study until a definite answer can be found. [36]

The pairing theory has had wide application to Fermi systems other than electrons in metals. For example, the theory has been used to account for

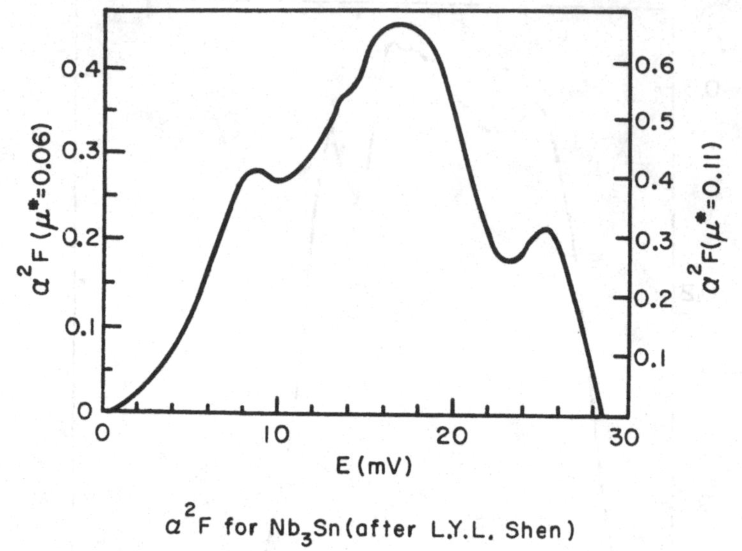

Fig. 8.
α^2F for Nb_3Sn (after Y. L. Y. Shen).

many aspects of nuclear structure. It is thought the nuclear matter in neutron stars is superfluid. Very recently, evidence has been found for a possible pairing transition in liquid He^3 at very low temperatures [37]. Some of the concepts, such as that of a degenerate vacuum, have been used in the theory of elementary particles. Thus pairing seems to be a general phenomenon in Fermi systems.

The field of superconductivity is still a very active one in both basic science and applications. I hope that these lectures have given you some feeling for the accomplishments and the methods used.

References

1. Shoenberg, D. Superconductivity, Cambridge Univ. Press, Cambridge (1938). Second edition, 1951.
2. London, F. Proc. Roy. Soc. (London) *152A*, 24 (1935).
3. Bardeen, J. Phys. Rev. *59*, 928A (1941).
4. Reynolds, C. A., Serin, B. Wright W. H. and Nesbitt, L. B. Phys. Rev. *78*, 487 (1950).
5. Maxwell, E., Phys. Rev. *78*, 477 (1950).
6. Fröhlich, H., Phys. Rev. *79*, 845 (1950); Proc. Roy. Soc. (London) Ser. A 213, 291 (1952).
7. London, F., Superfluids, New York, John Wiley and Sons, 1950.
8. For recent review articles with references, see the chapters by D. J. Scalapino and by W. L. McMillan and J. M. Rowell in Superconductivity, R. D. Parks, ed., New York, Marcel Bekker, Inc., 1969, Vol. 1. An excellent reference for the theory and earlier experimental work is J. R. Schrieffer, Superconductivity, New York, W. A. Benjamin, Inc., 1964. The present lecture is based in part on a chapter by the author in Cooperative Phenomena, H. Haken and M. Wagner, eds. to be published by Springer.

9. Gor'kov, L. P., Zh. Eksper i. teor. Fiz. *34*, 735 (1958). (English transl. Soviet Phys. — JETP *7*, 505 (1958)).
10. Kadanoff L. P. and Martin, P. C. Phys. Rev. *124*, 670 (1961).
11. Migdal, A. B., Zh. Eksper i. teor. Fiz. *34*, 1438 (1958). (English transl. Soviet Phys. — JETP *7*, 996 (1958)).
12. Eliashberg, G. M., Zh. Eksper i. teor. Fiz. *38*, 966 (1960). Soviet Phys. — JETP *11*, 696 (1960).
13. Nambu, Y., Phys. Rev. *117*, 648 (1960).
14. Bardeen, J., Phys. Rev. *79*, 167 (1950); *80*, 567 (1950); *81* 829 (1951).
15. Bardeen, J., Rev. Mod. Phys. *23*, 261 (1951).
16. Schafroth, M. R., Helv. Phys. Acta *24*, 645 (1951); Nuovo Cimento, *9*, 291 (1952).
17. For a review see Bardeen, J., Encyclopedia of Physics, S. Flugge, ed., Berlin, Springer-Verlag, (1956) Vol. XV, p. 274.
18. Bardeen, J., L. N. Cooper and J. R. Schrieffer, Phys. Rev. *108*, 1175 (1957).
19. For references, see the review article of M. A. Biondi, A. T. Forrester, M. B. Garfunkel and C. B. Satterthwaite, Rev. Mod. Phys. *30*, 1109 (1958).
20. Pippard, A. B., Proc. Roy. Soc. (London) *A216*, 547 (1954).
21. See N. N. Bogoliubov, V. V. Tolmachev and D. V. Shirkov, A New Method in the Theory of Superconductivity, New York, Consultants Bureau, Inc., 1959.
22. Morel P. and Anderson, P. W., Phys. Rev. *125*, 1263 (1962).
23. Giaever, I., Phys. Rev. Letters, *5*, 147; *5*, 464 (1960).
24. Giaever, I., Hart H. R., and Megerle K., Phys. Rev. *126*, 941 (1962).
25. Rowell, J. M., Anderson P. W. and Thomas D. E., Phys. Rev. Letters, *10*, 334 (1963).
26. Culler, G. J., Fried, B. D., Huff, R. W. and Schrieffer, J. R., Phys. Rev. Letters *8*, 339 (1962).
27. Schrieffer, J. R., Scalapino, D. J. and Wilkins, J. W., Phys. Rev. Letters *10*, 336 (1963); D. J. Scalapino, J. R. Schrieffer, and J. W. Wilkins, Phys. Rev. *148*, 263 (1966).
28. Scalapino, D. J., Wada, Y. and Swihart, J. C., Phys. Rev. Letters, *14*, 102 (1965); *14*, 106 (1965).
29. Eliashberg, G. M., Zh. Eksper i. teor. Fiz. *43*, 1005 (1962). English transl. Soviet Phys. — JETP *16*, 780 (1963).
30. Bardeen, J. and Stephen, M., Phys. Rev. *136*, A1485 (1964).
31. McMillan, W. L. and Rowell, J. M. in Reference 8.
32. Lou, L. F. and Tomasch, W. J., Phys. Rev. Lett. *29*, 858 (1972).
33. Shen, L. Y. L., Phys. Rev. Lett. *29*, 1082 (1972).
34. Carbotte, J. P., Superconductivity, P. R. Wallace, ed., New York, Gordon and Breach, 1969, Vol. 1, p. 491; J. P. Carbotte and R. C. Dynes, Phys. Rev. *172*, 476 (1968); C. R. Leavens and J. P. Carbotte, Can. Journ. Phys. *49*, 724 (1971).
35. Ashcroft N. W., Phys. Rev. Letters, *21*, 1748 (1968).
36. See V. L. Ginzburg, "The Problem of High Temperature Superconductivity," *Annual Review of Materials Science*, Vol. 2, p. 663 (1972).
37. Osheroff D. D., Gully W. J., Richardson R. C. and Lee, D. M. Phys. Rev. Lett. *29*, 1621 (1972).

NOTES AND REFERENCES

1. F. London, *Superfluids*, Vol. I, Wiley, New York, 1950.
2. C. J. Gorter and H. G. B. Casimir, *Phys. Z.*, **35**, 963 (1934); *Z. Tech. Phys.*, **15**, 539 (1934).
3. F. London, *Superfluids*, Vol. II, Wiley, New York, 1950.
4. D. Schoenberg, *Superconductivity*, Cambridge, New York, 1952.
5. F. London, *Phys. Rev.*, **74**, 562 (1948).
6. W. Meissner and R. Ochsenfeld, *Naturwiss.*, **21**, 787 (1933).
7. H. Kamerlingh Onnes, *Comm. Phys. Lab. Univ. Leiden*, Nos. 119, 120, 122 (1911).
8. J. Bardeen, L. N. Cooper, and J. R. Schrieffer, *Phys. Rev.*, **106**, 162 (1957); **108**, 1175 (1957).
9. J. Bardeen and J. R. Schrieffer, *Progr. Low Temp. Phys.*, Vol. III, North-Holland, Amsterdam, 1961.
10. H. Frohlich, *Phys. Rev.*, **79**, 845 (1950).
11. C. A. Reynolds, B. Serin, W. H. Wright, and L. B. Nesbitt, *Phys. Rev.*, **78**, 487 (1950).
12. E. Maxwell, *Phys. Rev.*, **78**, 477 (1950).
13. J. Bardeen, *Rev. Mod. Phys.*, **23**, 261 (1951).
14. M. R. Schafroth, *Helv. Phys. Acta*, **24**, 645 (1951).
15. A. B. Migdal, *Soviet Phys. JETP*, **1**, 996 (1958).
16. (a) See Ref. 4.
    (b) B. Serin, *Handbuch der Physik*, **15**, 210, Springer, Berlin, 1950.

(c) J. Bardeen, *Handbuch der Physik*, **15**, 274, Springer, Berlin, 1950.
(d) M. A. Biondi, A. T. Forrester, M. P. Garfunkel, and C. B. Satterthwaite, *Rev. Mod. Phys.*, **30**, 1109 (1958).
(e) E. A. Lynton, *Superconductivity*, Methuen, London, 1963.
(f) M. Tinkham, *Low Temperature Physics*, p. 149, Gordon and Breach, New York, 1962.
(g) D. H. Douglass and L. M. Falicov, *Progr. Low Temp. Phys.*, Vol. IV, C. J. Gorter (ed.), North-Holland, Amsterdam, 1963.

17. A. A. Abrikosov, *J. Exptl. Theoret. Phys. (USSR)*, **32**, 1442 (1957), translated as *Soviet Phys. JETP*, **5**, 1174 (1957).
18. (a) M. Tinkham, *Phys. Rev.*, **129**, 2413 (1963).
(b) P. W. Anderson, *Proc. Ravello Spring School*, 1963.
19. N. Byers and C. N. Yang, *Phys. Rev. Letters*, **7**, 46 (1961).
20. (a) B. D. Deaver, Jr. and W. M. Fairbank, *Phys. Rev. Letters*, **7**, 43 (1961).
(b) R. Doll and M. Näbauer, *Phys. Rev. Letters*, **7**, 51 (1961).
21. W. H. Keesom and J. H. van den Ende, *Commun. Phys. Lab. Univ. Leiden*, No. 2196 (1932); W. H. Keesom and J. A. Kok, *Physica*, **1**, 175 (1934).
22. D. Mapother, private communication.
23. T. H. Geballe, B. T. Matthias, G. W. Hull, Jr., and E. Corenzwit, *Phys. Rev. Letters*, **6**, 275 (1961); T. H. Geballe and B.T. Matthias, *IBM J. Res. Develop.*, **6**, 256 (1962).
24. J. W. Garland, *Phys. Rev. Letters*, **11**, 111, 114 (1963).
25. R. E. Glover, III, and M. Tinkham, *Phys. Rev.*, **108**, 243 (1957); M. A. Biondi and M. Garfunkel, *Phys. Rev.*, **116**, 853 (1959).
26. I. Giaever, *Phys. Rev. Letters*, **5**, 147, 464 (1960).
27. (a) R. W. Morse, *Progr. Cryog.*, Vol. I, p. 220, K. Mendelssohn (ed.), Heywood, London, 1959.
(b) B. T. Geilikman and V. Z. Kresin, *J. Exptl. Theoret. Phys. (USSR)*, **41**, 1142 (1961), translated in *Soviet Phys. JETP*, **14**, 816 (1961).
(c) V. L. Pokrovskii, *J. Exptl. Theoret. Phys. (USSR)*, **40**, 143 (1961), translated in *Soviet Phys. JETP*, **13**, 100 (1961).
28. L. C. Hebel and C. P. Slichter, *Phys. Rev.*, **113**, 1504 (1959); L. C. Hebel, *Phys. Rev.*, **116**, 79 (1959).

29. A. G. Redfield, *Phys. Rev. Letters*, **3**, 85 (1959); A. G. Redfield and A. G. Anderson, *Phys. Rev.*, **116**, 583 (1959).
30. J. Bardeen, G. Rickayzen, and L. Tewordt, *Phys. Rev.*, **113**, 982 (1959).
31. F. London and H. London, *Proc. Roy. Soc.* (*London*), **A149**, 71 (1935); *Physica*, **2**, 341 (1935).
32. L. Onsager, *Phys. Rev. Letters*, **7**, 50 (1961).
33. A. B. Pippard, *Proc. Roy. Soc.* (*London*), **A216**, 547 (1953).
34. R. G. Chambers, *Proc. Roy. Soc.* (*London*), **A65**, 458 (1952).
35. R. A. Ferrell and R. E. Glover, III, *Phys. Rev.*, **109**, 1398 (1958); M. Tinkham and R. A. Ferrell, *Phys. Rev. Letters*, **2**, 331 (1959).
36. V. L. Ginsburg and L. D. Landau, *J. Exptl. Theoret. Phys.* (*USSR*), **20**, 1064 (1950).
37. L. P. Gor'kov, *J. Exptl. Theoret. Phys.* (*USSR*), **36**, 1918 (1959), translated in *Soviet Phys. JETP*, **9**, 1364 (1959).
38. (a) N. R. Werthamer, *Phys. Rev.*, **132**, 663 (1963).
    (b) L. Tewordt, *Phys. Rev.*, **132**, 595 (1963).
39. C. N. Yang, *Rev. Mod. Phys.*, **34**, 694 (1962).
40. P. W. Anderson, *Phys. Chem. Solids*, **11**, 26 (1959).
41. L. N. Cooper, *Phys. Rev.*, **104**, 1189 (1956).
42. M. R. Schafroth, J. M. Blatt, and S. T. Butler, *Helv. Phys. Acta*, **30**, 93 (1957). Subsequent to the work of Bardeen, Cooper, and Schrieffer, Matsubara and Blatt attempted to improve the mathematical treatment of SBB. Owing to the complexity of their formalism, they did not carry through the calculation far enough to compare their results with the BCS theory. In particular, a well-defined, simple equation corresponding to the BCS energy-gap equation was not derived, nor was the quasi-particle spectrum obtained. See T. Marsubara and J. M. Blatt, *Progr. Theoret. Phys.* (*Kyoto*), **23**, 451 (1960). Later, this approach was extended to obtain the results of the pairing theory; see J. M. Blatt, *Progr. Theoret. Phys.* (*Kyoto*), **27**, 1137 (1962). See also M. Baranger, *Phys. Rev.*, **130**, 1244 (1963).
43. M. R. Schafroth, *Phys. Rev.*, **111**, 72 (1958).
44. D. Pines, *The Many-Body Problem*, Benjamin, New York, 1962.
45. L. D. Landau, *J. Exptl. Theoret. Phys.* (*USSR*), **30**, 1058 (1956).

46. J. M. Luttinger and P. Nozières, *Phys. Rev.*, **127**, 1423, 1431 (1962).

47. Anderson has exploited the formal similarity between the commutation relations (2-21) and those for a collection of fictitious Pauli spin operator $\mathbf{S}_k$. The connection between the two sets is

$$2S_{z\mathbf{k}} = 1 - (n_{\mathbf{k}\uparrow} + n_{-\mathbf{k}\downarrow})$$
$$S_{x\mathbf{k}} + iS_{y\mathbf{k}} = b_{\mathbf{k}}^{+}$$
$$S_{x\mathbf{k}} - iS_{y\mathbf{k}} = b_{\mathbf{k}}$$

He was able to obtain the results of the BCS theory by a semi-classical treatment of this fictitious spin system. See P. W. Anderson, *Phys. Rev.*, **110**, 827 (1958); **112**, 1900 (1958).

48. S. Tomonaga, *Progr. Theoret. Phys. (Kyoto)*, **2**, 6 (1947).

49. (a) T. D. Lee, F. Low, and D. Pines, *Phys. Rev.*, **90**, 297 (1953).
(b) T. D. Lee and D. Pines, *Phys. Rev.*, **92**, 883 (1953).

50. L. P. Kadanoff and P. C. Martin, *Phys. Rev.*, **124**, 670 (1961).

51. (a) J. M. Blatt, *Progr. Theoret. Phys. (Kyoto)*, **27**, 1137 (1962); *Proc. Superconductivity Conference*, Cambridge 1959, unpublished; *Theory of Superconductivity*, Academic, New York, 1964.
(b) M. Baranger, *Phys. Rev.*, **130**, 1244 (1963).
(c) F. Bloch and H. E. Rorschach, *Phys. Rev.*, **128**, 1697 (1962).

52. N. N. Bogoliubov, *Nuovo Cimento*, **7**, 6, 794 (1958); also see N. N. Bogoliubov, V. V. Tolmachev, and D. V. Shirkov, *A New Method in the Theory of Superconductivity*, Consultants Bureau, New York, 1959.

53. J. Valatin, *Nuovo Cimento*, **7**, 843 (1958).

54. (a) N. N. Bogoliubov, *Physica, Suppl.*, **26**, 1 (1960).
(b) J. Bardeen and G. Rickayzen, *Phys. Rev.*, **118**, 936 (1960).
(c) B. Mühlschlegel, *J. Math. Phys.*, **3**, 522 (1962).
(d) D. C. Mattis and E. Lieb, *J. Math. Phys.*, **2**, 602 (1961).
(e) R. Haag, private communication.

55. J. Valatin, private communication, 1957.

56. The operator R is particularly useful in tunneling calculations; see J. Bardeen, *Phys. Rev. Letters*, **9**, 147 (1962) and B. D. Josephson, *Phys. Letters*, **1**, 251 (1962).

57. (a) H. Suhl, *Bull. Am. Phys. Soc.*, **6**, 119 (1961).
(b) Y. Wada, *Rev. Mod. Phys.*, **36**, 253 (1964).

58. P. W. Anderson, *Phys. Rev.*, **112**, 1900 (1958).
59. (a) J. C. Fisher, private communication, 1959.
(b) K. A. Brueckner, T. Soda, P. W. Anderson, and P. Morel, *Phys. Rev.*, **118**, 1442 (1960).
(c) R. Balian and N. R. Werthamer, *Phys. Rev.*, **131**, 1553 (1963).
60. B. Bayman, *Nucl. Phys.*, **15**, 33 (1960).
61. B. R. Mattelson, *The N-Body Problem*, Wiley, New York, 1959.
62. The point of view taken here in calculating finite-temperature properties is due to J. Bardeen.
63. L. I. Schiff, *Quantum Mechanics*, Chap. 8, McGraw-Hill, New York, 1949.
64. In order that the limits on ϵ_p and $\epsilon_{p'}$ do not affect this argument, one must satisfy the condition $|\mathbf{q}_0|\xi_0 \gg 1$. If $|\mathbf{q}_0|\xi_0 \ll 1$, the term $\epsilon_p\epsilon_{p'} \to \epsilon_p{}^2$ and does not average to zero. In this case the average coherence factor is twice as large as that for $|\mathbf{q}_0|\xi_0 \gg 1$; however, only one-half as many states contribute to the sum for small $\mathbf{q}_0$. Therefore, the resultant expression for α is the same in the two cases.
65. (a) See Ref. 9.
(b) D. H. Douglass and L. M. Falicov, *Progr. Low Temp. Phys.*, Vol. IV, C. J. Gorter (ed.), North-Holland, Amsterdam (to be published).
(c) J. Bardeen, *Rev. Mod. Phys.*, **34**, 667 (1962).
(d) M. Tinkham, *Phys. Rev.*, **129**, 2413 (1963).
(e) B. T. Matthias, T. H. Geballe, V. B. Compton, *Rev. Mod. Phys.*, **35**, 1 (1963).
(f) E. A. Lynton, *Superconductivity*, Methuen, London, 1963.
66. S. B. Chandrasekhar and J. A. Rayne, *Phys. Rev.*, **124**, 1011 (1961). Anderson estimated a relative shift of the sound velocity of order $(m/M) \simeq 10^{-5}$, in agreement with the results of Chandrasekhar and Rayne [see *Phys. Rev.*, **112**, 1900 (1958)], while Ferrell [*Phys. Rev. Letters*, **6**, 541 (1961)] suggested that a phonon-frequency shift $\delta\omega_q/\omega_q \sim 0.1$ might be obtained for $q\xi_0 > 1$. Detailed calculations by Markowitz (private communication) and by Toxin and Liu (private communication) show that the shift is definitely less than 1 per cent over the entire frequency range. See also R. E. Prange, *Phys. Rev.*, **129**, 2495 (1963).

67. A. B. Pippard, *Phil. Mag.*, **46**, 1104 (1955); *Low Temperature Physics*, Gordon and Breach, New York, 1962; J. R. Liebowitz, *Bull. Am. Phys. Soc.*, Ser. II, **9**, 267 (1964).
68. C. P. Slichter, *Principles of Magnetic Resonance*, Harper, New York, 1963.
69. (a) D. C. Mattis and J. Bardeen, *Phys. Rev.*, **111**, 412 (1958).
(b) A. A. Abrikosov and L. P. Gor'kov, *Soviet Phys. JETP*, **35**, 1558 (1958), translated in **8**, 1090 (1959); **36**, 319 (1959), translated in **9**, 220 (1959).
(c) G. Rickayzen, in C. Fronsdal (ed.), *The Many-Body Problem*, Benjamin, New York, 1961.
70. P. B. Miller, *Phys. Rev.*, **113**, 1209 (1959).
71. D. M. Ginsberg and M. Tinkham, *Phys. Rev.*, **118**, 990 (1960). See also D. M. Ginsberg, P. L. Richards, and M. Tinkham, *Phys. Rev. Letters*, **3**, 337 (1959).
72. P. L. Richards and M. Tinkham, *Phys. Rev.*, **119**, 575 (1960).
73. D. M. Ginsberg and J. D. Leslie, *Rev. Mod. Phys.*, **36**, 198 (1964); J. Bardeen, *Rev. Mod. Phys.*, **36**, 198 (1964). See also D. J. Scalapino, J. R. Schrieffer, and J. W. Wilkins (to be published).
74. J. Bardeen, *Phys. Rev. Letters*, **6**, 57 (1961); **9**, 147 (1962).
75. M. H. Cohen, L. M. Falicov, and J. C. Phillips, *Phys. Rev. Letters*, **8**, 316 (1962).
76. R. E. Prange, *Phys. Rev.*, **131**, 1083 (1963).
77. W. A. Harrison, *Phys. Rev.*, **123**, 85 (1961).
78. (a) J. R. Schrieffer, D. J. Scalapino, and J. W. Wilkins, *Phys. Rev. Letters*, **10**, 336 (1963);
(b) *Phys. Rev.* (to be published).
79. (a) J. W. Wilkins, Ph.D. thesis, University of Illinois, 1963.
(b) J. R. Schrieffer, *Rev. Mod. Phys.*, **36**, 200 (1964).
(c) J. R. Schrieffer, in T. Bak (ed.), *Phonons and Phonon Interactions*, Benjamin, New York, 1964.
80. E. Burstein, D. N. Langenberg, and B. N. Taylor, *Phys. Rev. Letters*, **6**, 92 (1961); *Advances in Quantum Electronics*, J. R. Singer (ed.), Columbia Univ. Press, New York, 1961.
81. B. N. Taylor and E. Burstein, *Phys. Rev. Letters*, **10**, 14 (1963).
82. J. R. Schrieffer and J. W. Wilkins, *Phys. Rev. Letters*, **10**, 17 (1963).
83. B. D. Josephson, *Phys. Rev. Letters*, **1**, 251 (1962).
84. P. W. Anderson, *Proc. Ravello Spring School*, 1963. See also

R. A. Ferrell and R. E. Prange, *Phys. Rev. Letters*, **10**, 479 (1963) and V. Amkegaokar and A. Baratoff, *Phys. Rev. Letters*, **10**, 486 (1963).

85. P. W. Anderson and J. M. Rowell, *Phys. Rev. Letters*, **10**, 230 (1963).
86. (a) J. M. Rowell, *Phys. Rev. Letters*, **11**, 200 (1963); *Rev. Mod. Phys.*, **36**, 199 (1964).
    (b) B. D. Josephson, Thesis, Cambridge University, 1962; *Phys. Rev. Letters*, **1**, 251 (1962); *Rev. Mod. Phys.*, **36**, 216 (1964).
    (c) S. Shapiro et al., *Rev. Mod. Phys.*, **36**, 223 (1964).
87. I. Giaever, H. R. Hart, and K. Megerle, *Phys. Rev.*, **126**, 941 (1962).
88. J. M. Rowell, P. W. Anderson, and D. E. Thomas, *Phys. Rev. Letters*, **10**, 334 (1963).
89. R. A. Ferrell, *Phys. Rev. Letters*, **3**, 262 (1959).
90. P. W. Anderson, *Phys. Rev. Letters*, **3**, 325 (1959).
91. (a) D. J. Thouless, *The Quantum Mechanics of Many-Body Systems*, Academic, New York, 1961.
    (b) D. Pines, *The Many-Body Problem*, Benjamin, New York, 1962.
    (c) L. P. Kadanoff and G. Baym, *Quantum Statistical Mechanics*, Benjamin, New York, 1962.
    (d) P. Nozières, *Le problem du N corpes*, Dunod, Paris, 1963, translated as *The Theory of Interacting Fermi Systems*, Benjamin, New York, 1963.
    (e) T. D. Schultz, *Quantum Field Theory and the Many-Body Problem*, Gordon and Breach, New York, 1963.
    (f) V. L. Bonch-Bruevitch and S. V. Tyablikov, *Method Funcii Grina Statisticeskei Mexanike*, Moscow, 1960, translated by D. Ter Haar, North-Holland, Amsterdam, 1962.
    (g) A. A. Abrikosov, L. P. Gor'kov, and I. E. Dzyaloshinskii, *Methods of Quantum Field Theory in Statistical Mechanics*, Prentice-Hall, Englewood Cliffs, N.J., 1963.
    (h) *1962 Cargèse Lectures in Theoretical Physics*, M. Lévy (ed.), Benjamin, New York, 1962.
    (i) *The Many-Body Problem*, C. Fronsdal (ed.), Benjamin, New York, 1962.
    (j) *The Many-Body Problem* (Les Houches Notes), Dunod, Paris, 1959.

92. R. E. Peierls, *Quantum Theory of Solids*, Clarendon Press, Oxford, 1956.
93. J. Bardeen and D. Pines, *Phys. Rev.*, **99**, 1140 (1953).
94. J. Wilkins, Ph.D. thesis, University of Illinois, 1963.
95. F. Bassani, J. Robinson, B. Goodman, and J. R. Schrieffer, *Phys. Rev.*, **127**, 1969 (1962).
96. J. M. Ziman, *Electrons and Phonons*, Clarendon Press, Oxford, 1960. See also *The Fermi Surface*, W. A. Harrison and M. B. Webb (eds.), Wiley, New York, 1960.
97. L. J. Sham, *Proc. Phys. Soc.* (*London*), **78**, 895 (1961).
98. W. A. Harrison, *Phys. Rev.*, **129**, 2503 (1963); **129**, 2512 (1963); **131**, 2433 (1963).
99. (a) J. Schwinger, *Proc. Natl. Acad. Sci.* (*U.S.*), **37**, 452 (1951).
(b) P. C. Martin and J. Schwinger, *Phys. Rev.*, **115**, 1342 (1959).
100. V. M. Galitskii and A. B. Migdal, *Soviet Phys. JETP*, **7**, 96 (1958).
101. (a) L. P. Kadanoff and P. C. Martin, *Phys. Rev.*, **124**, 670 (1961).
(b) L. P. Kadanoff, *Proc. Ravello Spring School*, 1963. This time-dependent form of the instability was known to many workers in the field in 1957–1958, e.g., Goldstone, Anderson, Bogoliubov, etc.
102. (a) R. P. Feynman, *Phys. Rev.*, **76**, 769 (1949).
(b) S. Tomonaga, *Progr. Theoret. Phys.* (*Kyoto*), **1**, 27 (1946).
(c) F. J. Dyson, *Phys. Rev.*, **75**, 486 (1949).
103. (a) J. Hubbard, *Proc. Roy. Soc.* (*London*), **A240**, 539 (1957).
(b) N. N. Bogoliubov and D. V. Shirkov, *Introduction to the Theory of Quantized Fields*, Wiley-Interscience, New York, 1959.
104. L. I. Schiff, *Quantum Mechanics*, McGraw-Hill, New York, 1949.
105. D. Bohm and D. Pines, *Phys. Rev.*, **92**, 609 (1953).
106. M. Gell-Mann and K. Brueckner, *Phys. Rev.*, **106**, 364 (1957).
107. P. Nozières and D. Pines, *Phys. Rev.*, **111**, 442 (1958).
108. J. Lindhard, *Kgl. Danske Videnskab, Selskab. Mat. Fys. Medd.*, **28**, 8 (1954).
109. (a) J. S. Langer and S. H. Vosko, *Phys. Chem. Solids*, **12**, 196 (1959).
(b) W. Kohn and S. H. Vosko, *Phys. Rev.*, **119**, 912 (1960).

(c) J. Friedel, *Phil. Mag.*, **43**, 153 (1952); *Nuovo Cimento Suppl.*, **2**, 287 (1958).
110. T. J. Rowland, *Phys. Rev.*, **125**, 459 (1962).
111. R. P. Feynman and M. Cohen, *Phys. Rev.*, **102**, 1189 (1956).
112. D. Pines and J. R. Schrieffer, *Nuovo Cimento*, **10**, 496 (1958).
113. G. Rickayzen, *Phys. Rev.*, **115**, 765 (1959).
114. Y. Nambu, *Phys. Rev.*, **117**, 648 (1960).
115. J. J. Quinn and R. A. Ferrell, *Phys. Rev.*, **112**, 812 (1958).
116. M. Gell-Mann, *Phys. Rev.*, **106**, 359 (1957).
117. P. Nozières and D. Pines, *Phys. Rev.*, **111**, 442 (1958).
118. S. D. Silverstein, *Phys. Rev.*, **128**, 631 (1962).
119. T. Staver, Ph.D. thesis, Princeton University, 1952 (unpublished).
120. S. Engelsberg and J. R. Schrieffer, *Phys. Rev.*, **131**, 993 (1963).
121. L. P. Gor'kov, *J. Exptl. Theoret. Phys.* (*USSR*), **34**, 735 (1958), translated in *Soviet Phys. JETP*, **7**, 505 (1958).
122. G. M. Eliashberg, *J. Exptl. Theoret. Phys.* (*USSR*), **38**, 966 (1960), translated in *Soviet Phys. JETP*, **11**, 696 (1960).
123. See Ref. 78b.
124. P. Morel and P. W. Anderson, *Phys. Rev.*, **125**, 1263 (1962).
125. J. C. Swihart, *IBM J. Res. Develop.*, **6**, 14 (1962).
126. G. J. Culler, B. D. Fried, R. W. Huff, and J. R. Schrieffer, *Phys. Rev. Letters*, **8**, 399 (1962).
127. (a) D. J. Scalapino and P. W. Anderson, *Phys. Rev.*, **133**, A291 (1964).
(b) D. J. Scalapino, *Rev. Mod. Phys.*, **36**, 205 (1964).
128. A. A. Abrikosov, L. P. Gor'kov, and I. E. Dzyaloshinskii, *Soviet Phys. JETP*, **9**, 636 (1959).
129. D. M. Ginsberg and M. Tinkham, *Phys. Rev.*, **118**, 990 (1960).
130. (a) P. B. Miller, *Phys. Rev.*, **118**, 928 (1960).
(b) D. M. Ginsberg and J. D. Leslie, *Rev. Mod. Phys.*, **36**, 198 (1964).
(c) K. Maké and T. Tsuneto, *Progr. Theoret. Phys.* (*Kyoto*), **28**, 163 (1962).
131. J. Bardeen, *Nuovo Cimento*, **5**, 1766 (1957).
132. G. Rickayzen, *Phys. Rev.*, **115**, 795 (1959).
133. G. Baym and L. P. Kadanoff, *Phys. Rev.*, **124**, 287 (1961).
134. D. Pines and J. R. Schrieffer, *Nuovo Cimento*, **10**, 496 (1958).
135. R. P. Feynman and M. Cohen, *Phys. Rev.*, **102**, 1189 (1956).

136. (a) L. P. Kadanoff and V. Ambegaokar, *Nuovo Cimento*, **22**, 914 (1961).
(b) V. Ambegaokar and L. P. Kadanoff, *The Many-Body Problem*, C. Fronsdal (ed.), Benjamin, New York, 1962.
(c) J. C. Ward, *Phys. Rev.*, **78**, 182 (1950).
137. Z. Koba, *Progr. Theoret. Phys. (Kyoto)*, **6**, 322 (1951).
138. T. Tsuneto, *Phys. Rev.*, **118**, 1029 (1960).
139. A. Bardasis and J. R. Schrieffer, *Phys. Rev.*, **121**, 1050 (1961).
140. Subsequent to the experiments of Deaver and Fairbank and of Doll and Näbauer, Little and Parks performed a beautiful experiment in which they observed a periodic variation in the transition temperature of a superconducting cylinder as a function of the enclosed magnetic flux ϕ. Their results confirm the quantized flux unit $hc/2e$; see W. A. Little and R. D. Parks, *Phys. Rev. Letters*, **9**, 9 (1962); W. A. Little, *Rev. Mod. Phys.*, **36**, 264 (1964).
141. A. Bohr and B. R. Mottelson, *Phys. Rev.*, **125**, 495 (1962).
142. W. D. Knight, *Solid State Physics*, Vol. 2, Seitz and Turnbull (eds.), Academic, New York, 1956.
143. F. Reif, *Phys. Rev.*, **106**, 208 (1957).
144. G. M. Androes and W. D. Knight, *Phys. Rev. Letters*, **2**, 386 (1959).
145. R. A. Ferrell, *Phys. Rev. Letters*, **3**, 262 (1959).
146. P. W. Anderson, *Phys. Rev. Letters*, **3**, 325 (1959).
147. R. J. Noer and W. D. Knight, *Rev. Mod. Phys.*, **36**, 177 (1964).
148. V. Heine and A. B. Pippard, *Phil. Mag.*, **3**, 1046 (1958).
149. J. R. Schrieffer, *Phys. Rev. Letters*, **3**, 323 (1959). See also Ref. 9, p. 262. A similar suggestion was advanced by J. M. Blatt, *Proc. Superconductivity Conference*, Cambridge, 1959, unpublished.
150. L. N. Cooper, *Phys. Rev. Letters*, **8**, 367 (1962); B. B. Schwartz and L. N. Cooper, *Rev. Mod. Phys.*, **36**, 280 (1964).
151. A. M. Clogston, A. C. Gossard, V. Jaccarino, and Y. Yafet, *Phys. Rev. Letters*, **9**, 262 (1962).
152. R. Kubo and Y. Obata, *J. Phys. Soc. Japan*, **11**, 547 (1956).
153. V. L. Ginsburg, *J. Exptl. Theoret. Phys. (USSR)*, **29**, 748 (1955), translated in *Soviet Phys. JETP*, **2**, 589 (1956).
154. See Ref. 38a.
155. See Ref. 38b.
156. P. W. Anderson, *Proc. Ravello Spring School*, 1963.

157. (a) T. Kinsel et al., *Rev. Mod. Phys.*, **36**, 105 (1964).
(b) L. Dubeck et al., *Rev. Mod. Phys.*, **36**, 110 (1964).
158. (a) T. G. Berlincourt, *Rev. Mod. Phys.*, **36**, 19 (1964).
(b) P. W. Anderson and Y. B. Kim, *Rev. Mod. Phys.*, **36**, 39 (1964).
159. G. Chanin, E. A. Lynton, and B. Serin, *Phys. Rev.*, **114**, 719 (1959).
160. (a) B. T. Matthias et al., *Rev. Mod. Phys.*, **36**, 155 (1964).
(b) B. T. Matthias, T. H. Geballe, and V. B. Compton, *Rev. Mod. Phys.*, **35**, 1 (1963).
161. (a) T. Tsuneto, *Progr. Theoret. Phys.* (*Kyoto*), **28**, 857 (1962).
(b) P. G. de Gennes et al., *Phys. Condensed Matter*, **1**, 176 (1963).
(c) D. Markowitz and L. P. Kadanoff, *Phys. Rev.*, **131**, 563 (1963).
(d) L. Gruenberg (to be published).
162. H. Suhl and B. T. Matthias, *Phys. Rev.*, **114**, 977 (1959).
163. W. Baltenspenger, *Rev. Mod. Phys.*, **36**, 157 (1964).
164. J. Bardeen, *Rev. Mod. Phys.*, **34**, 667 (1962); D. H. Douglass, *Rev. Mod. Phys.*, **36**, 316 (1964).
165. P. G. de Gennes, *Rev. Mod. Phys.*, **36**, 225 (1964).
166. A. Bohr, B. R. Mottelson, and D. Pines, *Phys. Rev.*, **110**, 936 (1958).
167. (a) A. Bohr and B. R. Mottelson (book to be published).
(b) M. Baranger, *Cargèse Lectures 1962 in Theoretical Physics*, M. Lévy (ed.), Benjamin, New York, 1963.
168. Y. Nambu and G. Jona-Lasinio, *Phys. Rev.*, **122**, 345 (1961); **124**, 246 (1961).
169. J. C. Fisher, *Phys. Rev.*, **129**, 1414 (1963).
170. K. A. Brueckner, T. Soda, P. W. Anderson, and P. Morel, *Phys. Rev.*, **118**, 1442 (1960).
171. W. A. Little, *Rev. Mod. Phys.*, **36**, 264 (1964).
172. (a) A. A. Abrikosov and L. P. Gorkov, *J. Exptl. Theoret. Phys.* (*USSR*), **39**, 1781 (1960); translated in *Soviet Phys. JETP*, **12**, 1243 (1961).
(b) F. Reif and M. A. Woolf, *Phys. Rev. Letters*, **9**, 315 (1962).
(c) J. C. Phillips, *Phys. Rev. Letters*, **10**, 96 (1963).
(d) H. Suhl and D. R. Fredkin, *Phys. Rev. Letters*, **10**, 131, 268 (1963).

INDEX

Printed in the United States
by Baker & Taylor Publisher Services